ELECTRONIC TRANSFORMERS AND CIRCUITS

ELECTRONIC TRANSFORMERS AND CIRCUITS

THIRD EDITION

Reuben Lee
Leo Wilson
Charles E. Carter

WILEY

A WILEY-INTERSCIENCE PUBLICATION
JOHN WILEY & SONS
NEW YORK   CHICHESTER   BRISBANE   TORONTO   SINGAPORE

Library of Congress Cataloging in Publication Data:

Lee, Reuben.
  Electronic transformers and circuits.

  "A Wiley-Interscience publication."
  Includes index.
  1. Electronic transformers.   I. Wilson, Leo
II. Carter, Charles E.   III. Title.
TK7872.T7L4   1988   621.3815'3   87-34062
ISBN 0-471-81976-X

Printed in the United States of America

10 9 8 7 6 5 4 3 2 1

PREFACE TO THIRD EDITION

In the generation since the second edition of this book was published, the electronics industry has progressed at a fast and furious pace. The vacuum tube has been almost entirely replaced. Magnetic amplifiers have been almost forgotten, but their advantages are now being rediscovered and for that reason the chapter on this subject was not completely revised as originally planned. Transformers have in some ways followed the general trend. New and better core materials and insulations are available, and better winding equipment exists. Thirty years ago, most transformers were convection cooled. Now almost all airborne and many transformers for ground-based equipment are conduction cooled.

With all of the changes and advancements that have occurred, it was originally thought that almost the entire text would have to be rewritten. As the review and writing began it soon became apparent that the text was so fundamental and well designed that it could be used much as it was by changing an occasional word here and there to reflect the advancements in electronics.

One of the possibilities in the future is the use of computers in all phases of the design and manufacture of magnetic components. Computer programs are not presented, although there are places where the use of a digital computer can save much work. There are many programs available for all phases of design and analysis. Computers were used to develop some of the new curves presented in this edition as well as doing many of the line drawings. The broader use of computers in magnetic design is bound to come.

This revision has been a joint effort. The work of the revision was done by Charles Carter and Leo Wilson, who have both known and worked with Reuben Lee over the years. We are part of a generation that has grown up to appreciate the legacy left us in the earlier editions of this book. We hope to leave

v

a similar legacy to our successors. But as in any work, the work would not have been accomplished without encouragement and input from many sources. First and foremost are the contributions made by our colleagues at Westinghouse over the years.

In addition, we wish to thank our wives, both Jean, for their patience and help with the typing and reviewing of the manuscript. Most of the material in Chapters 1 to 7, 10, and 11 is from the second edition, with changes such as "amplifying device" replacing "vacuum tube" in the previous edition. Where necessary, information on bipolar devices and applications is presented. It is shown how the original curves can be used with the new devices by showing similarities and differences to vacuum-tube circuits. This will give young engineers the opportunity to expand their horizons into more types of devices. To go with advances in magnetic materials, some of the curves have been updated. Where the older materials are still in use, the information has been retained and new material added.

Some of the subjects discussed in the preceding edition are now primarily history or are of little interest. These have been either deleted or condensed. Chapter 2 on materials has been substantially rewritten. Chapter 14 is now a condensation of the material which was previously contained in Chapter 8. Chapter 9 on high voltage design is entirely new, as are Chapters 12 and 13 on inverter transformers and applications.

In a book of general coverage, there is room for only a brief treatment of any subject. It is hoped that the treatment given to the various subjects will inspire young engineers entering the field into a broader investigation of the possibilities in the future design of magnetic components.

REUBEN LEE
LEO WILSON
CHARLES E. CARTER

Baltimore, Maryland
April 1988

PREFACE TO SECOND EDITION

In the years since the first edition of this book was published, several new developments have taken place. This second edition encompasses such new material as will afford acquaintance with advances in the art. Some old topics which were inadequately presented have received fuller treatment. Several sections, especially those on electronic amplifiers and wave filters, have been deleted because more thorough treatments of these subjects are available in current literature. Thus the original objectives of a useful book on electronic transformers and related devices, with a minimum of unnecessary material, have been pursued in the second edition. Wherever the old material appeared adequate, it has been left unchanged, and the general arrangement is still the same, except for the addition of new Chapters 9 and 11. More information in chart form, but few mathematical proofs, are included.

In a book of general coverage, there is room only for a brief treatment of any phase of the subject. Thus the new chapter on magnetic amplifiers is a condensed outline of the more common components and circuits of this rapidly growing field. It is hoped that this chapter will be helpful as a general introduction to circuit and transformer designers alike. Recent circuit developments are reported in the *AIEE Transactions*.

In response to inquiry it should be stated that, where a mathematical basis is given, graphical performance is always calculated. There has been good general correspondence between the graphs and experimental tests. This correspondence is quite close in all cases except pulse transformers; for these, the graphs presented in this book predict wave shape with fair accuracy, but to predict exactly all the superposed ripples would be impracticable. This is pointed out in Chapter 10.

Although technical words usually have the same meaning as in the first edition, there are several new magnetic terms in the second edition. These terms conform with ASTM Standard A127-48.

Pascal said that an author should always use the word "our" rather than "my" in referring to his work, because there is in it usually more of other people's than his own. Never was this more true than of the present volume. Acknowledgment is due many Westinghouse engineers, especially R. M. Baker, L. F. Deise, H. L. Jessup, J. W. Ogden, G. F. Pittman, R. A. Ramey, T. F. Saffold, and D. S. Stephens, all of whom assisted immeasurably by their constructive comments on the manuscript. D. G. Little's continued interest was most encouraging.

Helpful comment has been received from men outside Westinghouse. Mr. P. Fenoglio of the General Electric Co. kindly pointed out an omission in the first edition. Output wave shapes given for the front or leading edge of a pulse transformer were accurate for a hard-tube modulator, but not for a line-type modulator. The missing information is included in the second edition.

Finally, to my wife Margaret, my heartfelt thanks not only for her understanding of the long disruption of normal social life but also for her patience in checking proofs.

Reuben Lee

Baltimore, Maryland
August, 1955

PREFACE TO FIRST EDITION

The purpose of this book is twofold: first, to provide a reference book on the design of transformers for electronic apparatus and, second, to furnish electronic equipment engineers with an understanding of the effects of transformer characteristics on electronic circuits. Familiarity with basic circuit theory and transformer principles is assumed. Conventional transformer design is treated adequately in existing books, so only such phases of it as are pertinent to electronic transformers are included here. The same can be said of circuit theory; only that which is necessary to an understanding of transformer operation is given. It is intended that in this way the book will be encumbered with a minimum of unnecessary material. Mathematical proofs as such are kept to a minimum, but the bases for quantitative results are indicated. The A.I.E.E. "American Standard Definitions of Electrical Terms" gives the meaning of technical words used. Circuit symbols conform to A.S.A. Standards Z32.5—1944 and Z32.10—1944.

Chapter headings, except for the first two, are related to general types of apparatus. This arrangement should make the book more useful. Design data are included which would make tedious reading if grouped together. For instance, the design of an inductor depends on whether it is for power or wave filter work, and the factors peculiar to each are best studied in connection with their respective apparatus.

Parts of the book are based on material already published in the *Proceedings of the Institute of Radio Engineers, Electronics,* and *Communications.* Much of it leans heavily upon work done by fellow engineers of the Westinghouse Electric Corporation, the warmth of whose friendship I am privileged to enjoy. To list all their names would be a difficult and inadequate expression of gratitude, but I should be guilty of a gross omission if I did not mention the encouragement given me by Mr. D. G. Little, at whose suggestion this book was written.

REUBEN LEE

July, 1947

CONTENTS

LIST OF SYMBOLS

Page numbers are those on which the corresponding symbol first appears. A symbol formed from one of the tabulated letters, with a subscript or prime added, is not listed unless it is frequently and prominently used in the book. Sometimes the same symbol denotes more than one property; the meaning is then determined by the context. Units are given wherever symbols are used. Lower-case letters indicate instantaneous or varying electrical quantities, and upper-case letters indicate steady, effective, or scalar values.

a	Coil radius, 222
a	Coil winding height, 72
a	N_2/N_1, 142
A	Area, 163
A_c	Core area, 12
A_n	Ripple amplitude, 108
AWG	American Wire Gauge, 34
b	Winding traverse, 72
B	X_C/R_1 at frequency f_r, 146
B	Core flux density, 12
B_{ac}	Ac flux density, 91
B_{dc}	Dc flux density, 91
B_m, B_{max}	Maximum operating flux density, 91
B_r	Residual flux density, 27
B_s	Saturating flux density, 246
c	Insulation thickness, 72
c	Specific heat, 52

C	Capacitance, 58
C_1, C_p	Primary capacitance, 289, 142
C_2, C_s	Secondary capacitance, 289, 142
C_e	Effective capacitance, 164
C_g	Capacitance of winding to ground, 307
C_1	Capacitance per layer, 164
C_w	Capacitance across winding, 307
C_N	PFN capacitance, 327
cm/A	Current density, 47
d	Core tongue width, 34
d	Toroid diameter, 257
D	Core strip width, 47
D	Winding height, 34
D	X_C/R_2 at frequency f_r, 151
e	Voltage (instantaneous value), 7
e_i	Alternating input voltage, 134
e_o	Alternating output voltage, 134
E	Core build, 47
E	Emissivity, 52
E	Voltage (effective value), 8
E_a	Voltage at top of pulse, 291
E_{cc}	Dc power supply voltage, 178
E_{dc}	Dc voltage, 105
E_f	Fundamental voltage amplitude, 154
E_o	Output voltage, 173
E_{pk}	Peak value of alternating voltage, 105
E_1	Primary voltage, 9
E_1	Input voltage, 391
E_2	Secondary voltage, 9
E_2	Output voltage, 390
E_B	Applied dc voltage, 135
E_L	Secondary full-load voltage, 7
E_C	Collector voltage, 137
E_H	Harmonic distortion in the output voltage, 173
E_H'	Harmonic distortion with feedback, 173
E_R	Ripple distortion in the output voltage, 173
E_R'	Ripple distortion with feedback, 173
E_S	Secondary no-load voltage, 9
f	Frequency, 8
f_m	Midband frequency, 186
f_r	Resonance frequency, 146
f_c	Cutoff frequency, 182
$f(\ )$	Function of, 108
F	Core window height, 47

F	Factor, 223
G	Core window length, 47
G_x	Electric stress, 273
H	Magnetizing force, 12
H_c	Coercive force, 27
i	Current (instantaneous value), 12
$I, \lvert I\rvert, I_{rms}$	Current (effective value), 8, 18
$\bar{I}, I_{av}$	Average value of current, 18
I_{dc}	Direct component of current, 17
I_f	Fundamental current amplitude, 154
$\hat{I}, I_{pk}$	Peak value of current, 17
I_C	Collector current, 137
I_E	Loss component of exciting current, 12
I_G	Grid current (dc), 138
I_H	Harmonic current amplitude, 154
I_L	Load current, 9
I_M	Magnetizing current, 11
I_N	Exciting current, 11
I_B	Plate current (dc), 138
I_1	Primary current, 342
I_2	Secondary current, 342
j	$\sqrt{-1}$ (vector operator), 143
J	Low frequency permeability/pulse permeability, 328
k	Thermal conductivity, 52
k	Coefficient of coupling, 219
k	1/2 ratio of impedance/circuit resistance $= \sqrt{L/C}/2R$, 97
K	Constant, 76
K_c	Core space factor, 47
K_g	Gap loss constant, 187
K_w	Winding space factor, 47
l_c	Mean length of core (or magnetic path), 10
l_g	Air gap, 83
L	Inductance, 85
L_e	Open circuit inductance (OCL), 29
L_m	Mutual inductance, 219
L_s	Short circuit inductance, 72
m	Decrement, 98
m	Order of harmonic, 107
M	Modulation factor, 17
MT	Mean turn length, 40
n	Number (e.g., of dielectrics), 72
N	Turns, 7
N_L	Number of layers (of wire in coil), 164
N_1	Primary turns, 7

N_2	Secondary turns, 7
OCL	Open-circuit inductance, 100
p	Densitx, 29
p	Ratio of voltages (in autotransformer), 387
p	Rectifier ripple frequency/line frequency (number of phases), 107
P_a	Volt-amperes per pound, 29
P_c	Core loss, 29
P_A	Ripple amplitude/E_{dc} (in rectifier), 108
P_R	Ripple amplitude/E_{dc} (across load), 108
PFN	Pulse-forming network, 326
PRF	Pulse repetition frequency, 331
Q	$\omega L/R$ = coil reactance/coil ac resistance, 100
r	Radius, 34
r_e	Equivalent radius, 52
R	Resistance, 8
R_1	Source resistance, 143
R_2	Load resistance, 143
R_E, R_{sh}	Equivalent core loss (shunt) resistance, 10
R_L	Load resistance, 10
R_S	Equivalent secondary resistance, 61
R_{ser}	Equivalent series core loss resistance, 185
S	Secondary winding, 68
S	Core window width, 96
t	Time (independent variable), 7
t	Thickness of insulation, 163
T	Period of a wave, 18
T	$2\pi\sqrt{L_s C_2}$ (undamped period of oscillatory wave), 290
V	Commutation voltage, 112
V	Volume (of core), 85
VA	Volt-Amperes, 46
w	Core-stacking dimension, 34
W_e	Core loss, 76
W_g	Gap loss, 187
W_s	Copper loss, 76
X	Reactance, 8
X_C	Capacitive reactance = $1/(2\pi f C)$, 105
X_L	Inductive reactance = $2\pi f L$, 105
X_N	Open-circuit reactance = $2\pi f L_e$, 11
Z	Impedance, 8
Z_G	Source impedance, 135
Z_L	Load impedance, 135
Z_N	PFN impedance, 326
Z_0	Characteristic impedance, 141
Z_1	Series-arm impedance, 180
Z_2	Shunt-arm impedance, 180

GREEK SYMBOLS

α	Amplifier gain, 173	
α'	Amplifier gain with feedback, 173	
α	$\sqrt{C_g/C_w}$, 307	
α	Damping factor, 314	
β	Feedback constant, 298	
β	Natural angular frequency, 178	
δ	Small interval of time, 18	
Δ	Increment (e.g., of flux), 28	
Δ	Exciting current/load current, 295	
ϵ	Base of natural logarithms ($=2.718$), 52	
ϵ	Dielectric constant of insulation, 163	
ρ	Volume resistivity, 274	
η	Efficiency, 16	
θ	Temperature, 52	
θ	Phase angle, 112	
μ	Amplification factor, 135	
μ	Permeability, 27	
μ_0	Initial pdrmeability, 27	
μ_Δ	Incremental permeability, 28	
π	3.1416, 8	
ϕ	Phase angle, 384	
ϕ	Flux (varying), 8	
Φ_{max}	Peak value of flux, 8	
$\sum$	Summation (of a series of elements), 40	
τ	Pulse duration, 293	
ω	$2\pi f$ (angular frequency), 8	

1 INTRODUCTION

1.1 WHAT IS A TRANSFORMER?

In its most elementary form, a transformer consists of two coils wound of wire
and inductively coupled to each other. When alternating current at a given
frequency flows in either coil, an alternating voltage of the same frequency is
induced in the other coil. The value of this voltage depends on the degree of
coupling and the flux linkages in the two coils. The coil connected to a source of
alternating voltage is usually called the primary coil, and the voltage across this
coil is the primary voltage. Voltage induced in the secondary coil may be greater
than or less than the primary voltage, depending on the ratio of primary to
secondary turns. A transformer is termed a step-up or a step-down transformer
accordingly.

Most transformers have stationary iron cores, around which the primary and
secondary coils are placed. Because of the high permeability of iron, most of the
flux is confined to the core, and a greater degree of coupling between the coils is
thereby obtained. So tight is the coupling between the coils in some transformers
that the primary and secondary voltages bear almost exactly the same ratio to
each other as the turns in the respective coils or windings. Thus the *turns ratio* of
a transformer is a common index of its function in raising or lowering voltage.
This function makes the transformer an important adjunct of modern electrical
power systems. Raising the voltage makes possible the economical transmission
of power over long distances; lowering the voltage again makes this power
available in useful form. It is safe to say that, without transformers, modern
industry could not have reached its present state of development.

1.2 ELECTRONIC TRANSFORMERS

No exact line of demarcation can be drawn between power transformers and electronic transformers except that electronic transformers are usually smaller. The source of power on a 60-Hz network is extremely large and may be the combined generating capacity of half a continent. Power in electronic equipment is limited to the capabilities of the rectifying, amplifying, and other components of the system, of which even the largest is small compared to the electric generators in a power utility system.

Transformers are needed in electronic apparatus to provide the various levels of voltage required for proper system operation, to provide electrical isolation between parts of the system which operate at different potentials, to furnish high impedance to alternating current but low impedance to direct current, and to maintain or modify wave shape and frequency response at different potentials. The very concept of impedance, so characteristic of electronics, almost necessarily presupposes a means of changing from one impedance level to another; that means is most often a transformer.

Impedance levels in electronic equipment are usually much higher than those in power systems. For example, the connected load on an 11,000 V power line may easily total 1,000,000 kVA. Compare this with a large airborne radar transmitter operating at 90,000 V and drawing 135 kVA. The currents in the two cases are 90,000 and 1.50 A, respectively. For the power line the load impedance is 11,000/90,000, or slightly more than 0.1 Ω; for the transmitter it is 90,000/1.50, or about 60,000 Ω. Source impedances are approximately proportional to these load impedances. In low-power electronic circuits the source impedance often exceeds the load impedance, which further influences the performance of the transformer.

Weight and space are usually at a premium in electronic equipment; efficiency, safety, and reliability are of primary importance. Transformers can account for a significant percentage of a system's volume and weight and are also a prime element in its reliability.

These and other differences of application render many power transformers unsuitable for electronic circuit use. The design, construction, and testing of electronic transformers have become separate arts, directed toward the most effective use of materials for electronic applications.

1.3 NEW TECHNOLOGY

Equipment that utilizes electronic circuits is in a continual state of evolution and change; these changes bring about corresponding changes in transformers. As new or modified materials are introduced, as more sophisticated manufacturing, processing, and test equipment becomes available, and as new design and analytical aids are developed, changes in the electronic transformer naturally follow.

Probably the single factor that has had the most profound impact on electronic equipment and therefore on the electronic transformer has been the change from an electron-tube technology to a semiconductor technology. This has led in turn to the replacement of many analog circuit applications with digital circuits. The net result of these changes in electronic circuits has been the need for processed power at lower voltages, higher currents, and at higher frequencies than had been the case with electron-tube circuits. This aspect of transformer design is discussed in Chapters 12 and 13.

Some of the new materials, processes, and design technologies that have resulted in changes of some type in electronic transformers are listed below.

1. Magnetic core materials
   a. Magnetically annealed cobalt iron
   b. High-flux, grain-oriented silicon steel
   c. Amorphous magnetic materials
   d. Improved ferrite materials
2. Insulating materials
   a. Film insulations: Mylar, polyimide, polypropelene
   b. Sheet insulation: polyamide
   c. Plastic laminates: epoxy glass
   d. Gas dielectrics: SF_6
   e. Liquid dielectrics: silicone fluids, fluorinated hydrocarbons, polyester-modified silicone fluids
   f. Improved wire enamels: polyimide
   g. Improved thermoplastic embedding resins
3. Manufacturing and test equipment
   a. Microprocessor-controlled, multifunction winding equipment
   b. Programmable, multifunction, automatic test equipment
   c. Improved partial discharge detection equipment
   d. Microprocessor-controlled resin embedment equipment
4. Transformer design
   a. Digital computer design programs
   b. Programmable calculators
   c. Electrostatic and electromagnetic field plot programs
   d. Circuit analysis programs
   e. Interactive graphic design aids

Through the application of the foregoing, it has been possible to do the following:

1. Reduce the size and weight of audio and power transformers and inductors.

2. Reduce the size of high-voltage transformers and inductors.

3. Design more efficient transformers for nonsinusoidal wave shapes such as are encountered in inverters and in pulse, video, and sweep amplifiers.

4. Extend the operating frequency range of power transformers into the hundreds of kilohertz.

5. Optimize transformers for any desired parameter: weight, efficiency, cost, and so forth.

6. Permit high-volume manufacture at lower unit cost.

With the computational power available to designers through digital computer programs and multifunction calculators the question is often asked: Why not let the computer do all the work? A computer design program can be very useful in designing a transformer and, depending on the simplicity of the transformer and the sophistication of the computer program, satisfactory designs can be produced. However, as the transformer requirements become more demanding, the design program must become more complex and must be more interactive with the designer to produce optimized designs. Figure 1.1 is a flowchart of the design of a typical electronic transformer. It can be seen that there are several steps in the design process that may require a decision based on judgment, a process that is difficult to program.

Many of the transformer programs that are available are extensions of design curves or charts which show a relation between volts, turns, wire size, and power rating and can be helpful when designing the simpler transformers. However, the charts and design programs may not be universally applicable, for the following reasons:

1. *Regulation.* This property is rarely negligible in electronic circuits. It often requires care and thought to use the most advantageous winding arrangement in order to obtain the proper *IX* and *IR* voltage drops. Sometimes the size is dictated by such considerations.

2. *Frequency Range.* The low-frequency end of a wideband transformer operating range in a given circuit is determined by the transformer open-circuit inductance. The high-frequency end is governed by the leakage inductance and distributed capacitance. Juggling the various factors, such as core size, number of turns, interleaving, and insulation, in order to obtain the optimum design constitutes a technical problem too complex to solve on charts and would require an extremely complex computer program to produce an optimum design.

3. *Voltage.* It would be exceedingly difficult, if not impossible, to reduce to chart form the use of high voltages in the restricted space of a transformer. Circuit considerations are very important here, and the transformer designer must be thoroughly familiar with the function of the transformer to ensure reliable operation, low cost, and small size. However, there are field plot

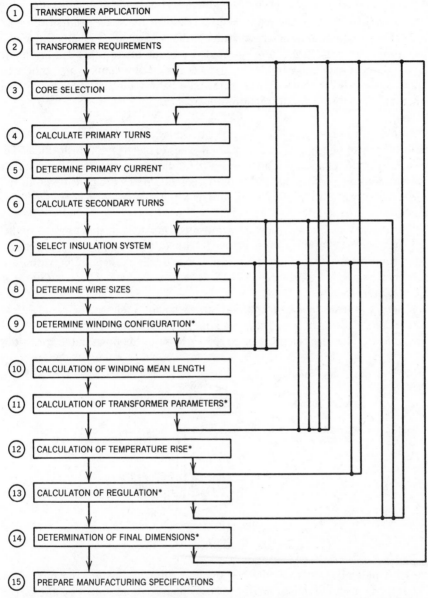

1. TRANSFORMER APPLICATION
2. TRANSFORMER REQUIREMENTS
3. CORE SELECTION
4. CALCULATE PRIMARY TURNS
5. DETERMINE PRIMARY CURRENT
6. CALCULATE SECONDARY TURNS
7. SELECT INSULATION SYSTEM
8. DETERMINE WIRE SIZES
9. DETERMINE WINDING CONFIGURATION*
10. CALCULATION OF WINDING MEAN LENGTH
11. CALCULATION OF TRANSFORMER PARAMETERS*
12. CALCULATION OF TEMPERATURE RISE*
13. CALCULATON OF REGULATION*
14. DETERMINATION OF FINAL DIMENSIONS*
15. PREPARE MANUFACTURING SPECIFICATIONS

*STEPS 9, 11, 12, 13 AND 14 ARE "DECISION" STEPS, WHERE THE DESIGN CONTINUES OR RETURNS TO A PREVIOUS STEP FOR CHANGE.

Figure 1.1 Power transformer design flowchart.

programs that will permit the designer to determine if a given design overstresses any of the insulating materials by providing plots of gradient and electric fields.

4. *Size.* Much electronic equipment is cramped for space, and since transformers often constitute the largest items in the equipment, it is imperative that they, too, be of small size. An open-minded attitude toward this condition and good judgment may make it possible to meet the requirements, which otherwise might not be fulfilled. New materials, too, can be instrumental in reducing size, sometimes down to a small fraction of former size.

In succeeding chapters the foregoing considerations will be applied to the performance and design of several general types of electronic transformers. The remainder of this chapter is a brief review of fundamental transformer principles. Only iron-core transformers with closed magnetic paths are considered in this introduction. Air-core transformers, with or without slugs of powdered iron, are discussed in Chapter 7.

While many electronic power transformers are now being used in higher frequency inverter applications, most still operate at power frequencies; the basic transformer theory is most easily explained by beginning with low-frequency principles. These principles will be modified for other conditions in later chapters.

A simple transformer coil and core arrangement is shown in Fig. 1.2. The primary and secondary coils are wound one over the other on an insulating coil tube or form. The core is laminated to reduce losses. Flux flows in the core along

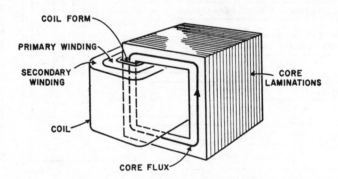

Figure 1.2. Transformer coil and core,

the path indicated, so that all the core flux threads through or links both windings. In a circuit diagram the transformer is represented by the circuit symbol of Fig. 1.3.

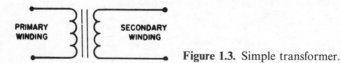

Figure 1.3. Simple transformer.

1.4 TRANSFORMER FUNDAMENTALS

The simple transformer of Fig. 1.3 has two windings. The left-hand winding is assumed to be connected to a voltage source and is called the primary winding. The right-hand winding is connected to a load and is called the secondary. The transformer merely delivers to the load a voltage similar to that impressed across its primary, except that it may be smaller or greater in amplitude.

For a transformer to perform this function, the voltage across it must vary with respect to time. A dc voltage such as that of a storage battery produces no voltage in the secondary winding or power in the load. If both varying and dc voltages are impressed across the primary, only the varying part is delivered to the load. This comes about because the voltage e in the secondary is induced in that winding by the core flux ϕ according to the law

$$e = -\frac{N\,d\phi}{dt} \times 10^{-8} \tag{1.1}$$

This law may be stated in words as follows: The voltage induced in a coil is proportional to the number of turns and to the time rate of change of magnetic flux in the coil. This rate of change of flux may be large or small. For a given voltage, if the rate of change of flux is small, many turns must be used. Conversely, if a small number of turns is used, a large rate of change of flux is necessary to produce a given voltage. The rate of change of flux can be made large in two ways, by increasing the maximum value of flux and by decreasing the period of time over which the flux change takes place. At low frequencies, the flux changes over a relatively large interval of time, and therefore a large number of turns is required for a given voltage, even though moderately large fluxes are used. As the frequency increases, the time interval between voltage changes is decreased, and for a given flux fewer turns are needed to produce a given voltage. And so it is that low-frequency transformers are characterized by the use of a large number of turns, whereas high-frequency transformers have but few turns.

If the flux ϕ did not vary with time, the induced voltage would be zero. Equation 1.1 is thus the fundamental transformer equation. The voltage variation with time may be of any kind: sinusoidal, exponential, sawtooth, or impulse. The essential condition for inducing a voltage in the secondary is that there be a flux *variation*. Only that part of the flux which links both coils induces a secondary voltage.

In equation 1.1, if ϕ denotes *maxwells* of flux and t time in seconds, e denotes *volts* induced.

If all the flux links both windings, equation 1.1 shows that equal volts per turn are induced in the primary and secondary, or

$$e_1/e_2 = N_1/N_2 \tag{1.2}$$

where e_1 = primary voltage N_1 = primary turns
$\qquad e_2$ = secondary voltage N_2 = secondary turns

1.5 SINUSOIDAL VOLTAGE

If the flux variation is sinusoidal,

$$\phi = \Phi_{max} \sin \omega t$$

where Φ_{max} is the peak value of flux, ω the angular frequency, and t the time. Equation 1.1 becomes

$$e = -N\Phi_{max}\omega \cos \omega t \times 10^{-8} \tag{1.3}$$

or the induced voltage also is sinusoidal. This voltage has an effective value

$$E = 0.707 \times 2\pi f N \Phi_{max} \times 10^{-8}$$
$$= 4.44 f N \Phi_{max} \times 10^{-8} \tag{1.4}$$

where f is the frequency of the sine wave. Equation 1.4 is the relation between voltage and flux for sinusoidal voltage.

Sufficient current is drawn by the primary winding to produce the flux required to maintain the winding voltage. The primary *induced* voltage in an unloaded transformer is just enough lower than the *impressed* voltage to allow this current to flow into the primary winding. If a load is connected across the secondary terminals, the primary induced voltage decreases further, to allow more current to flow into the winding in order that there may be a load current. Thus the primary of a loaded transformer carries both an exciting current and a load current, but only the load part is transformed into secondary load current.

Primary induced voltage would exactly equal primary impressed voltage if there were no resistance and reactance in the winding. Primary current flowing through the winding causes a voltage drop IR, the product of primary current I and winding resistance R. The winding also presents a reactance C, which causes an IX drop. Reactance X is caused by the *leakage* flux or flux which does not link both primary and secondary windings. There is at least a small percentage of the flux which is not common to both windings. Leakage flux flows in the air spaces adjacent to the windings. Because the primary turns link leakage flux an inductance is thereby introduced into the winding, producing leakage reactance X at the line frequency. The larger the primary current, the greater the leakage flux and the greater the reactance drop IX. Thus the leakage reactance drop is a series effect, proportional to primary current.

1.6 EQUIVALENT CIRCUIT AND VECTOR DIAGRAM

For purposes of analysis the transformer may be represented by a 1:1 turns ratio *equivalent circuit*. This circuit is based on the following assumptions:

1. Primary and secondary turns are equal in number. One winding is chosen as the *reference* winding; the other is the *referred* winding. The voltage in the referred winding is multiplied by the actual turns ratio after it is computed from the equivalent circuit. The choice between primary and secondary for the reference winding is a matter of convenience.
2. Core loss may be represented by a resistance across the terminals of the reference winding.
3. Core flux reactance may be represented by a reactance across the terminals of the reference winding.
4. Primary and secondary IR and IX voltage drops may be lumped together; the voltage drops in the referred winding are multiplied by a factor derived at the end of this section, to give them the correct equivalent value.
5. Equivalent reactances and resistances are linear.
6. Capacitances are neglected.

As will be shown later, some of these assumptions are approximate, and the analysis based on them is accurate only as far as the assumptions are justified. With proper attention to this fact, practical use can be made of the equivalent circuit.

With many sine-wave electronic transformers, the transformer load is resistive. A tube filament heating load, for example, has a 100% power factor. Under this condition the relations between voltages and currents become appreciably simplified in comparison with the same relations for reactive loads. In what follows, the secondary winding will be chosen as the reference winding. At low frequencies such a transformer may be represented by Fig. 1.4(*a*). The transformer equivalent circuit is approximated by Fig. 1.4(*b*), and its vector diagram for a 100% power factor load by Fig. 1.4(*c*). Secondary load voltage E_L and load current I_L are in phase. Secondary induced voltage E_S is greater than E_L because it must compensate for the winding resistances and leakage reactances. The winding resistance and leakage reactance voltage drops are shown in Fig. 1.4(*c*) as IR and IX, which are respectively in phase and in quadrature with I_L and E_L. These voltage drops are the sum of secondary and primary winding voltage drops, but the primary values are multiplied by a factor to be derived later. If voltage drops and losses are temporarily forgotten, the same power is delivered to the load as is taken from the line. Let subscripts 1 and 2 denote the respective primary and secondary quantities:

$$E_1 I_1 = E_2 I_2 \qquad (1.5)$$

or

$$E_1/E_2 = I_2/I_1 \qquad (1.6)$$

so that the voltages are inversely proportional to the currents. Also, from equation 1.2, they are directly proportional to their respective turns.

$$E_1/E_2 = N_1/N_2 \tag{1.2a}$$

Now the transformer may be replaced by an impedance Z_1 drawing the same current from the line, so that

$$I_1 = E_1/Z_1$$

Similarly,

$$I_2 = E_2/Z_2$$

where Z_2 is the secondary load impedance, in this case R_L. If these expressions for current are substituted in equation 1.6,

$$\frac{Z_1}{Z_2} = \left(\frac{E_1}{E_2}\right)^2 = \left(\frac{N_1}{N_2}\right)^2 \tag{1.7}$$

Equation 1.7 is strictly true only for negligible voltage drops and losses. It is approximately true for voltage drops up to about 10% of the winding voltage or for losses less than 20% of the power delivered, but it is not true when the voltage drops approach in value the winding voltage or when the losses constitute most of the primary load.

Not only does the load impedance bear the relation of equation 1.7 to the equivalent primary load impedance; the winding reactance and resistance may also be referred from one winding to the other by the same ratio. This can be seen if the secondary winding resistance and reactance are considered part of the load, across which the secondary induced voltage E_S appears. Thus the factor by

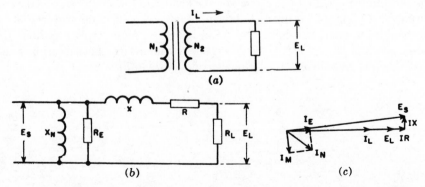

Figure 1.4. (*a*) Transformer with resistive load; (*b*) equivalent circuit; (*c*) vector diagram.

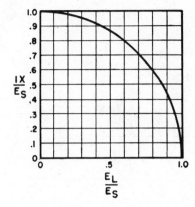

Figure 1.5. Relation between reactive voltage drop and load voltage.

which the primary reactance and resistance are multiplied, to refer them to the secondary for addition to the secondary drops, is $(N_2/N_1)^2$. If the primary had been the reference winding, the secondary reactance and resistance would have been multiplied by $(N_1/N_2)^2$.

In Fig. 1.4(c) the IR voltage drop subtracts directly from the terminal voltage across the resistive load, but the IX drop makes virtually no difference. How much the IX drop may be before it becomes appreciable is shown in Fig. 1.5. If the IX drop is 30% of the induced voltage, a 4% reduction in load voltage results; a 15% IX drop causes only a 1% reduction.

1.7 MAGNETIZING CURRENT

In addition to the current entering the primary because of the secondary load, there is the core exciting current I_N which flows in the primary whether the secondary load is connected or not. This current is drawn by the primary core reactance X_N and equivalent core-loss resistance R_E and is multiplied by N_1/N_2 when it is referred to the secondary side. It has two components: I_M, the magnetizing component, which flows 90° lagging behind induced voltage E_S; and I_E, the core-loss current, which is in phase with E_S. Ordinarily, this current is small and produces negligible voltage drop in the winding.

Core loss current is often divided into two components: eddy current and hysteresis. Eddy current loss is caused by current circulating in the core laminations. Hysteresis loss is the *power* required to magnetize the core first in one direction and then in the other on alternating half-cycles. Hysteresis loss and magnetization are intimately connected, as can be seen from Fig. 1.6. Here induced voltage e is plotted against time, and core flux ϕ lags e by 90°, in accordance with equation 1.3. This flux is also plotted against magnetizing current in the loop at the right. This loop has the same shape as the *B-H* loop for

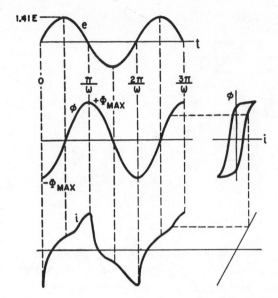

Figure 1.6. Transformer voltage, flux, and exciting current.

the grade of iron used in the core, but the scales are changed so that

$$\left.\begin{array}{r}\phi = BA_c \\ i = Hl_c/0.4\pi N\end{array}\right\} \tag{1.8}$$

where B = core flux density, G
$\quad A_c$ = core cross-sectional area, cm^2
$\quad H$ = core magnetizing force, Oe
$\quad l_c$ = core flux path length, cm

Current is projected from the ϕ-i loop to obtain the alternating current i at the bottom of Fig. 1.6. This current contains both the magnetizing and the hysteresis loss components of current. In core material research it is important to separate these components, for it is mainly through reduction of the B-H loop area (and hence hysteresis loss) that core materials have been improved. Techniques have been developed to separate the exciting current components, but it is evident that these components cannot be separated by current measurement only. It is nevertheless convenient for analysis of measurements to add the loss components and call their sum I_E, and to regard the magnetizing component I_M as a separate lagging current, as in Fig. 1.4. As long as the core reactance is large, the vector sum I_N of I_M and I_E is small, and the nonsinusoidal shape of I_N does not seriously affect the accuracy of Fig. 1.4.

Core flux reactance may be found by measuring the magnetizing current, that is, the current component which lags the applied voltage 90° with the secondary

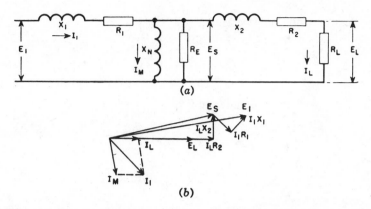

Figure 1.7. (*a*) Equivalent circuit and (*b*) vector diagram for transformer with high magnetizing current.

circuit open. Because of the method of measurement, this is often called the *open-circuit* reactance, and this reactance divided by the angular frequency is called the open-circuit inductance. The secondary and primary winding leakage reactances are found by short-circuiting the secondary winding and measuring the primary voltage with rated current flowing. The component of primary voltage which leads the current by 90° is divided by the current; this is the sum of the leakage reactances, the secondary reactance being multiplied by the (turns ratio)^2, and is called the *short-circuit* reactance.

Practical cases sometimes arise where the magnetizing component becomes of the same order of magnitude as I_L. Because current I_M flows only in the primary, a different equivalent circuit and vector diagram are necessary, as shown in Fig. 1.7. Note that the leakage reactance voltage drop has a marked effect on the load voltage, and this effect is larger as I_M increases relative to I_L. Therefore, the statement that IX voltage drop causes negligible difference between secondary induced and terminal voltages in transformers with resistive loads is true only for small values of exciting current. Moreover, the total primary current I_1 has a largely distorted shape, so that treating the IR and IX voltage drops as vectors is a rough approximation. For accurate calculation of load voltage with large core exciting current, a point-by-point analysis would be necessary.

1.8   FLUX AND AVERAGE VOLTAGE

If the variables are separated in equation 1.1, thus

$$e\,dt = -N \times 10^{-8}\,d\phi$$

an expression for flux may be found:

$$\int e \, dt = -N \times 10^{-8} \int d\phi$$

Now if we consider the time interval 0 to π/ω, we have

$$\int_0^{\pi/\omega} e \, dt = -N \times 10^{-8} \int_{-\Phi_{max}}^{+\Phi_{max}} d\phi$$
$$= -2N\Phi_{max} \times 10^{-8} \qquad (1.9)$$

Equation 1.9 gives the relation between maximum flux and the time integral of voltage. The left side of the equation is the area under the voltage-time wave. For a given frequency, it is proportional to the *average* voltage value. This is perfectly general and holds true regardless of waveform. If the voltage waveform is alternating, the average value of the time integral over a long period of time is zero. If the voltage waveform is sinusoidal, the flux waveform is also sinusoidal but is displaced 90° as in Fig. 1.6, and the integral over a half-cycle is

$$-1.41E\left[\frac{\cos \omega t}{\omega}\right]_0^{\pi/\omega} = \frac{2.82E}{\omega}$$

whence

$$\Phi_{max} = \frac{1.41 \times 10^8 E}{\omega N} \qquad (1.10)$$

Equation 1.10 is the relation between maximum flux, effective voltage, frequency, and turns. It is a transposed form of equation 1.4.

1.9   IDEAL TRANSFORMER

The use of equivalent circuits enables an engineer to calculate many transformer problems with comparative ease. It is always necessary to multiply properties in the referred winding by the proper ratio. This has led to the interposition of a transformer of the right turns ratio somewhere in the equivalent circuit, usually across the load. The transformer thus used must introduce no additional losses or voltage drops in the circuit. It is called an *ideal transformer* (see MIT, 1943), and it has negligibly small winding resistances, leakage flux, core loss, magnetizing current, and winding capacitances. Some power and audio transformers very nearly approach the ideal transformer at some frequencies. For example, in a typical 50-kVA plate transformer, the winding resistance IR drops a total of 1%

and the leakage reactance IX drops 3% of rated voltage, the core loss 0.6% of output power, and the magnetizing current 2% of rated primary current. When the term *ideal transformer* is used, it should be borne in mind that *negligibly small* is not *zero*. Particularly in electronic work, where frequency may vary, a limiting frequency may be reached at which the transformer is no longer ideal. Moreover, even if the limiting frequency is very low, it is never zero. There must be voltage variation if transformation is to take place. The assumptions of equations 1.5 to 1.7 were the same as for an ideal transformer.

1.10 POLARITY

Let turns from equation 1.2a be substituted in equation 1.5. Then we have

$$N_1 I_1 = N_2 I_2 \tag{1.11}$$

or the primary and secondary ampere-turns are equal and opposite. This equality holds for only the load component of I_1; that is, exciting current has been regarded as negligibly small. If there is a direct current in the load but not in the primary, or vice versa, equation 1.11 is true for only the ac components.

A 1:1 turns ratio transformer is shown diagrammatically in Fig. 1.8. The impressed voltage is E_1, and the primary current is I_1. Induced voltage E_i is slightly less than E_1 and is the same in magnitude and direction for both windings. The secondary current I_2 flows in the opposite direction to I_1. Instantaneous polarities are indicated by $+$ and $-$ signs. That is, when E_1 reaches a positive maximum, so do E_i and E_2. Dots are conventionally used to indicate terminals of the same polarity; dots in the circuit symbol at the right of Fig. 1.8 are used to indicate the same winding directions as in the left-hand figure.

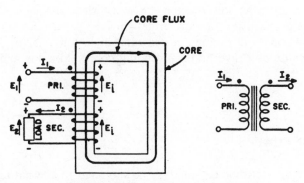

Figure 1.8. Transformer polarity.

1.11   REGULATION, EFFICIENCY, AND POWER FACTOR

Transformer regulation is the difference in the secondary terminal voltage at full load and at no load, expressed as a percentage of the full-load voltage. For the resistive load of Fig. 1.4,

$$\text{percent regulation} = 100\left(\frac{E_S - E_L}{E_L}\right) \tag{1.12}$$

Since with low values of leakage reactance $E_S - E_L = IR$,

$$\text{percent regulation} = 100IR/E_L \tag{1.13}$$

provided that R includes the primary winding resistance multiplied by the factor $(N_2/N_1)^2$ as well as the secondary winding resistance. If the leakage reactance is not negligibly small, approximately

$$\text{percent regulation} = 100\left[\frac{IR}{E_L} + \frac{1}{2}\left(\frac{IX}{E_L}\right)^2\right] \tag{1.14}$$

The efficiency is the ratio

$$\eta = \frac{\text{output power}}{\text{output power plus losses}} \tag{1.15}$$

where losses include both core and winding losses.

A convenient way of expressing the power factor is

$$\text{power factor} = \frac{\text{output power plus losses}}{\text{input volt-amperes}} \tag{1.16}$$

Equation 1.16 gives the power factor of a transformer plus its load.

One of the problems of transformer design is the proper choice of induction to obtain low values of exciting current and high power factor. Low power factor may cause excessive primary winding copper loss, low efficiency, and overheating.

1.12   WAVE SHAPES

Transformers in electronic circuits may be subjected to alternating and direct currents simultaneously, to modified sine waves, or to other nonsinusoidal waves. Although there is a relation between current and voltage wave shapes in a transformer, the two are frequently not the same, as already seen in Fig. 1.6. Dc components of primary voltage are not transformed; only the varying ac component is transformed. Secondary current may be determined by the connection of the load. For example, if the load is a rectifier, the current will be some form of rectified wave; if the load is a modulator, the secondary current may be the superposition of two waves. If the primary voltage is nonsinusoidal, the secondary current almost certainly will be nonsinusoidal.

If the primary voltage comes from an alternating source only and the load is a half-wave rectifier, the secondary current has a dc component, but the primary current has no dc component except under changing conditions. That is, in the steady state there is no primary dc component resulting from secondary dc component alone. This is true because any direct current in the primary requires a dc source. But by the initial assumption there is no direct current present in the primary. Under these conditions, the core flux may be very much distorted because the flux excursions go into saturation in one direction only.

In succeeding chapters, two values of current will be of interest in circuits with nonsinusoidal waves, the average and the rms. Average current causes core saturation unless there is an air gap. Rms current determines the heating of the windings and is limited by the permissible temperature rise. Voltage waveform will be dealt with in subsequent chapters. Common current waveforms are shown in Table 1.1.

TABLE 1.1   Nonsinusoidal Current Waveforms

Current Wave Shape	Description	I_{rms}	I_{av}
	Direct current with superposed sine wave	$I_{dc}\sqrt{1+\dfrac{M^2}{2}}$	I_{dc}
	Half-sine loops of T duration and f repetition rate	$I_{pk}\sqrt{\dfrac{fT}{2}}$	$\dfrac{2I_{pk}fT}{\pi}$
	Square waves of T duration and f repetition frequency	$I_{pk}\sqrt{fT}$	$I_{pk}fT$
	Sawtooth wave of T duration and f repetition frequency	$I_{pk}\sqrt{\dfrac{fT}{3}}$	$\dfrac{I_{pk}fT}{2}$
	Trapezoidal wave of f repetition frequency	$I_{pk}\sqrt{\dfrac{f(2\delta+3T)}{3}}$	$I_{pk}f(\delta+T)$

There are several different voltage waveforms which are used in the succeeding chapters. Figure 1.9 shows the most commonly used of these waveforms. The equations showing the voltage-flux-turns relationships are based on equation 1.10 with the constants corrected so that turns and flux density can be calculated directly using the normally state values of voltage, time and frequency.

SINE WAVE VOLTAGE

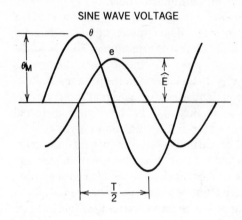

$$(1)\quad e' = -N\frac{d\phi}{dt} \times 10^{-8} = -N\frac{\Delta\phi}{\Delta t} \times 10^{-8}$$

$$(2)\quad e' = -N\frac{2\phi_m}{T/2} \times 10^{-8} = -N\frac{2\phi_m}{1/2f} \times 10^{-8}$$

(3) $e' = 4fN\theta m \times 10^{-8} = 4fNAcBm \times 10^{-8}$
Where B is in Maxwells per In^2
For B in Gauss

(4) $e' = 25.8fNAcBm \times 10^{-8}$
Where $e' = $ average induced volts

(5) $Erms = 28.6fNAcBm \times 10^{-8}$

$$(6)\quad N = \frac{E \times 10^8}{28.6fAcBm} = \frac{3.49E \times 10^6}{fAcBm}$$

SQUARE WAVE VOLTAGE

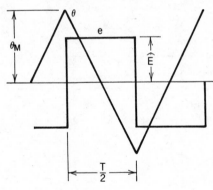

EQUATIONS (1) THRU (4) ABOVE

$$E(pk) = e'$$

(5) $E = 25.8fNAcBm \times 10^{-8}$

$$(6)\quad N = \frac{E \times 10^8}{25.8fAcBm} = \frac{3.88E \times 10^6}{fAcBm}$$

PULSE VOLTAGE

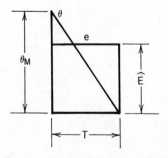

EQUATION (1) ABOVE

$$(2)\quad e' = -N\frac{\phi_m}{T} \times 10^{-8} = -N\frac{AcBm}{T} \times 10^{-8}$$

$E(pk) = e'$ and for B in Gauss

$$(3)\quad E = \frac{6.45NAcBm}{T} \times 10^{-8}$$

$$(4)\quad N = \frac{ET \times 10^8}{6.45AcBm}$$

Figure 1.9 Voltage-flux-turns relationship.

Root-mean-square (rms) current values are based on the equation

$$I_{rms} = \sqrt{f \int_0^T i^2 \, dt}$$ (1.17)

where i = current at any instant
$\quad f$ = frequency of repetition of current waves per second
$\quad T$ = duration of current waves, s
$\quad t$ = time, s

Average current values are

$$I_{av} = f \int_0^T i \, dt$$ (1.18)

In the first wave shape, $T = 1/f$. In the fifth wave shape, $T + 2\delta$ is the current wave duration.

In both equations 1.17 and 1.18, T refers to a full period. This is in contrast to steady-state sinusoidal alternating currents, the rms and average values of which are developed over a half-period because of the symmetry of such currents about the zero axis.

2 TRANSFORMER CONSTRUCTION, MATERIALS, AND RATINGS

2.1 CONFIGURATION

Transformers used in modern electronic circuits may be fabricated in a variety of configurations and from a variety of materials whose choice may be influenced by many factors: cost, circuit application, environment, efficiency, weight, volume, and others. Regardless of the configuration, all transformers consist of two or more electrically conducting coils coupled through magnetic induction. The more common transformer configurations are shown in Fig. 2.1, but many variations of these exist. Usually, the configuration used in a given application is not of particular importance insofar as the transformer is concerned; however, each may have characteristics that may lead to its use in a particular application rather than some other configuration. Table 2.1 lists some of the characteristics of those configurations shown in Fig. 2.1, together with applications for each.

2.2 TRANSFORMER CORES

The component that has the greatest effect on the size and weight of the transformer is the magnetic core. Figures 2.2, 2.3, and 2.4 show the shapes of various types of magnetic cores: laminations, tape-wound cores, and common cast or machined shapes, respectively. The laminations and tape-wound shapes are, respectively, punched from thin sheets of crystalline magnetic alloys or wound from narrow tape of the same material. It can be seen that these cores may be assembled to produce the configurations in Fig. 2.1.

20

TABLE 2.1   Transformer Configuration

Type	Characteristic	Uses
Simple	One coil loop, one core loop: large mean turn	Power transformers and inductors
Core	Two coil loops, one core loop: permits winding balance, lower winding capacitance, reduces stray magnetic fields	Power, wide-band, inverter, and pulse transformers; inductors
Shell	One coil loop, two core loops: smaller mean turn and better winding efficiency than simple	Power, wide-band, and inverter transformers; inductors
Three phase	E core has higher VA/weight ratio than three single-phase	Power transformers
Toroid	Low leakage reactance, dc may cause saturation, difficult to wind and to cool	Wide-band, power, pulse and inverter transformers; inductors
Cup core	Available only in ferrite and iron powder, fragile, should not be potted, low external magnetic field	Wide-band, power, pulse and inverter transformers; inductors
Solenoid	Can be ferromagnetic or air-core devices	IF and RF coils
Pie winding	Lowest capacitance	IF and RF coils

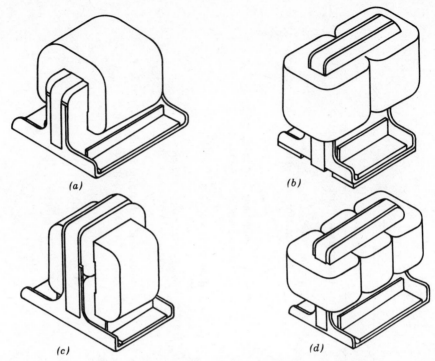

Figure 2.1. Typical transformer configurations. (*a*) Single phase, simple type (one core loop, one coil); (*b*) single phase, core type (one core loop, two coils), (*c*) single phase, shell type (two core loops, one coil); (*d*) three phase, core type (three core legs, three coils).

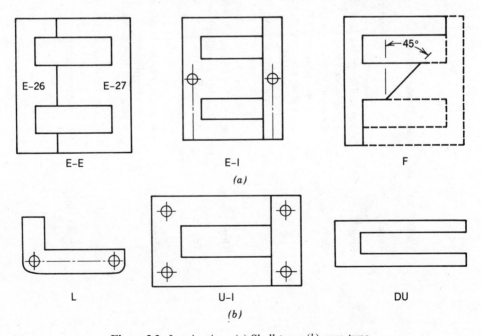

Figure 2.2. Laminations (*a*) Shell type; (*b*) core type.

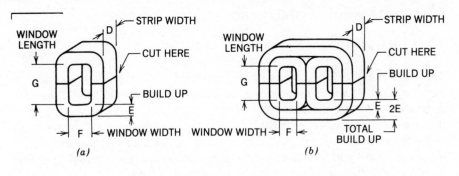

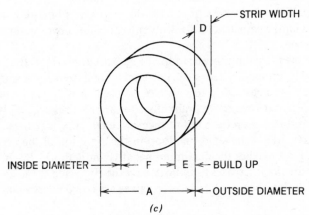

Figure 2.3. Tape-wound cores. (*a*) Cut "C" core; (*b*) cut "E" core; (*c*) toroidal core.

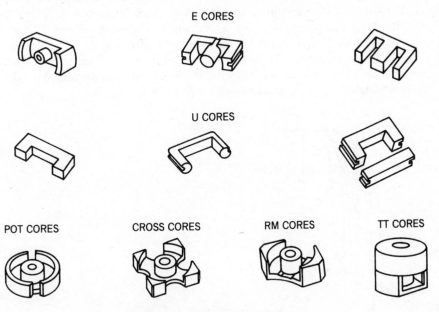

Figure 2.4. Ferrite core shapes.

2.3   CORE MATERIALS

There are many different types of core material available to the transformer designer, most of which have been available for many years. Table 2.2 lists the properties of several different types of magnetic materials. Most new magnetic materials are improvements in existing alloys or modifications in the processing of the materials which improve one or more of the magnetic properties. The most significant addition to magnetic material metallurgy in recent years has been the introduction of the amorphous magnetic alloys.

In this chapter the properties of several different types of core material are presented. To guard against possible ambiguity, definitions of magnetic terms are first reviewed.

Referring to the typical hysteresis loop of Fig. 2.5, curve OB_m is the manner in which completely unmagnetized steel becomes magnetized by a magnetizing force H gradually increasing up to value H_m. Flux density or induction is not proportional to H but rises more gradually as it approaches H_m, B_m. Once the material reaches this state, it does not retrace curve OB_m if H is reduced. Instead, it follows the left side of the solid-line loop in the direction of the arrow until, with negative H_m, it reaches the maximum negative induction $-B_m$. If H is now reversed, the induction increases as indicated by the right side of the loop, which is symmetrical in that the upper and lower halves are equal in area and have the same shape.

In laboratory tests of magnetic material, the changes in H are made slowly by means of a permeameter. The solid curve of Fig. 2.5 is then called the *dc hysteresis loop*. If the changes are made more rapidly, for example, at a 60-Hz rate, the loop is wider, as shown by the dashed lines. If a higher frequency is used, the loop becomes still wider, as shown by the dot-dash lines. At any frequency, energy is expended in changing induction from B_m to $-B_m$ and back to B_m; this energy is called the *hysteresis loss* and is proportional to the area of the B-H loop. Increase in loop width with frequency is usually attributed to eddy currents which flow, even in laminated cores, to some degree.

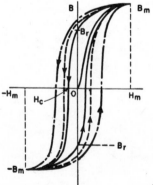

Figure 2.5. Ac and dc hysteresis loops.

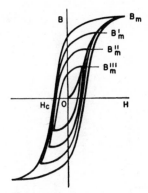

Figure 2.6. Normal induction.

If a closed magnetic core is magnetized to induction B_m and then the magnetizing force completely removed, induction decreases to *residual* induction B_r and remains at this value in the absence of magnetizing force, or for $H = 0$. The value of H required to reduce B to zero is called the *coercive force* (H_c). From Fig. 2.5 it is evident that B_r and H_c may change with frequency for the same B_m and grade of core material, and the design of transformers and inductors may be affected by the influence of frequency on core steel properties.

According to equation 1.10, the core flux is proportional to effective alternating voltage for a given frequency and number of turns, and so is the flux density in a given core. Therefore, the largest loop of Fig. 2.6 corresponds to a definite effective voltage and frequency, applied across a coil linking a definite core, and magnetizing it to maximum flux density B_m. If effective voltage is reduced 20% a smaller B-H loop results, with lower maximum flux density B'_m. If effective voltage is reduced further, still lower maximum flux density B''_m is reached. The locus of points B_m, B'_m, B''_m, and so on, is drawn in Fig. 2.6, and is called the *normal induction* curve. It is similar in shape to, but not identical with, the *virgin curve* OB_m of Fig. 2.5. Each time the maximum flux density is lowered, a short time elapses before the new loop is traced each cycle. Thus the loops of Fig. 2.6 represent symmetrical steady-state or cyclic magnetization at different levels of maximum induction.

A normal induction curve is drawn in Fig. 2.7. The ratio of B to H at any point on the curve is the *normal permeability* for that value of B. For the maximum flux density B_m, the normal permeability is

$$\mu = B_m/H_m \qquad (2.1)$$

It is the slope of a straight line drawn through the origin and B_m. A similar line drawn tangent to the curve at its "knee" is called the *maximum* permeability and is the ratio $\mu_m = B'/H'$. The slope B_0/H_0 of normal induction at the origin (enlarged in Fig. 2.7) is the permeability for very low induction B_0; it is called *initial* permeability and is usually much less than μ_m.

TABLE 2.2 Core Materials

Approximate Description	Trade Name	Typical Maximum Permeability	Maximum Operating Flux Density (kG)	Chief Uses
Silicon-iron				
Unoriented	AISI M19	8,500	15.0	Small-power and voice-frequency audio transformers; 50–4000 Hz
	AISI M15	8,500	15.0	
Oriented	Hipersil A	30,000	13.0	Larger sizes of power and audio transformers; saturable inductors; 50–12,000 Hz
	Hipersil Z	30,000	17.6	
50% Nickel-iron				
Round loop	Hipernik	50,000	12.0	Wide-band audio transformers and inductors; 100–20,000 Hz
Square loop	Deltamax	50,000	13.5	Saturable inductors and inverter transformers; to 50,000 Hz
80% Nickel-iron				
Round loop	Permalloy	100,000	6.5	Wide-band transformers, no dc current; 100–30,000 Hz
Square loop	Square Permalloy	20,000	7.0	Saturable inductors and inverter transformers; to 50,000 Hz
Special anneal	Supermalloy	300,000	7.0	Very wide-band and current transformers; 10–100,000 Hz

Cobalt-iron: magnetic anneal	Supermendur	50,000	21.0	Power transformers; 50–3500 Hz
Amorphous alloys	Metglas	[a]	[a]	Power and inverter transformers; 50 Hz–1.0 MHz
Powder cores				
MPP powder	Moly-Permalloy	14–550	2.0	Filter inductors, loading coils; 10 Hz–100 kHz
Iron powder	Polyiron	5–75	9.0	Coils; 2 kHz–300 MHz
Ferrites				
MnZn	Ferroxcube	7,000	3.5	Power transformers and inductors, inverter transformers; to 1 MHz
NiZn	Ferroxcube	3,000	1.5	High-frequency transformers; 20 kHz–20 MHz

[a]The properties of the amorphous cores have not been standardized to the extent of the crystalline alloys; therefore, neither maximum nor typical values can be given.

27

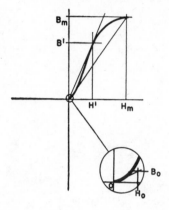

Figure 2.7. Normal permeabilities.

Maximum permeability as here defined is really the average slope of the normal induction curve up to induction B'. The actual slope from O to B' is greater at some points than maximum permeability because the curve is steepest below B'. The slope at any induction is called *differential* permeability.

From inspection of Fig. 2.6 it will be noticed that for $H = 0$, the sides of the B-H loop are steeper than any part of the normal induction curve and hence the slopes exceed μ_m. This fact has practical significance in the design of magnetic amplifiers.

In the foregoing, symmetrical magnetization has been assumed. If a core is magnetized with dc magnetizing force H_{dc} as in Fig. 2.8, and ac magnetization ΔH is superimposed, the cyclic magnetization follows a minor loop AB_m. Decreasing induction follows the left side of a major loop whose maximum induction is B_m, down to induction $A = B_m - \Delta B$. Increasing induction follows a line that joins the right side of the major loop. The area of this loop is small, but so is the average slope, or *incremental* permeability. This permeability is important in inductor design. It is defined by

$$\mu_\Delta = \Delta B/\Delta H \qquad (2.2)$$

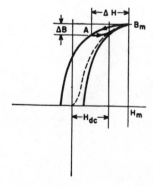

Figure 2.8. Incremental permeability.

and is generally smaller than μ_m. The dashed line in Fig. 2.8 is the normal induction curve, the locus of the tops of minor loops as H_{dc} is decreased.

Returning now to Fig. 2.6, if H_m is increased, an induction is finally reached at which unit increase of H produces only unit increase in B_m. This is known as *saturation induction* B_s. The value of H at which B_s is first reached is very large compared to H_c for most core materials. A striking development has been the production of core materials with *rectangular* hysteresis loops. In such materials B_s is reached at small values of H, as shown in Fig. 2.9. Core material having a rectangular hysteresis loop is especially useful in magnetic amplifiers and is discussed in Chapter 8.

The *volt-amperes per pound* or apparent core loss (P_a) of a magnetic material is the product of rms induced voltage and rms exciting current drawn from the source when a pound of the material is subjected to sinusoidally varying induction of a specified maximum value B_m and of a specified frequency f. Exciting current is nonsinusoidal, as can be seen from Fig. 1.6. The power component of P_a is the core loss P_c. The reactive component is usually the larger and is called VARs per pound. It is related to permeability in the following way: Let it be assumed that for conditions B_m, H_m in a core the magnetizing current is approximately sinusoidal, of effective value I_M, drawn from a supply of frequency f and effective voltage E. If we combine

1. Open-circuit inductance $L_e = E/2\pi f I_M$ $\qquad\qquad$ (2.3)

$$= B_m A_c N / \sqrt{2} I_M \times 10^8 \qquad\qquad (2.4)$$

2. Magnetizing force $H_m = 0.4\pi N I_M \sqrt{2}/l_c$ $\qquad\qquad$ (2.5)

3. VARs/lb $= EI_M p/A_c l_c$ $\qquad\qquad$ (2.6)

convert to inches, and put density $p = 0.27 \, \text{lb/in.}^3$, then

$$\mu = \frac{152 f B_m^2}{\text{VARs/lb} \times 10^8} \qquad\qquad (2.7)$$

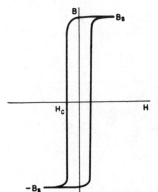

Figure 2.9. Rectangular hysteresis loop.

At 60 Hz, $\mu = B_m^2/11{,}000$ VARs/lb. Because of the nonlinearity of I_M, this equation is approximate. Moreover, there is no allowance for core gap.

In usual electronic transformer practice, it is necessary to avoid reaching saturation flux densities, because high exciting currents produce high winding IR drops, high losses, low efficiency, and large size. Curves of induction and core loss are available from manufacturers of laminations and tape-wound cores.

2.4 PROPERTIES OF IMPORTANT FERROMAGNETIC ELEMENTS

Of the three main ferromagnetic elements, cobalt, iron, and nickel, iron is the most important, although all three are rarely used in their pure form.

2.4.1 Iron-silicon Alloys

Silicon was first added to steel to make it harder; later it was discovered that it reduced the loss in transformer steel by a factor of 6. Even more important is the lack of magnetic aging in silicon steel. Silicon reduces both eddy current loss and hysteresis losses; the silicon-iron alloys are the least expensive magnetic steels available. A wide variety of silicon steel magnetic materials in the form of laminations and tape-wound core shapes are available to the electronic transformer designer: laminations in thicknesses of 0.006 to 0.025 in. and wound cut cores with tape thicknesses ranging from 0.001 to 0.012 in.

A significant improvement in the properties of silicon iron core material can be accomplished by "grain orienting" the core material. The orientation is accomplished by cold rolling the strip in the direction shown in Fig. 2.10. Magnified sections of laminations are shown in the figure; part (a) shows the random directions of "easy" magnetization in magnetic domains of nonoriented silicon steel. When magnetic flux is established in the lamination, the domains must be aligned in the same direction, as in Fig. 2.10(b). If the domains are already oriented in this direction during the rolling process, a much smaller magnetizing force is required to produce the desired flux. Coercive force and hysteresis loss are less than in non-oriented steel, and permeability is greater, as is B_r, so the rectangular loop of Fig. 2.9 is approached in grain-oriented steel. In

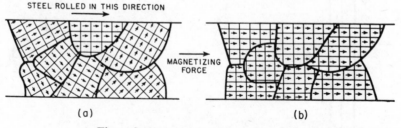

STEEL ROLLED IN THIS DIRECTION

MAGNETIZING FORCE

(a)   (b)

Figure 2.10. Oriented magnetic domains.

grain-oriented silicon-steel the maximum permeability is increased from about 12,000 to 52,000 and the practical limit of operating flux density is increased from 13 to more than 18 kG.

2.4.2 Iron-nickel Alloys

Alloys of 45 to 80% nickel are the most useful as ferromagnetic materials, with the most common alloys at 50 and 80%. The properties of these alloys are markedly affected by the method of reduction to the final lamination or strip thickness and subsequent heat treatment and cooling. Extremely square loop materials can be obtained with proper processing of the 50% nickel alloys, and special treatment will produce very high permeability near 78% nickel.

2.4.3 Iron-cobalt Alloys

Supermendur is a very pure alloy containing 2% vanadium, 49% iron, and 49% cobalt. Grain orientation is obtained by annealing the core in a magnetic field. Supermendur has the highest permeability above 15,000 G (100,000 at 20,000 G) and the highest residual induction (21,000 G) of any usable magnetic material known today. The hysteresis loop for this material is very rectangular with low coercive force (< 0.3 Oe).

2.4.4 Ferrites

Ferrites are nonmetallic ferromagnetic oxides whose electrical resistivities are normally 10^6 times that of the metallic magnetic alloys. Mechanically, ferrites resemble high-grade porcelain or ceramic materials.

   Because of their high resistivities, ferrite materials have relatively low specific core loss and therefore can be used at higher frequencies than can most of the metallic alloys. However, the operating frequency range is determined by the formulation of the material. Ferrites are limited by their low Curie temperatures and low saturating flux density, which decreases with increasing temperature and by low permeability. In many applications it is necessary to operate ferrite cores at less than 50% of room temperature B_{max} so that the core will not saturate at its operating temperature. However, even with these limitations ferrite cores are widely used in higher frequency transformers.

2.4.5 Powdered Cores

There are two main types of cores manufactured from powdered or pulverized magnetic materials: Molybdenum Permalloy powdered cores and the iron powder cores.

2.4.5.1 Molybdenum Permalloy Powder Cores. The outstanding feature of these cores is the stability of permeability over wide ranges of induction; they also have high resistivity and low hysteresis and eddy current losses. Standard permeabilities of these cores range from 14 to more than 500. Maximum flux densities are less than 5 kG. These cores are most often used in toroidal shapes.

2.4.5.2 *Iron Powder Cores.*

The maximum permeability of the iron powder cores is less than that of the Molybdenum Permalloy powder core, ranging from values of 3 up to 85. These cores are used from the power frequencies to above 250 MHz and are available in a wide variety of shapes.

2.4.6 Amorphous Alloys

A new type of magnetic alloy has been developed (ca. 1970–1980) which has significantly lower core losses than those for comparable crystalline alloys and because of this can be operated at higher frequencies. They can also be operated at much higher flux densities than can ferrite materials.

These alloys are produced by liquid quenching at a rate (10^6 deg/s) so rapid that the alloy passes through the crystallization temperature before crystallization can occur.

The amorphous magnetic alloys are normally composed of a ferromagnetic transition element and one or more metalloid elements at a ratio of about 80/20, respectively. Because of the metalloid elements the amorphous metals are lower in saturating flux density than are their crystalline counterparts.

The major advantages of the amorphous alloys are:

1. There is significantly lower loss at higher frequencies.
2. These alloys have higher operating flux densities than 80% Ni-Fe alloys.
3. The μ/f ratio is better than for Ni/Fe alloys.
4. The magnetic properties can be modified by the annealing process and by changing alloying elements.

The major disadvantages of amorphous alloys are

1. The material is extremely strain sensitive.
2. The material after anneal is very brittle.
3. Cut cores are not available; this requires special coil winding techniques if either uncut rectangular cores or toroidal cores are used.

2.5 WINDINGS

The size of the winding conductors can be influenced by regulation and efficiency requirements, by temperature rise, and by the method of winding. However, the effect of wire size on regulation, efficiency, and temperature rise can only be determined after the basic transformer design is completed.

Regulation and efficiency are calculated as in Section 1.11 and temperature rise as in Sections 2.8 and 2.9.

Often the wire size is determined from a set current density, 500 circular mils per ampere (cm/A) as an example. These are useful guides in picking out a first choice of wire size for a given current requirement but should not be regarded as final. Such an approach could result in excessive regulation and I^2R loss in high-VA transformers and an inefficient utilization of conductor size in lower-VA transformers. In Fig. 2.11 the current density in circular mils per ampere is plotted for several types of transformer. It can be seen that for low power, enclosed, dry-type transformers with Hipersil C cores and a winding temperature rise of 55°C the current density varies appreciably. In 400 Hz, low power, single-phase transformers, up to 500 VA, current densities may range from 350 to 900 cm/A.

A more accurate method of determining wire size in a power transformer is to apportion a part of the winding area in the core window to each winding based on that winding's percentage of the total transformer VA. The total VA is the load VA plus the VA supplied by the primary, which includes all transformer losses.

The winding area is the gross window area minus the space occupied by insulation, margins, the coil form, and clearance. When the area available for each winding is determined, the turns per square inch for that winding can be calculated and the wire size picked from an appropriate winding chart such as Table 2.5.

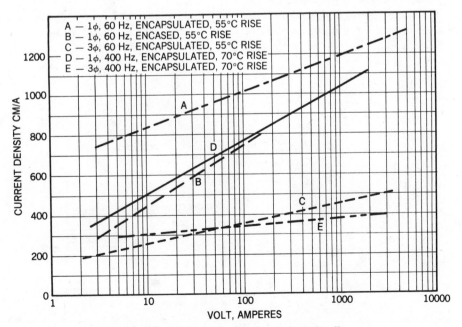

Figure 2.11. Current density, layer wound coils.

The space occupied by the wire depends on the wire insulation as well as on the wire section. This is especially noticeable in small wire sizes. Table 2.3 gives the bare and insulated dimensions for single-build, film-coated copper wire, and Table 2.4 gives the turns per square inch of winding space for several common types of insulated magnet wire. The minimum insulation increase on heavy-build magnet wire is about twice that of single-build wire and that for triple-build magnet wire is about three times that of single-build wire.

Present-day manufacturers still provide magnet wire insulated with some of the natural nonconductive fibers, cotton, glass, and paper as well as polyester fibers. The insulated diameters of fiber-served magnet wires are usually greater than those of film-insulated wires of the same AWG size.

The more recent advances in magnet wire insulation are the result of the application of organic polymer materials such as polyurethanes, epoxy, nylon, acrylics, modified polyesters, polytetrafluoroethylene, and polyimide as insulating films. Some of these insulations are supplied in bondable and solderable forms and some are used as coatings with fiber-served wires.

Winding space usually can be saved by avoiding the fibrous insulated wires and using enameled wire with sheet layer and winding insulations such as kraft or aramid paper, polyester film, or polyimide film. Figure 2.12 shows a paper-insulated coil. The thickness of layer insulation may be governed by the layer voltage. However, in coils where layer voltages are low, the thickness is determined by the mechanical strength necessary to produce even layers and still produce a tightly wound coil. Table 2.5 gives the minimum paper thickness based on this consideration.

"Space factor" may refer to linear spacing across a layer or to the total coil section area. It is more convenient to use as linear space factor in designing layer-wound coils and an area space factor in random-wound coils. The values

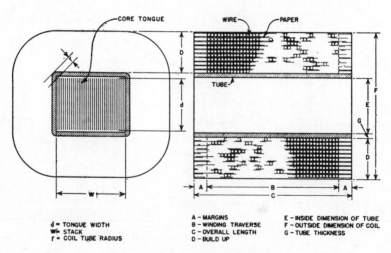

d = TONGUE WIDTH
w = STACK
r = COIL TUBE RADIUS

A – MARGINS
B – WINDING TRAVERSE
C – OVERALL LENGTH
D – BUILD UP

E – INSIDE DIMENSION OF TUBE
F – OUTSIDE DIMENSION OF COIL
G – TUBE THICKNESS

Figure 2.12. Paper-insulated coil.

in each case depend largely on the method of winding. For example, it is possible to wind AWG No. 30 enameled wire with a 97% linear space factor by hand, but with only an 89% factor on an automatic multiple-coil winding machine. Moreover, values of space factor may vary from winding machine to winding machine and from operator to operator. An average for multiple winding machines is given in Table 2.5.

The mean length of turn must be calculated for a coil in order to find its resistance in ohms. This may be found by referring to the side view of Fig. 2.12. Note that there is a small clearance space between core and coil form or tube. Let d be the core tongue and w the stack. Suppose that there are several concentric windings. The length of mean turn of a winding V at distance r from the core and having height D is

$$MT = 2w + 2d + 2\pi(r + D/2)$$

$$= 2(w + d) + \pi(2\sum D + D) \qquad (2.8)$$

where $\sum D$ is the sum of all winding heights and insulation thicknesses between winding V and the core.

The mean turn of the winding U just below V ordinarily is calculated before that of winding V. This fact simplifies the calculation of winding V, the mean turn of which is

$$MT_V = MT_U + \pi(D_U + D_V + 2c) \qquad (2.9)$$

where c is the thickness of insulation between U and V. Allowance must be made, with many coil leads, for bulging of the coil at the ends and consequent increase of mean turn length.

The placement, insulation, and soldering of leads constitute perhaps the most important steps in the manufacture of a coil. When coils are wound one at a time, the leads can be placed in the coil while it is being wound. The start lead may be placed on the coil form, suitable insulation may be placed over it, and coil turns may be wound over the insulation. Tap leads can be arranged in the same way. Finish leads must be anchored by means of tape, string, or yarn, because there are no turns of wire to wind over them. Typical lead anchoring is shown in Fig. 2.13.

In multiple-wound coils, the leads must be attached after the coils are wound. Extra wire on the start turn is pulled out of the coil and run up the side as shown in Fig. 2.14, with separator insulation between wire extension and coil. Outer insulation covers the wire extension up to the lead joint. A pad of insulation is placed under the joint, and one or more layers of insulation, which insulate and anchor the joint, are wound over the entire coil and the lead insulation. Electrical-grade adhesive tape is widely used for anchoring leads. It is important to avoid corrosive adhesives.

Leads should be large enough to introduce only a small amount of voltage

TABLE 2.3 Single-Build Enameled Copper Magnet Wire

AWG size	Bare wire diameter, nom. (in.)	Minimum insulation increase (in.)	Diameter (in.)			Weight		Resistance	
			Min.	Nom.	Max.	lb/1000 ft	ft/lb	Ω/1000 ft	Ω/lb
14	0.0641	0.0016	0.0651	0.0659	0.0666	12.51	79.94	2.524	0.2016
15	0.0571	0.0015	0.0580	0.0587	0.0594	9.962	100.4	3.181	0.3193
16	0.0508	0.0014	0.0517	0.0524	0.0531	7.888	126.8	4.018	0.5101
17	0.0453	0.0014	0.0462	0.0469	0.0475	6.275	159.4	5.054	0.8050
18	0.0403	0.0013	0.0412	0.0418	0.0424	4.974	201.1	6.386	1.285
19	0.0359	0.0012	0.0367	0.0373	0.0379	3.949	253.2	8.046	2.038
20	0.0320	0.0012	0.0329	0.0334	0.0339	3.141	318.4	10.13	3.225
21	0.0285	0.0011	0.0293	0.0298	0.0303	2.496	400.6	12.77	5.116
22	0.0253	0.0011	0.0261	0.0266	0.0270	1.972	507.1	16.20	8.215
23	0.0226	0.0010	0.0234	0.0239	0.0243	1.576	634.5	20.30	12.88
24	0.0201	0.0010	0.0209	0.0213	0.0217	1.242	805.2	25.67	20.67
25	0.0179	0.0009	0.0186	0.0190	0.0194	0.9896	1,010	32.37	32.69
26	0.0159	0.0009	0.0166	0.0170	0.0173	0.7810	1,280	41.02	52.74
27	0.0142	0.0008	0.0149	0.0153	0.0156	0.6244	1,603	51.44	82.46
28	0.0126	0.0008	0.0133	0.0137	0.0140	0.4922	2,028	65.31	132.5
29	0.0113	0.0007	0.0119	0.0123	0.0126	0.3971	2,519	81.21	204.6
30	0.0100	0.0007	0.0106	0.0109	0.0112	0.3108	3,215	103.7	333.4
31	0.0089	0.0006	0.0094	0.0097	0.0100	0.2464	4,049	130.9	530.0
32	0.0080	0.0006	0.0085	0.0088	0.0091	0.1997	5,000	162.0	810.0
33	0.0071	0.0005	0.0075	0.0078	0.0081	0.1576	6,345	205.7	1,305
34	0.0063	0.0005	0.0067	0.0070	0.0072	0.1241	8,058	261.3	2,106
35	0.0056	0.0004	0.0059	0.0062	0.0064	0.0980	10,204	330.7	3,375
36	0.0050	0.0004	0.0053	0.0056	0.0058	0.0785	12,739	414.8	5,284

37	0.0045	0.0003	0.0047	0.0050	0.0052	0.0634	15,773	512.1	8,077
38	0.0040	0.0003	0.0042	0.0045	0.0047	0.0503	19,881	648.2	12,887
39	0.0035	0.0002	0.0036	0.0039	0.0041	0.0384	26,042	846.6	22,047
40	0.0031	0.0002	0.0032	0.0035	0.0037	0.0303	33,003	1,079	35,610
41	0.0028	0.0002	0.0029	0.0031	0.0033	0.02448	40,850	1,323	54,045
42	0.0025	0.0002	0.0026	0.0028	0.0030	0.01960	51,020	1,659	84,642
43	0.0022	0.0002	0.0023	0.0025	0.0026	0.01532	65,274	2,143	139,882
44	0.0020	0.0001	0.0020	0.0022	0.0024	0.01247	80,192	2,593	207,938

TABLE 2.4 Turns per Square Inch of Insulated Wire[a]

AWG Size	Single Enamel Wire	Double Enamel Wire	Triple Enamel Wire	Single Cotton Enamel Wire	Single Glass-Enamel Wire	Double Glass-Enamel Wire	Single Polyester Glass Wire	Double Polyester Glass Wire
8		56						51
9		70						64
10		88						81
11		111						100
12		139						124
13		174						153
14	225	215	204	198	190	181	195	186
15	283	269	254	244	233	223	241	230
16	355	336	316	303	286	273	267	282
17	443	419	393	368	349	333	362	345
18	556	523	489	455	426	404	445	421
19	696	654	606	555	518	491	543	514
20	870	811	754	675	628	592	660	621
21	1,090	1,010	940	846	758	714	802	754
22	1,370	1,260	1,160	1,045	918	860	1,096	912
23	1,695	1,560	1,430	1,260	1,089	1,020	1,164	1,089
24	2,125	1,940	1,760	1,530	1,514	1,420	1,402	1,301
25	2,655	2,420	2,180	1,820	1,826	1,693	1,679	1,562
26	3,340	3,010	2,680	2,270	2,047	2,029	2,010	1,857
27	4,110	3,710	3,340	2,720	2,603	2,402	2,356	2,183
28	5,100	4,620	4,110	3,280	3,086	2,859	2,770	2,576
29	6,300	5,650	4,960	3,810	3,629	3,341	3,228	2,986

AWG								
30	7,970	7,060	6,100	4,580	4,328	3,955	3,810	3,501
31	10,000	8,570		5,330				
32	12,000	10,400		6,210				
33	15,200	12,900		7,280				
34	19,200	16,400		8,430				
35	24,400	20,400		9,600				
36	29,700	25,100		11,100				
37	36,900	30,700						
38	45,200	38,400						
39	59,400	49,300						
40	73,000	62,500						
41	91,800	77,100						
42	111,100	97,600						
43	147,900	118,900						
44	173,600	137,100						

[a]Wires per square inch based on maximum insulated diameter.

39

TABLE 2.5   Layer-Wound Coil Data

AWG Size[a]	Turns per inch	Turns per square inch	Layer Insulation (in.)	Thickness of Layer (in.)	Space Factor (%)
10	8	65	0.010	0.1290	90
11	9	82	0.010	0.1164	90
12	10	102	0.010	0.1052	90
13	12	126	0.010	0.0952	90
14	13	157	0.010	0.0851	90
15	15	194	0.010	0.0771	90
16	17	242	0.010	0.0701	90
17	19	297	0.010	0.0639	90
18	21	396	0.005	0.0527	90
19	24	509	0.005	0.0471	91
20	27	632	0.005	0.0427	91
21	30	781	0.005	0.0384	92
22	34	1,043	0.003	0.0326	92
23	38	1,279	0.003	0.0297	92
24	43	1,616	0.003	0.0266	93
25	47	2,017	0.002	0.0233	92
26	53	2,524	0.002	0.0210	92
27	59	3,089	0.002	0.0191	92
28	66	3,793	0.002	0.0174	92
29	73	4,771	0.0015	0.0153	92
30	82	5,857	0.0015	0.0140	91
31	92	7,301	0.0015	0.0126	91
32	101	8,707	0.0015	0.0116	91
33	113	10,561	0.0015	0.0107	90
34	127	13,789	0.001	0.00921	89
35	144	17,122	0.001	0.00841	88
36	158	20,440	0.001	0.00773	88
37	177	25,106	0.001	0.00705	88
38	196	29,924	0.001	0.00655	87
39	224	40,215	0.00075	0.00557	87
40	246	47,582	0.00075	0.00517	86
41	278	58,403	0.00075	0.00476	85
42	307	74,514	0.0005	0.00412	85
43	354	96,986	0.0005	0.00365	85
44	383	112,317	0.0005	0.00341	85

[a]AWG 10–13; heavy-build enamel; AWG 14–44, single-build enamel.

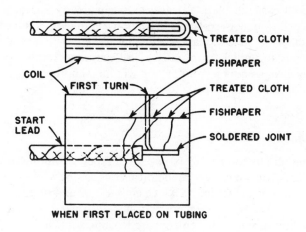

TREATED CLOTH

FISHPAPER

COIL

FIRST TURN

TREATED CLOTH

FISHPAPER

START LEAD

SOLDERED JOINT

WHEN FIRST PLACED ON TUBING

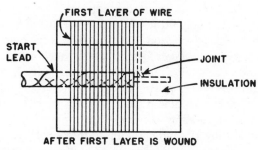

FIRST LAYER OF WIRE

START LEAD

JOINT

INSULATION

AFTER FIRST LAYER IS WOUND

Figure 2.13. Start-lead insulation in hand-wound coils.

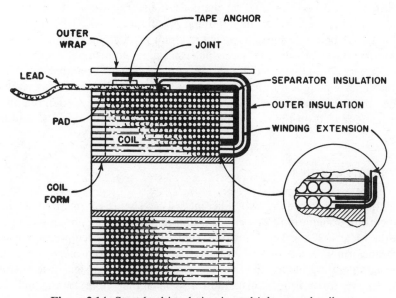

TAPE ANCHOR

OUTER WRAP

JOINT

LEAD

SEPARATOR INSULATION

OUTER INSULATION

PAD

WINDING EXTENSION

COIL

COIL FORM

Figure 2.14. Start-lead insulation in multiple-wound coils.

41

drop and should have insulation clearances adequate for the test voltage. These clearances can be found as explained in Section 9.9. In high-voltage transformers it would often be possible to seal the windings if there were no leads; hence lead placement calls for much care and skill. Leads and joints should also be mechanically strong enough to withstand winding, impregnating, and handling stresses without breakage.

2.5.1 Winding Machines

The preceding discussion considers two basic types of winding machinery, the multiple machine on which several identical coils can be wound simultaneously and the hand winding machine on which coils are wound singly. Over the years many improvements have been incorporated into winding machinery; some increase winding speed, some control wire tension, some improve the accuracy of the winding and simplify the operator's job, and some machines are so automatic that there is virtually no need for an operator. Regardless of all of these improvements, the basic function of the machine is the placement of turns of magnet wire on some type of coil form in a controlled manner.

Most modern winding machines are microprocessor controlled; these may be programmed to wind precisely very complex coils with repeatable results from coil to coil. These machines eliminate much of the variability in coil winding which results from hand winding and from operator to operator. It is also possible to wind "perfect layer" windings without the use of layer insulation (i.e., multilayer windings without any turn crossovers). Some of the machines will also attach leads, place winding insulation, apply the finish wrap of insulation, mark leads, and perform tests on the coils. Generally, the more sophisticated machines are geared to very high volume production, in the 100,000's of units, and may not be economically applied to small production runs.

Another type of winding for which special equipment is required is for transformers and inductors designed on toroidal cores. These are wound singly on machines having a shuttle that is passed through the toroid. Magnet wire is loaded onto the shuttle, which is then rotated through the toroid, one turn of wire being placed on the core with each complete revolution of the shuttle.

2.6 INSULATION

Insulation is classified in two different ways. The IEEE method assigns a temperature designation to types of insulating materials, as shown in Table 2.6, and the MIL-T-27 method, shown in Table 2.7, assigns a temperature class to the transformer based on the hottest spot in the transformer to provide a specified operating life for the transformer; in a sense it assigns a temperature class to the transformer's insulation system rather than to the individual materials that make up the system. Regardless of whether a transformer is

TABLE 2.6   IEEE Insulation Classifications

Class	Maximum Temperature	Insulating Materials[a]
O	90	Unimpregnated cotton, silk, paper, etc.
A	105	Impregnated cotton, silk, paper, etc.
B	130	Mica, glass fibers, asbestos, etc., with suitable bonding substances
F	155	Mica, glass fibers, asbestos, etc., with suitable bonding substances
H	180	Silicone elastomer, mica, glass fibers, asbestos, etc., with suitable bonding substances such as silicone resins
C	220	Materials that have been shown to have required life at 220°C
>C	>220	Mica, quartz, porcelain, glass, and similar inorganic materials

[a]Other materials or combinations may be included in each class if by experience or tests they can be shown to have comparable thermal life at the temperature of the class.
Source: IEEE, *Standard Dictionary of Electrical and Electronic Terms*, IEEE Std 100-1977.

TABLE 2.7   MIL-T-27 Temperature Classes

Symbol	Maximum operating temperature (°C)
Q — — — — — — —	85
R — — — — — — —	105
S — — — — — — —	130
V — — — — — — —	155
T — — — — — — —	170
U — — — — — — —	>170, as specified

Source: MIL-T-27, *Transformers and Inductors (Audio, Power, and High Power Pulse), General Specification for.*

manufactured for a military or a commercial application it is usually designed to operate with the lowest practicable temperature rise.

In general, the vital difference between these classes of insulation is one of operating temperature. Class B materials or class S transformers are more expensive than class A insulations or class R transformers and are used only when other advantages outweigh the cost.

Life test data are plotted in Fig. 2.15 for class A and class B insulation. The temperature scale is based on Dakin's (1948) data, showing that insulation life is proportional to the reciprocal of absolute temperature. The two lines indicate

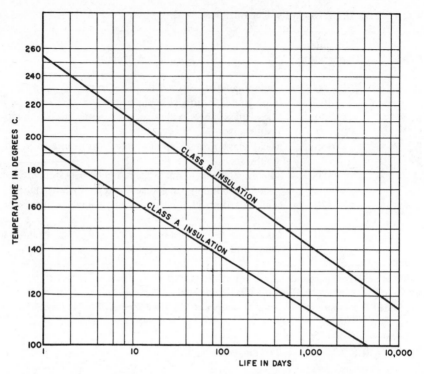

Figure 2.15. Approximate life expectancy of electrical insulation.

how operating temperature may be increased for a given life when class B insulation is used. Equal life is obtained when class A insulation is operated at 105°C maximum (40°C ambient, 55°C rise, 10°C hottest spot gradient), and when class B insulation is operated at 130°C maximum (40°C ambient, 80°C rise, 10°C hottest spot gradient). Intermittent load temperatures may be high for short periods. These periods are additive. For example, class A insulation has approximately the same life whether it is operated at 115°C continuously or half the time at 123°C and half the time at 25°C. Figure 2.15 shows only the influence of temperature on insulation life. Life is further reduced by moisture, vibration, on-off operating cycles, and corona. It is therefore important that insulation be protected against damage caused by all these factors, although there can be very little control over the on-off operation of the transformer.

The need for small size and light weight in aerospace applications or in mobile apparatus is continually increasing the tendency to use materials at their fullest capabilities. As size decreases, the ability of a transformer to radiate a given number of watts loss also decreases. Hence it operates at a higher temperature. Transformers for 400-Hz power supplies can be made in smaller overall dimensions by using class B insulation rather than class A. This results in a 30 to 50% decrease in size. The higher temperature classes of insulation are very important in aerospace applications.

The high-temperature classes, above 170°C, are rarely operated at their maximum thermal rating because other factors usually limit the transformer hot-spot temperature to something well below these temperatures. Transformer regulation and changes in the magnetic properties of the core material are often factors that limit the maximum operating temperature.

2.6.1 Dielectric Properties of Insulation

The dielectric properties of insulation which are of importance in a transformer are its dielectric strength, dielectric constant, loss tangent, and volume resistivity. These, together with other properties, are discussed at greater length in Chapter 9.

2.7 IMPREGNATION

After a coil is wound the best practice is to impregnate it with a material in liquid form which solidifies, either by the removal of a solvent or by a chemical reaction. This is done for several reasons. First, it protects the magnet wire from movement and possible mechanical damage. Second, it prevents the entrance of moisture and foreign matter which might cause wire corrosion or insulation deterioration. Third, it increases the dielectric strength of fibrous and porous sheet insulation. Fourth, it helps the flow of heat from the coil to its surface. Single-layer coils are often dipped in the liquid, drained and dried, but thicker coils must be evacuated to remove the air and admit the liquid to the interior. The best mechanical result is obtained when the coils are assembled with the cores and mounting bracket before the impregnation treatment.

Coils having little or no temperature rise in normal use can be impregnated with a chemically neutral mineral wax. The wax is melted in a sealed tank and drawn into another tank in which preheated coils have been placed and a vacuum is maintained. Coils are then removed from the tank, drained, and allowed to cool. Wax treatment provides good dielectric properties and moisture protection. It is a quick, simple process.

Transformers that operate at temperatures of 65°C or higher are normally impregnated with a varnish or a resin of some type. For many years oleoresinous varnishes diluted with a solvent to lower the viscosity had been used. When the coils are baked the varnish polymerizes and the solvent is driven off. The process of impregnation involves drying the coils, drawing a vacuum, and very close control of the curing temperature. Most varnishes now used to impregnate coils are synthetic products, primarily polyesters or epoxies which are diluted with a solvent. The impregnation process is essentially the same as for the oleoresinous varnishes; however, the curing often involves two or three baking temperatures to optimize the electrical and mechanical properties of the varnish.

Most varnishes are diluted with a solvent and as a consequence leave voids in the coil when the varnish polymerizes and the solvent is driven off. If a void-free coil is required, a solventless resin must be used. These materials change from a liquid to a solid state, usually by heat polymerization. There are some resins that are polymerized by ultraviolet radiation and some by other activating media. However, heat is the main method of curing the resin systems used as transformer coil impregnants. The first solventless resin systems were polyesters such as the Fosterite process, which was introduced in the early 1940s. Many resin systems are now in use: polyurethanes, epoxies, silicones, and modified polyesters, to name a few. Epoxies or modified epoxy systems are probably the most widely used.

The development of resin systems continues, with much effort being directed toward resins whose properties are tailored to very unique applications.

2.8   SIZE VERSUS RATING

The core cross-sectional area depends on voltage, induction, frequency, and the number of turns. For a given frequency and grade of core material, core area depends on the applied voltage. The window area depends on the coil size or the number of turns wound into the window. Since the window area and the core area determine the core size, there is a relation between the transformer size and its VA rating.

Usually, the first thing the transformer designer must do is to select the core to be used. Many factors can influence the selection of the core and, by implication, the core material; these factors can include operating frequency, ambient temperature, size limitations, efficiency, acoustic noise, cost, and even the use of standardized cores.

Often the best selection is a scaled size based on the designer's experience with similar transformers. Table 2.10 lists some theoretical relations involving transformer characteristics which are based on similitude in design. Those relating VA to core dimensions follow directly from equation 2.10. The effect of insulation, cooling technique, voltage, and flux density is to modify these factors in actual practice. In addition these factors may be modified appreciably when applied beyond two orders of magnitude of VA.

In applying these relationships it is assumed that related core dimensions are changed in the same proportion, that is, if the core width (D) is increased 50% the core build (E) is also increased 50%. Hence VA would be increased by 1.5^2 or 2.25 times. To determine the change in weight of a transformer resulting from an increase in VA from 500 to 2000, a fourfold increase. The weight would increase by $4.0^{3/4}$ or 2.83 times.

Whatever method is used, the core size for a given VA rating is based on the cross-sectional area of the core and the area of the core window. The following equation relates core cross-sectional area and core window area to transformer VA.

$$VA = \frac{36.5 \times 10^{-2} \times B \times f \times D \times E \times F \times G \times K_w \times K_c}{CM/A} \qquad (2.10)$$

where B = maximum flux density, G

f = frequency, Hz

D = core strip width or lamination stack height, in.

E = core build, in.

F = core window height, in.

G = core window length, in.

CM/A = current density, cmils/A

K_w = window space factor

K_c = core space factor

This equation assumes that all of the core window can be filled with conductor and that the core cross-sectional area is all magnetic material, neither of which is the case. Modifications must be made which reflect the type of transformer, type of winding, insulation, margins, clearances, and the various space factors that are involved.

The current density is taken from Fig. 2.11, and the core space and winding factors are taken from Tables 2.8 and 2.9, respectively. The winding space factors

TABLE 2.8   Core Space Factors

Core Type	Strip Thickness	Space Factor (%)
Tape-wound cut core	0.0005 in.	80
	0.001 in.	83
	0.002 in.	89
	0.004 in.	90
	0.012 in.	95
Laminations	0.004 in., butt stack	90
	0.004 in., 1 × 1	80
	0.006 in., butt stack	90
	0.006 in., 1 × 1	85
	0.014 in., butt stack	95
	0.014 in., 1 × 1	90
	0.0185 in., butt stack	95
	0.0185 in., 1 × 1	90
Tape-wound toroid	0.000125 in.	25
	0.00025 in.	37.5
	0.0005 in.	50
	0.001 in.	75
	0.002 in.	85
	0.004–0.009 in.	90
	0.010–0.014 in.	95

TABLE 2.9   Winding Space Factors

Winding Type	Space Factor (%)
C core	
Simple configuration	20.5
Core configuration	20
Shell configuration	21
Three-phase E core	9
Cup core	24
E-E laminations, single-phase	21

TABLE 2.10   Theoretical Relations

If the core proportions are maintained reasonably closely the following relations give the theoretical variations of various transformer factors.

VA varies as the frequency, f
VA varies as the area of the core, D × E
VA varies as the area of the core window, F × G
VA varies as the transformer core dimensions, D × E × F × G
Dimensions vary as the $VA^{1/4}$
Weight varies as the $VA^{3/4}$
Volume varies as the $VA^{3/4}$
Cost varies as the $VA^{3/4}$
Total losses vary as the $VA^{3/4}$
Volts per turn vary as the $VA^{1/2}$
Area of the core varies as the reciprocal of frequency, 1/f

in Table 2.8 are based on a single secondary transformer; for each additional secondary, reduce these factors by 1%. If the transformer is conductively cooled, increase the calculated VA rating by 20%. There are many other factors in the design of a transformer that could affect the practical VA rating of a core; these must be considered by the designer.

With other factors, such as operating frequency and core material, unchanged, the larger, higher VA transformers dissipate less heat per unit volume than do the lower VA units. This is true because the dissipation area increases as the square of the equivalent spherical radius, whereas the volume increases as its cube. Therefore, larger units are more commonly of the open type, whereas smaller units are often totally enclosed. Where enclosure is feasible, it tends to cause size to increase by limiting heat dissipation. Figure 2.16 shows the relation between size and rating for small, enclosed, low-voltage, two-winding, 60-Hz transformers having grain-oriented silicon-iron and class A insulation and operating continuously in a 40°C ambient. Figure 2.17 shows the same relationship for larger, open, low-voltage, two-winding, 60-Hz transformers

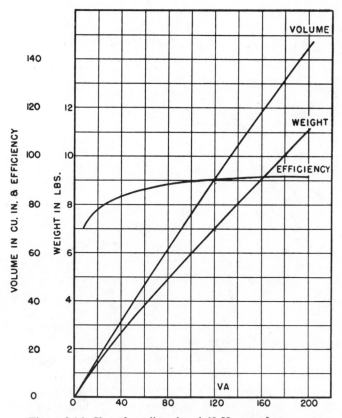

Figure 2.16. Size of small enclosed 60-Hz transformers.

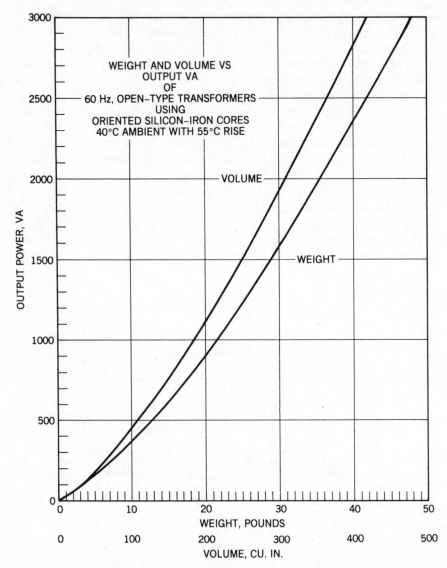

Figure 2.17. Size of 60-Hz open-type transformers.

having grain-oriented silicon-iron cores and which are impregnated with class A insulation and operate continuously in a 40°C ambient.

As noted above, open-type transformers have better heat dissipation than enclosed units. The lamination stacking dimension can be made to suit the rating, so that one size of lamination may cover a range of VA ratings. Heat dissipation from the end cases is independent of stacking dimension, but that from the laminations is directly proportional to this dimension. This is shown in

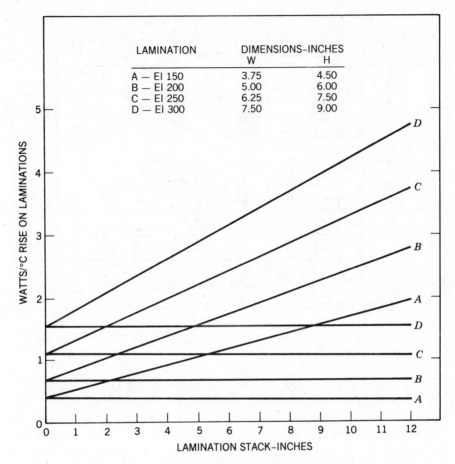

Figure 2.18. Heat dissipation from open-type transformers with end cases.

Fig. 2.18 for several lamination sizes. For each size the horizontal line represents heat dissipation from the end cases; the sloping line represents dissipation from the end cases plus that from the lamination edges. At ordinary working temperatures heat is dissipated at the rate of 0.008 W/in.^{2} per degree Celsius rise. In Fig. 2.18 the watts per degree Celsius of temperature rise are given as a function of lamination stack height. This refers to temperature rise at the core surface only. In addition, there is a temperature gradient between the core and the coil which is given in a similar manner in Fig. 2.19.

To find the average coil temperature rise, divide the winding loss by the watts per degree Celsius from the sloping line of Fig. 2.19. To this add the total of winding and core losses divided by the appropriate ordinate from Fig. 2.18. That is, the total coil temperature rise is equal to the sum of the temperature drop

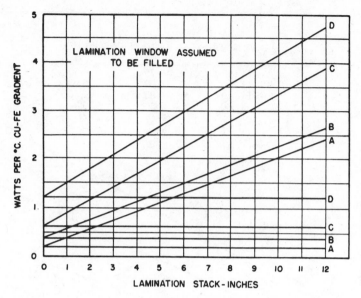

Figure 2.19. Winding-to-core gradient for open-type transformers.

across the insulation (marked "Cu-Fe gradient" in Fig. 2.19) and the temperature drop from the core to the ambient air. Data like those in Figs. 2.18 and 2.19 can be established for any lamination by making a heat run on two transformers, one having a core stack near the minimum and one near the maximum that is likely to be used. Usually, stacking dimensions lie between the extremes of one-half to three times the lamination tongue width; poor use of space results from stacking outside these limits. If the end cases are omitted, coil dissipation may be improved by as much as 50%.

The same method can be used for figuring tape-wound cut C core designs; here the strip width takes the place of the lamination stack height and the core buildup corresponds to the lamination tongue width. When a shell configuration is formed by using two C cores the temperature rise can be approximated by using data for the lamination whose dimensions are closest to those of the two C cores.

For irregular or unknown heat dissipation surfaces, an approximation of the temperature rise can be found from the transformer weight as derived in the next section. More accurate determination of the temperature rise in a transformer can be had by digital computer programs designed specifically for this purpose. However, the transformer designer must be able to determine the temperature rise in the transformer reasonably closely when it is being designed; the more-time-consuming heat-flow programs can be used to confirm the accuracy of the design.

2.9   INTERMITTENT RATINGS

It often happens that electronic equipment is operated for repeated short lengths of time, between which the power is off. In such cases the average power determines the heating and size. Transformers operating intermittently can be built smaller than if they were operated continuously at full rating.

Intermittent operation affects size only if the "on" periods are short compared to the thermal time constant of the transformer; that is, small transformers have less heat storage capacity and hence rise to final temperature more quickly than do large ones. It is important, therefore, to know the relation between size and thermal time constant, or the time that would be required to bring a transformer to 63% of the temperature to which it would finally rise if the power were applied continuously.

The exact determination of temperature rise time in objects such as transformers, having irregular shapes and non-homogeneous materials, has not yet been attempted. Even in simple shapes of homogeneous material, and after further simplifying assumptions have been made, the solution is too complicated (see Ingersoll and Zobel, 1913, p. 142) for rapid calculation. However, under certain conditions, a spherical object can be shown to cool according to the simple law (Ingersoll and Zobel, 1913, p. 143)

$$\theta = \theta_0 \epsilon^{-3Et/pcr} \tag{2.11}$$

where θ = temperature above ambient at any instant t
θ_0 = initial temperature above ambient
E = emissivity, cal/s per degree Celsius per square centimeter
p = density of material
c = specific heat of material
r = radius of sphere
ϵ = 2.718

The conditions involved in this formula are that the sphere is so small or the cooling so slow that the temperature at any time is sensibly uniform throughout the whole volume. Mathematically, this is fulfilled when the expression Er/k (where k is the thermal conductivity of the material) is small compared to unity. Knowing the various properties of the transformer material, we can tell (1) whether the required conditions are met, and (2) what the thermal time constant is. The latter is arrived at by the relation

$$t_c = pcr_e/3E \tag{2.12}$$

where r_e is the radius of the equivalent sphere.

To convert the nonhomogeneous transformer into a homogeneous sphere the average product of density and specific heat pc is found. Figures on widely different transformers show a variation from 0.862 to 0.879 in this product;

hence an average value of 0.87 can be taken, with only 1% deviation in any individual case.

Since the densities of iron and copper do not differ greatly, and insulation brings the coil density closer to that of iron, it may be further assumed that the transformer has material of uniform density 7.8 throughout. The equivalent spherical radius can then be found from

$$r_e = (\text{weight}/1.073)^{1/3} \qquad (2.13)$$

where r_e is in inches and weight is in pounds. The time constant is plotted from equations 2.12 and 2.13 in terms of weight in Fig. 2.20.

The condition that Er/k be small compared to unity is approximated by assuming that k is the conductivity of iron—a safe assumption, because the conductivity of copper is 7 to 10 times that of iron. A transformer weighing as much as 60 lb has $r_e = 5.45$ in., $E = 0.00028$ cal/s per square centimeter per degree Celsius, and $k = 0.11$. Changing r_e to metric units gives $Er/k = (0.00028 \times 5.45 \times 2.54)/0.11 = 0.34$, which is small enough to meet the necessary condition of equation 2.11.

It will be noticed that equation 2.11 is a law for cooling, not temperature rise. But if the source of heat is steady (as it nearly is) the equation can be inverted to the form $\theta_0 - \theta$ for temperature rise, and θ_0 becomes the final temperature.

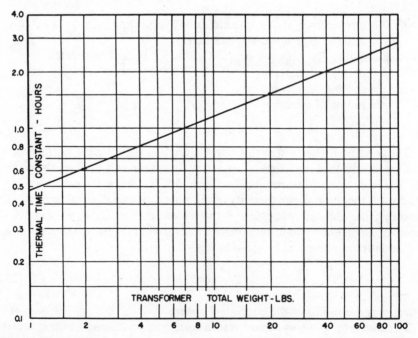

Figure 2.20. Transformer time constant, or time required to reach 63% of the final temperature.

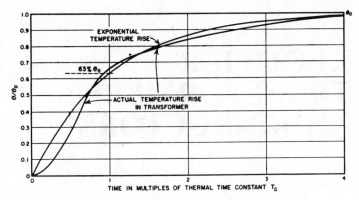

Figure 2.21. Transformer temperature rise time.

Temperature rise of a typical transformer is shown in Fig. 2.21, together with the exponential law which is $\theta_0 - \theta$, where θ is the temperature of equation 2.11. The actual rise is less at first than that of the foregoing simplified theory, then more rapid, and with a more pronounced "knee." The 63% of final temperature is reached in about 70% of the theoretical time constant t_c for transformers weighing between 5 and 200 lb. This average correction factor is included in Fig. 2.20 also.

If a transformer is operated for a short time and then allowed to cool to room temperature before operating again, the temperature rise can be found from Figs. 2.20 and 2.21. As an example, suppose that the continuously operated final coil temperature rise is 100°C, the total weight is 5 lb, and the operating duty is infrequent periods of 2 h. From Fig. 2.20, the transformer has a thermal time constant of 0.85 h. This corresponds to $t_c = 1$ in Fig. 2.21. Two hours are therefore $2 \div 0.85 = 2.35$ times t_c, and the transformer rises to 90% of final temperature, or a coil temperature rise of 90°C in 2 h.

If, on the other hand, the transformer has regular off and on intervals, the average watts dissipated over a long period of time govern the temperature rise. A transformer is never so small that it heats up more in the first operating interval than at the end of many intervals.

From equation 2.13 can be found a relation between weight, losses, and final temperature rise. For, since heat is dissipated at 0.008 W/in.^{2} per degree Celsius rise, and the area A_S of the equivalent sphere is $4\pi r_e^2$,

$$\theta_0 = \frac{\text{total watts loss}}{0.008 A_S} = \frac{\text{total watts loss}}{0.1(\text{total weight in pounds}/1.073)^{2/3}} \qquad (2.14)$$

where θ_0 is the final temperature rise in degrees Celsius. This equation is subject to the same approximations as equation 2.11; test results show that it is most reliable for transformers weighing 20 lb or more, with 55°C temperature rise at 40°C ambient.

3 RECTIFIER TRANSFORMERS AND INDUCTORS

Rectifiers are used to convert alternating current into direct current. While semiconductor rectifiers have replaced electron tubes in many applications, a comparison and discussion of both types are included. Most rectifiers have two electrodes, the cathode and the anode. Current flows only when the anode is positive with respect to the cathode: the forward resistance is low when the applied voltage has this polarity. When the anode is negative with respect to the cathode it exhibits high resistance and current does not flow. The reverse resistance of electron tube rectifiers is essentially infinite, while that of semiconductors is finite, about 1000 times the forward resistance. A high-vacuum rectifier tube characteristic voltage-current curve is shown in Fig. 3.1.

Semiconductor rectifiers are usually much smaller than tube rectifiers and require no filament power; however, in high-power applications the semiconductors must be mounted on heat sinks to limit the rise in junction temperature. Figure 3.2 is a typical characteristic curve of instantaneous forward current versus instantaneous forward voltage. The curve may vary depending on the type of rectifying element: silicon, germanium, or selenium.

In this chapter the rectifier circuits are first summarized and then the rectifier transformers and inductors are discussed.

Chapter 3.1. High-vacuum rectifier voltage-current curve.

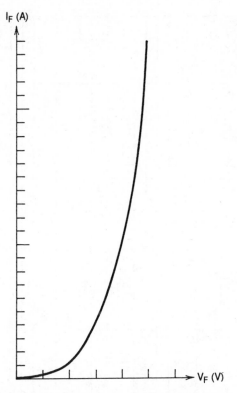

I_F (A)

V_F (V)

Figure 3.2. Typical semiconductor rectifier voltage-current curve.

3.1   RECTIFIERS WITH INDUCTOR INPUT FILTERS

Table 3.1 shows the more commonly used rectifier circuits, together with current and voltage relations in the associated transformers. This table is based on the use of an inductor input filter to reduce ripple. The inductance of the inductor is assumed to be great enough to keep the output direct-current constant. With any finite inductance there is always some superposed ripple current which is neglected in the table and which is considered further in Chapter 4.

Unbalanced direct current in the half-wave rectifiers requires larger transformers than in the full-wave rectifiers. This is partly overcome in three-phase transformers by using the zigzag connection; it can also be overcome by using an E type of three-phase transformer core. The transformer in the three-phase full-wave rectifier can be delta connected in both primary and secondary if desired; the secondary current is multiplied by 0.577 and the secondary voltage by 1.732. For a given load current a delta-connected winding has more turns of smaller wire than does a wye-connected winding. Single-phase bridge and three-phase full-wave rectifiers require notably low ac voltage for a given dc output, and low inverse voltage on the rectifiers and small transformers.

TABLE 3.1

TYPE	SINGLE PHASE HALF WAVE	SINGLE PHASE FULL WAVE	SINGLE PHASE BRIDGE CIRCUIT	3 PHASE HALF WAVE	3 PHASE HALF WAVE ZIG ZAG	DOUBLE 3 PHASE WITH BALANCE COIL ‡	3 PHASE FULL WAVE (SEC. MAY BE Δ)	3 PHASE FULL WAVE ZIG ZAG *	SIX PHASE HALF WAVE
CIRCUITS									
RECTIFIER PHASES AND NUMBER OF TUBES	1	2	2-4	3	3	6	6	6	6
PHASES OF A-C SUPPLY	1	1	1	3	3	3	3	3	3
SECONDARY VOLT PER LEG	2.22	1.11 (HALF SECT.)	1.11 (WHOLE)	0.855	0.985 (HALF LEG .493)	0.855	0.428	0.493 HALF LEG .247	0.74
PRIMARY VOLTAGE	2.22	1.11	1.11	0.855	0.855	0.855	0.428	0.428	0.74
SECONDARY CURRENT PER LEG	1.57	0.707	1.00	0.577	0.577	0.289	0.816	0.816	0.408
PRIMARY CURRENT PER LEG	1.21	1.000	1.00	0.471	0.408	0.408	0.816	0.816	0.577
SECONDARY K V A	3.48	1.57	1.11	1.48	1.71	1.48	1.05	1.21	1.81
PRIMARY K V A	2.68	1.11	1.11	1.21	1.05	1.05	1.05	1.05	1.28
AVERAGE OF PRIMARY AND SECONDARY K V A ∮	3.08	1.34	1.11	1.35	1.38	1.26	1.05	1.13	1.55
INVERSE PEAK VOLTAGE	3.14	3.14	1.57	2.09	2.09	2.09	1.05	1.05	2.09 —
RMS CURRENT PER TUBE	1.57	0.707	0.707	0.577	0.577	0.289	0.577	0.577	0.408
PEAK CURRENT PER TUBE	3.14	1.00	1.00	1.00	1.00	0.500	1.00	1.00	1.00
AVERAGE CURRENT PER TUBE	1.00	0.50	0.50	0.33	0.33	0.167	0.33	0.33	0.167
RIPPLE FREQUENCY	f	2f	2f	3f	3f	6f	6f	6f	6f
RMS RIPPLE VOLTAGE	1.11	0.472	0.472	0.177	0.177	0.04	0.04	0.04	0.042
RIPPLE PEAKS	+2.14 / -1.00	+0.57 / -1.00	+0.57 / -1.00	+0.209 / -0.291	+0.209 / -0.291	+0.057 / -0.077	+0.057 / -0.077	+0.057 / -0.077	+0.057 / -0.077
LINE POWER FACTOR	0.373	0.90	0.90	0.826	0.955	0.955	0.955	0.955	0.955

NOTE: THE VALUES OF VOLTAGE AND CURRENT ARE EFFECTIVE OR R.M.S. UNLESS OTHERWISE STATED; THEY ARE GIVEN IN TERMS OF THE AVERAGE DC VALUES, AND THE KILOVOLT-AMPERES IN TERMS OF DC. KILOWATT OUTPUT. PERFECT TRANSFORMERS, RECTIFIERS AND DC CHOKE ARE ASSUMED, AND $N_1/N_2 = 1$, EXCEPT IN ZIG-ZAG CIRCUITS. SECONDARY AND PRIMARY REFER TO ANODE TRANSFORMER.
‡ FREQUENCY OF BALANCE COIL VOLTAGE 3f. BALANCE COIL VOLTAGE 0.356. PEAK BALANCE COIL VOLTAGE 0.605. BALANCE COIL K V A 0.173.
* USE COLUMN 5 VALUES FOR HALF VOLTAGE TAP.   ∮ MAGNETIZING CURRENT ASSUMED NEGLIGIBLE IN ALL CASES.

Source: Based mostly on R. W. Armstrong, "Polyphase Rectification Special Connections," *Proc. IRE, 19* (January 1931).

3.2 RECTIFIERS WITH CAPACITOR INPUT FILTERS

When the filter has no inductor intervening between the rectifier and the first capacitor, the rectifier current is not continuous throughout each cycle and the rectified waveform changes. During the voltage peaks of each cycle the capacitor charges and draws current from the rectifier. During the rest of the time no current is drawn from the rectifier and the capacitor discharges into the load.

Comparison between the rectified voltage of inductor input and capacitor input filters in a single-phase, full-wave rectifier may be seen in Fig. 3.3(a) and (b), respectively. The two rectifier currents, I_1 and I_2 in part (a), add to a constant dc

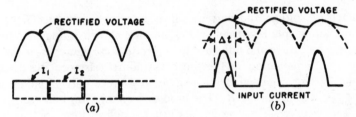

Figure 3.3. Voltage and current comparisons in inductor-input and capacitor-input circuits.

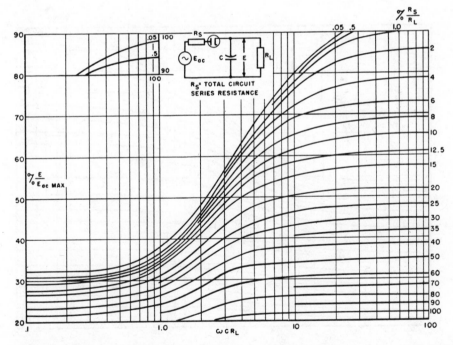

Figure 3.4. Relation of peak sine voltage to dc voltage in half-wave capacitor-input circuits.

output, whereas in part (b) the high-peaked rectifier currents flow only while the rectified voltage is higher than the average dc voltage. Average current per diode in both cases is half the rectifier output. With large values of capacitance the rectified voltage in Fig. 3.3(b) increases to within a few percent of the peak voltage. Ripple, average rectified voltage output, and rectifier current are dependent on the capacitance, the supply line frequency, and the load resistance. They are also dependent on rectifier internal resistance because it affects the peak value of current which the filter capacitor can draw during the charging interval, Δt. Analysis of this charge-discharge action involves complicated Fourier series which typically require lengthy solutions (see Waidelich, 1941); however, such analyses certainly can be programmed for solution on a digital computer.

Satisfactory voltage and current values have been obtained from experimental measurements by Schade (1943) and are shown in Figs. 3.4 to 3.6 for single-phase half-wave and full-wave rectifiers. In these figures R_S is the rectifier series resistance, including the transformer winding resistance. Results accurate to within 5% are obtained if the rectifier resistance corresponding to peak current $\hat{I}_P$ is used in finding R_S. The process is cut-and-try because $\hat{I}_P$ depends on R_S, and vice versa, but two trials usually suffice. Resistance is in ohms, capacitance in farads, and ω is 2π times the supply frequency.

In Fig. 3.6 the peak current indicates whether the current rating of a given

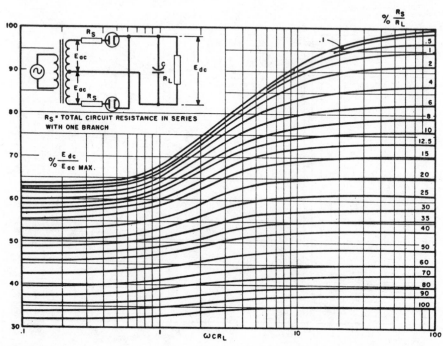

Figure 3.5. Relation of peak sine voltage to dc voltage in full-wave capacitor-input circuits.

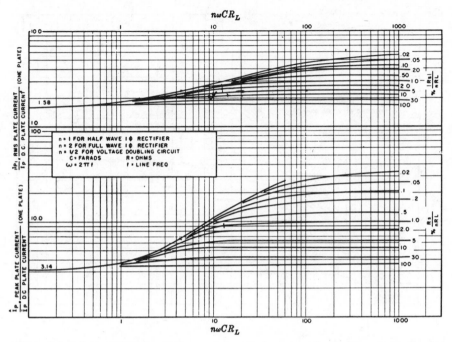

Figure 3.6. Relation of peak, average, and rms diode current in capacitor-input circuits.

rectifier is exceeded and the rms current determines the transformer secondary heating. The VA ratings are greater but the *ratios* of primary to secondary VA rating given in Table 3.1 hold for capacitor input transformers also.

3.3   VOLTAGE DOUBLERS

To obtain more dc output voltage from a rectifier, the circuit of Fig. 3.7 is often used. With proper values of circuit elements the output is nearly double the ac peak voltage. Rectifier peak inverse voltage is little more than the dc output and no dc unbalance exists in the transformer. Current output available from this circuit is less than from a single-phase, full-wave circuit for a given rectifier. Current relations are given in Fig. 3.6. Voltage tripling, quadrupling, and higher multiplying circuits are used either to increase the dc voltage or to avoid the use of a transformer (see Waidelich and Taskin, 1945).

3.4   THREE-PHASE CAPACITOR INPUT FILTERS

Three-phase capacitor input filters are not used too often, primarily because three-phase power supplies normally handle more power than do single-phase

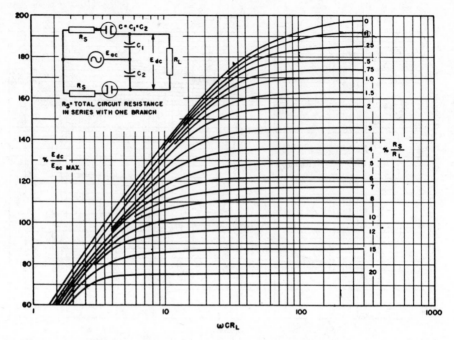

Figure 3.7. Relation of peak sine voltage to dc voltage in voltage-doubling circuits.

supplies. However, at higher frequencies there are some advantages to this type of filter; they eliminate the filter inductor, eliminate voltage transients during turn-on and load switching, make turn-off protection unnecessary, and eliminate the resonant peaks in the output impedance. The chief disadvantages are that they have poor regulation, low efficiency, and require large capacitors.

The circuit for the three-phase RC filter is shown in Fig. 3.8. R_S represents not only the external resistance but also that of two rectifiers, the secondary resistance of the transformer in series with the two rectifiers, and the resistance of the primary and the source reflected to the secondary. If R_S/R_L is 0.01 or greater, the current flow into C will be continuous and the effective current in R_S will be 1.1 times the dc output of the power supply and the power factor of the primary will be 90%. If R_S is less than 10% of R_L, the current will be discontinuous and the effective current will increase, as shown in Table 3.2.

The output voltage, E_L, is

$$E_L = (1.35E_{\text{rms}} - 2E_R)\frac{R_L}{R_S + R_L} \tag{3.1}$$

where E_{rms} = effective line-to-line voltage from the transformer
$\qquad E_R$ = forward drop across one leg of the rectifier
$\qquad R_L$ = load resistance
$\qquad R_S$ = series resistance, as explained above

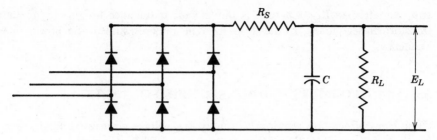

Figure 3.8. Three-phase RC filter circuit.

**TABLE 3.2   Effect of R_S/R_L
Ratio on Current**

Percent R_S/R_L	I_{rms}/I_{dc}
0.15	1.03
0.10	1.10
0.05[a]	1.24
0.025	1.36
0.01	1.49

[a]For $R_S/R_L = 0.05$ the current flows about 80% of the time.

In computing the rectifier forward resistance, use E/I for a small increment about the operating current. This resistance is usually negligible.

The output ripple is computed from the following:

$$\% \text{ ripple rms} = \frac{4X_C}{X_C^2 + R_S^2} \tag{3.2}$$

where $X_C = 1/12\pi fC$
  f = supply frequency
  C = filter capacity, F

If tantalum capacitors are used, this formula should be changed as follows: % ripple rms = $4Z_C/R_S$, where Z_C is the impedance at $3f$. The impedance Z_C for the worst operating condition must be used.

If R_S is less than $0.05R_L$, the voltage will be slightly higher than shown in formula 3.1; however, the error will be negligible in most cases.

For moderate amounts of leakage inductance the output voltage will be approximately $4\pi f L I_L$. where L is the leakage inductance for one leg of the transformer secondary, I_L the dc current in amperes, and f the power line frequency.

3.5   TRANSFORMERS WITH UNRECTIFIED OUTPUTS

Often transformers are required to supply low-voltage ac resistive loads, tube filaments for example. Sometimes these ac loads are operated at a high dc voltage where the electrical isolation of the high voltage from ground is provided by the insulation between the primary and secondary windings of the transformer. Examples of where the transformer provides this isolation are rectifier tube filaments, which operate at high dc voltages, and radar pulse circuitry, which often must operate at the tube cathode voltage. Usually, it is not feasible to combine high-voltage windings with low-voltage windings when the high voltage is greater than 3000 V dc, primarily because of difficulty in insulating leads. However, liquid and gaseous insulating fluids allow higher-voltage designs than are normally achievable in dry-type high-voltage transformers.

Low-capacitance windings are sometimes required for high-frequency circuits or in circuits where frequency response must be extended into the tens or hundreds of kilohertz. The problem is not particularly difficult to resolve in small VA ratings and at relatively low voltages. Here air occupies most of the space between windings. In higher ratings the problem is more difficult because the capacitance increases directly as the coil mean turn length for a given space between windings. As the voltage to ground increases, there comes a point beyond which creepage voltage effects necessitate a change to liquid or gaseous insulated windings. With liquid insulation the capacitance increases by the ratio of the dielectric constant of the liquid to the dry type of insulation, as much as 2.5:1 for a given size and spacing. There is a value of capacitance below which it is impossible to go because of space limitations in the transformer. The value in any given case can be estimated from the fact that the capacitance in picofarads of a body in free space is roughly equal to one-half its largest dimension in centimeters.

Except for high voltage isolation, the design of most low-voltage, low-frequency, low-power transformers with unrectified loads is essentially the same. The load is constant and of unity power factor. Leakage reactance plays practically no part because of its quadrature relationship to the load. Output voltage may therefore be figured as in Fig. 1.4.

When power is first applied to some resistive loads, tube filaments, for example, the resistance is a small value of its operating value. In these instances it is often necessary to protect the load from the high initial current they would draw at rated voltage. This is done by automatically reducing the starting voltage through the use of a current-limiting transformer having magnetic shunts between the primary and secondary windings. The design of these transformers is somewhat special and is included in Chapter 14.

3.6 LOW-VOLTAGE POWER TRANSFORMER DESIGN

It is important that design work be done systematically to save the designer's time and to afford a ready means of reviewing calculations at a later date. Calculation forms such as that in Fig. 3.9 are often used. These follow the design flow chart (Fig. 1.1); the form can be modified to meet the unique needs of any specific type of magnetic device.

Suppose that a transformer is required to supply power for tube filaments: 12.6 V at 1.5 A and 6.3 V at 7.0 A. In addition, the transformer must supply a 28.0 V rms load at 2.0 A.

  Primary: 117 V at 60 Hz
  Three secondaries:
    12.6 V rms at 1.5 A
    6.3 V rms at 7.0 A
    28.0 V rms at 2.0 A
  Ambient temperature: 40°C

First comes the choice of a core. Data such as that in Fig. 2.16 is helpful in this, but most often experience will guide the designer in the selection of the proper core. The core used in this transformer is a 1.5-in. stack of laminations, EI 150 (Fig. 2.17), which is described more fully in Fig. 3.10; it has enough heat dissipation surface for the 118-VA rating of this transformer. For AISI M-19, 26-gauge silicon steel, an induction of 10,000 G is practical. The primary turns are calculated using the maximum applied voltage, the minimum operating frequency, and the minimum effective core area at the maximum flux density for the core material selected. Using equation 1.4, making the substitution $\phi = BA_C$, and transposing yields

$$N_1 = \frac{E \times 10^8}{4.44 f A_C B} \tag{3.3}$$

where A_C is the core cross-sectional area, the product of the core tongue width and the stack height, and B is the core induction in lines/in.[2]. Since many core data are given in gauss, equation 3.3 will be changed for convenience to

$$N_1 = \frac{3.49 E \times 10^6}{f A_C B} \tag{3.4}$$

where the dimensions are inches and B is in gauss. In this transformer, with a 90% stacking factor, $A_c = 1.50 \times 1.50 \times 0.90 = 2.025$ in.[2] and the primary turns are 336.

Below this calculation are entered the voltage, current, volt-amperes, and insulation voltage for all secondary windings. These are designated S1 to S3 for identification. The transformer rating is the sum of the VA of the individual

REF.

1.50 in. of punchings or core   EI 150, Fig. 2.18   flux density   10000   gauss

$$N_p = \frac{3.49 \times E \times 10^6}{f \times A_c \times B} = \frac{3.49 \times 117 \times 10^6}{60 \times 2.025 \times 10000} = \underline{\quad 336 \quad}$$ pri. turns 21 wire. $A_c =$ 2.025   in.^{2}

Core wgt. =   5.5   lb

PRIMARY: 117.0 VOLTS 60 Hz   INS

							N_p/V_p		V_s		K		TURNS		WIRE
S1	6.3	V	7.0	A	44.1	VA	1500 V.	336/117 ×	6.3	×	1.05	=	19	;	13
S2	28.0	V	2.0	A	56.0	VA	1500 V.	336/117 ×	28.0	×	1.05	=	85	;	19
S3	12.6	V	1.5	A	18.9	VA	1500 V.	336/117 ×	12.6	×	1.05	=	38	;	20
S4		V		A		VA	V.	×		×		=		;	
S5		V		A		VA	V.	×		×		=		;	
S6		V		A		VA	V.	×		×		=		;	
S7		V		A		VA	V.	×		×		=		;	
S8		V		A		VA	V.	×		×		=		;	

TOTAL   119.0   VA. +   12.0   est. losses = 131.0 + j 4.7 =   1.12   PRI. A

117.0 PRI. V

Winding Area, A_w = 0.917 in.2

Winding	%VA	Wdg. A_w	N/in.^{2}	AWG
S1	17.6	0.162	117	13
S2	22.4	0.205	415	19
S3	7.6	0.069	551	20
P	52.4	0.481	698	21

Figure 3.9. Low-voltage transformer design calculations.

Wdg.	Wire Size	Wdg. Width	Turns/Layer	Layers	Volts/Layer	Volts 2 Lay.	$V_iN_p \times N_s - IR = V_T$ @ 25°C	V_T @ 85°C
P	21	1.75	53	7	17	34		
S3	20	1.75	47	1	12.6	—	113.964/336 × 38 − .371 = 12.52	12.80 − .46 = 12.34
S2	19	1.75	43	2	14	28	113.964/336 × 85 − 1.036 = 27.79	28.64 − 1.30 = 27.34
S1	0.003	1.75	1	19	0.33	0.66	113.964/336 × 19 − .172 = 6.27	6.40 − .22 = 6.18
		Second trial with				S3	114.0/332 × 38 − .371 = 12.68	12.96 − .46 = 12.50
		P turns = 332				S2	114.0/332 × 85 − 1.036 = 28.15	28.99 − 1.30 = 27.69
						S1	114.0/332 × 19 − .172 = 6.35	6.48 − .22 = 6.26

Wdg.	Layers(OD + INS)/SF	D	Coil M.T. (in.)	N × MT × OHMS/in.	DCR ohms	I_{rms}	IR (V)	I^2R Watts	Wgt. lbs.
P	7 × 0.0384	.2688	6.75 × π × .2688 (7.58)	336 × 7.58 × 12.77 / 12000	2.71	1.12	3.036	3.40	.530
S3	1 × 0.0368	0.368	6.75 × 2π × .1528 (7.71)	38 × 7.71 × 10.13 / 12000	.247	1.50	.371	.56	.077
S2	2 × 0.0471	.0942	6.75 × 2π × .3727 (9.09)	85 × 9.09 × 8.046 / 12000	.518	2.0	1.036	2.07	.254
S1	19 × (.003 + .002)/0.8	.095	6.75 × 2π × .4773 (9.75)	19 × 9.75 × 1.588 / 12000	.025	7.0	.172	1.20	.309
P* 2nd	7 × 0.0384	.2688	(7.58)	332 × 7.58 × 12.77 / 12000	2.68	1.12	3.00	3.36	.524

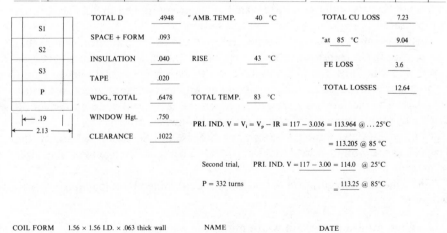

TOTAL D   .4948   "   AMB. TEMP.   40 °C   TOTAL CU LOSS   7.23

SPACE + FORM   .093   "at   85 °C   9.04

INSULATION   .040   RISE   43 °C

TAPE   .020   FE LOSS   3.6

WDG., TOTAL   .6478   TOTAL TEMP.   83 °C   TOTAL LOSSES   12.64

WINDOW Hgt.   .750

PRI. IND. $V = V_i = V_p - IR = 117 - 3.036 = 113.964$ @ ... 25°C

CLEARANCE   .1022

$= 113.205$ @ 85°C

.19   2.13

Second trial,   PRI. IND. $V = 117 - 3.00 = 114.0$ @ 25°C

P = 332 turns $= 113.25$ @ 85°C

COIL FORM   1.56 × 1.56 I.D. × .063 thick wall   NAME   DATE

Figure 3.9. (Cont.)

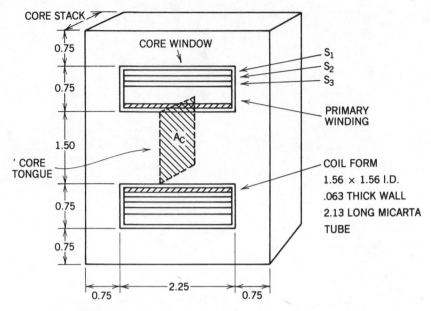

Figure 3.10. Dimensions and coil section of power transformer.

secondary windings. The primary current is determined from the sum of the total load VA and the estimated losses plus the reactive VA of the core and the core gap, which is in quadrature with the load and losses; the reactive VA of M-19 steel at 10,000 G is 0.85 VA/lb, 4.7 VA for a 5.5-lb core. This vector sum divided by the primary voltage gives the primary current.

The secondary turns are calculated from the product of the primary turns per volt and the secondary voltage. This is then multiplied by a factor that compensates for regulation, 5.0% in this design.

At this point in the design it is necessary to select the insulation system that will be used in the transformer. Kraft paper sheet insulation will be used between layers and windings, round magnet wire with polyester enamel and an epoxy impregnant will be used. Until induced or applied voltages affect insulation thickness, the mechanical stresses placed on the layer and winding insulation during winding determine the thickness of this insulation.

Methods of determining wire size are discussed in Section 2.5. Usually, the large wires are wound closest to the coil form to prevent damage to smaller sizes of magnet wire.

The winding configuration or coil build is determined by calculating for each winding:

1. Winding width, the coil width minus the margins
2. Turns per layer
3. Number of layers

4. Thickness of each layer

5. Thickness of each winding

6. Total winding build, including insulation and the coil form

All of this is included on the transformer calculation sheet (Fig. 3.9). It can now be determined if the coil is too big to fit in the core window or if there is excessive space remaining in the core window.

The larger sizes of magnet wire tend to bow outward when wound on relatively small coil forms, which can result in a greater than expected coil build; for that reason 0.003-in.-thick copper foil 1.75 in. wide was substituted for the AWG No. 13 round magnet wire in S1; the insulation between adjacent turns of foil is 0.002-in.-thick Kraft paper. The mean length of turn is calculated for each winding using formulas 2.22 and 2.23. With the mean turn values, winding resistances, and winding weights, the IR and I^2R for each winding can be calculated. The total copper loss is multiplied by 1.25 to correct for the estimated 85°C operating temperature. The core weight is 5.5 lb and the M-19 steel used has 0.66 W loss per pound at 10,000 G. This gives a core loss of 3.6 W and a total loss at 85°C of 12.64 W. Dividing these losses by the appropriate ordinates from Figs. 2.18 and 2.19, the coil temperature rise is figured to be 43°C, a coil hot-spot temperature of 83°C which is well below the maximum temperature for the insulation being used.

We know that the design is satisfactory insofar as temperature rise and physical size are concerned, but the secondary voltages must be checked. The method of equation 1.13 is used. In the initial design, at room temperature, the secondary voltages are within 1.0% of nominal voltage and at 85°C they are within 2.5%. This is a marginal condition and a design change is necessary. By decreasing the number of primary turns from 336 to 332, the secondary voltages are increased by almost 1% and the voltages are well within 2% of the nominal value.

If the design were deficient in any respect, even down to the last things figured, some change would have to be made which would require recalculation of all or part of the transformer; hence the importance of good estimating all the way along.

The filament transformer outlined above had a center tap (C.T.) in each filament winding. Such taps are used with directly heated cathodes, especially when plate current is large, to prevent uneven distribution of filament emission. In windings for supplying filaments of small tubes, center taps are sometimes omitted. Ripple in the rectified output then increases, and transformer core flux density becomes asymmetrical. Whether these effects are permissible depends on operating conditions. Usually, plate current is much smaller than filament current, so that center-tap leads may be smaller in copper section than start and finish leads. A certain amount of space is required for these leads; rectifier wiring is also more time consuming when there are center taps. Nevertheless, the extra work and size may be justified by improved performance.

An even number of layers, such as were used in the transformer windings described in this section, results in center-tap placement on the same coil end as the start and finish leads; if there were an odd number of layers, the tap lead would be at the opposite end. In a single-core, single-coil design, an odd number of turns cannot be center-tapped exactly. Usually, the unbalance caused by the tap being a half-turn off center is not serious, but it should not be disregarded without calculation.

3.7   ANODE TRANSFORMERS

Anode transformers differ from filament transformers in several respects.

1. *Currents* are nonsinusoidal. In a single-phase full-wave rectifier, for instance, current flows through one half of the secondary during each positive voltage excursion and through the other half during each negative excursion. For half of the time each half-secondary winding is idle.

2. *Leakage inductance* not only determines output voltage but also affects rectifier regulation in an entirely different manner than with a straight ac load. This is discussed in Chapter 4.

3. *Half-wave rectifiers* carry unbalanced direct current; this may necessitate less ac flux density, hence larger transformers, than full-wave rectifiers. Unbalance in the three-phase half-wave type can be avoided by the use of zigzag connections, but an increase in size over full-wave results because of the out-of-phase voltages. These connections are desirable in full-wave rectifiers when half voltage is obtained from a center tap (see Table 3.1).

4. *Single-phase full-wave rectifiers with two anodes* have higher secondary volt-amperes for a given primary VA rating than a filament transformer. Bridge-type (four-anode) rectifiers have equal primary and secondary volt-amperes, as well as balanced direct current, and plate transformers for these rectifiers are smaller than for other types. Three-phase rectifier transformers are smaller in total size but require more coils. The three-phase full-wave type has equal primary and secondary VA ratings.

5. *Induced secondary voltage* is much higher. Filament transformers are insulated for this voltage but have a few secondary turns of large wire, whereas anode transformers have many turns of small wire. For this reason the volts per layer are higher in anode transformers, and core windows having proportionately greater height and less width than those in Fig. 3.10 are often preferable. This trend runs counter to the conditions for low leakage inductance and makes it necessary to interleave the windings. Figure 3.11 shows the windings of a single-phase full-wave rectifier transformer with the primary interleaved between halves of the secondary. This arrangement is especially adaptable to transformers with grounded center tap. The primary-secondary insulation can be reduced to the amount suitable for primary to ground. This is called *graded* insulation.

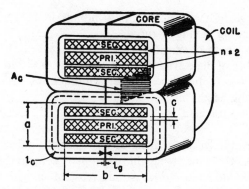

Figure 3.11. Dimensions and coil section of anode transformer.

In large power rectifiers of the gas-filled or pool types, anode current under short-circuit conditions may be very great, and anode transformer windings must be braced to prevent damage. If the conductors are small, solventless varnish is useful for solidly embedding the conductors.

3.8   LEAKAGE INDUCTANCE

Flux set up by the primary winding which does not link the secondary, or vice versa, gives rise to leakage or self-inductance in each winding without contributing to the mutual flux. The greater this leakage flux, the greater the leakage inductance, because the inductance of a winding equals the flux linkages with unit current in the winding. In Fig. 3.11, all flux that follows the core path l_c is mutual flux. Leakage flux is the relatively small flux which threads the secondary winding sections, enters the core, and returns to the other side of the secondaries, without linking the primary. The same is true of flux linking only the primary winding. But it is almost impossible for flux to leave the primary winding, enter the core, and reenter the primary without linking part of the secondary also. The more the primary and secondary windings are interleaved, the less leakage flux there is, up to the limit imposed by flux in the spaces c between sections. These spaces contain leakage flux also; indeed, if there is much interleaving or if the spaces c are large, most of the leakage flux flows in them. Large coil mean turn length, short winding traverse b, and tall window height a all increase leakage flux.

Several equations have been derived for the calculation of leakage inductance. That originated by Fortescue is generally accurate, and errs, if at all, on the conservative side (see Snelling, 1969):

$$L_S = \frac{10.6 N^2 (\text{MT})(2nc + a)}{10^9 n^2 b} \tag{3.5}$$

where L_S = leakage inductance of both windings in henrys, referred to the
       winding having N turns
    MT = mean length of turn for whole coil, in.
    n = number of dielectrics between windings ($n = 2$ in Fig. 3.11)
    c = thickness of dielectric between windings, in.
    a = winding height, in.
    b = winding traverse, in.

The greatest gain from interleaving comes when the dielectric thickness c is
small compared to the window height; when nc is comparable to the window
height, the leakage inductance does not decrease much as n is increased. It is
often difficult to reduce the leakage inductance which occurs in high-voltage
transformers because of leakage flux in spaces c. A small number of turns, short
mean turn, and low, wide-core windows all contribute to a low value of leakage
inductance.

3.9   ANODE TRANSFORMER DESIGN

Let the requirements of a rectifier be

1200-V 115-mA rectifier dc output
Single-phase full-wave circuit with 866 tubes
Primary 115 V, 60 Hz
Rectifier regulation 5% maximum
Ambient 55°C

To fulfill these requirements, an inductor input filter must be used. If 1% is
allowed for inductor IR drop, a maximum of 4% regulation is left in the anode
transformer. The approximate secondary output voltage is $1200 \times 2.22 = 2660$,
say 2700 V. The center tap may be grounded. Suppose that a transformer like
the one in Fig. 3.11 is used. The calculations are given in Fig. 3.12. The various
steps are performed in the same order as in filament transformers. The grain-
oriented type C core is worked at 38% higher induction, with but 60% of the core
loss of Fig. 3.9; its strip width is 2.25 in., build-up 0.625 in., and window 1.0 in. by
3.0 in. for each core loop. Note the difference in primary and secondary volt-
amperes and winding heights. Since the primary and secondary are symmetrical
about the primary horizontal centerline, they have the same mean turn length.
Losses and temperature rise are low. Regulation governs size. Secondary layer
voltage is high enough to require unusually thick layer paper. This coil is wound
on a multiple-coil machine. Winding height is figured on the basis of layer paper
adequate for the voltage instead of from Table 2.5, but turns per layer are taken
from this table. Since adjacent layers are wound with opposite directions of
traverse, the highest voltage across the layer insulation is twice the volts per

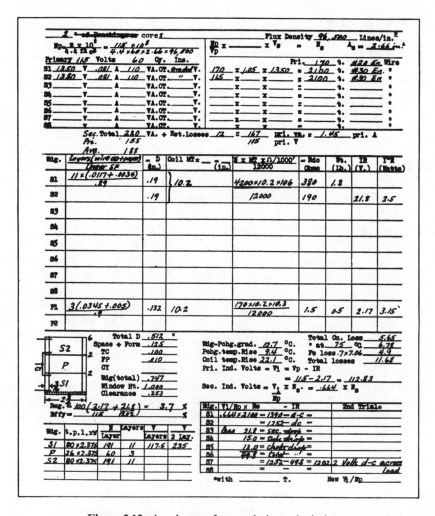

Figure 3.12. Anode transformer design calculations.

layer. Layer insulation is used at 46 V/mil in the secondary; this counts the 1.7 mils of double enamel, which must withstand impregnation without damage. Anode leads and margins withstand 5-kV rms test voltage. Since the secondary center tap is grounded, two thicknesses of 0.010-in. insulation between windings are sufficient. Clearance of 0.253 in. allows room for in-and-out coil taping.

Secondary leakage inductance, from equation 3.5, is

$$\frac{10.6 \times 4200^2 \times 10.2(4 \times 0.020 + 0.747)}{4 \times 2.375 \ 10^9} = 0.166 \, \text{H}$$

At 60 Hz this is $6.28 \times 60 \times 0.166 = 63 \, \Omega$, which would be $240 \, \Omega$ if the

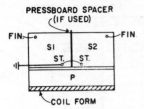

Figure 3.13. Anode transformer with center tap grounded.

secondary were a single section, and which would increase regulation as set forth in Chapter 4. The regulation calculated in Fig. 3.12 is that due to primary IR calculated in the normal manner, plus I_{dc} times one-half the secondary winding resistance.

When high voltage is induced in a winding, the layer insulation and coil size may often be reduced by using the scheme shown in Fig. 3.13. This is applicable to a plate transformer of the single-phase full-wave type with center tap grounded. It then becomes practical to make the secondary in two separately wound vertical halves or part coils. One of the part coils is assembled with the turns in the same direction as those of the primary, and the other part coil is reversed so that the turns are in the opposite direction. The two start leads are connected together and to ground as in Fig. 3.13. It is necessary then to provide only sufficient insulation between windings to withstand the primary test voltage. Channels may be used to insulate the secondaries from the core. With higher voltages, it may be necessary to provide pressboard spacers between the secondary part coils, or to tape the secondary coils separately, but margins must be provided sufficient to prevent creepage across the edges of the spacers.

3.10 POWER SUPPLY FREQUENCY

The foregoing design examples were based on a 60-Hz power source. For a given induction, 50-Hz core losses are somewhat lower than those at 60 Hz. It follows that for equal losses and magnetizing VA the induction at 50 Hz can be increased slightly. For equal VA a 50-Hz transformer will have about 15% greater volume than a 60-Hz unit. Otherwise, 50-Hz transformers are not appreciably different from 60-Hz transformers. In fact, many transformers are designed to operate at both frequencies.

Higher power supply frequencies, primarily 400 Hz, are used in aircraft and portable electronic equipment to reduce weight and volume. Thinner-gauge core materials are used at these higher frequencies to reduce eddy current losses; 0.004- and 0.006-in.-thick magnetic materials are used extensively in 400-Hz transformer applications. Losses at 400 Hz for several different types of core material are shown in Fig. 3.14. As the operating frequency increases, the core losses can become a controlling factor in determining transformer size. While the saturating flux density for a given core material is nearly the same at 60 and

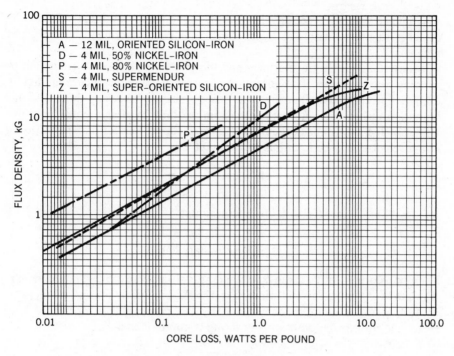

Figure 3.14. 400 Hz core loss, cut cores.

400 Hz, the core loss may limit the induction at which the core can be operated. Even with this increase in loss at higher frequencies the use of higher-temperature-rated insulation systems will permit a substantial reduction in transformer weight and volume. A 400-Hz transformer can be 25% the size of a comparably rated 60-Hz transformer.

Typical combinations of grain-oriented material and the temperature class are as follows:

Material	Frequency	Strip Thickness	B (G)	Temperature Class
AISI M15	60	0.012	15,000	105°C
Hipersil H	400	0.004	15,000	135°C
Hipersil Z	400	0.004	17,600	135°C
Supermendur	400	0.004	21,500	135°C

In very small units where regulation may limit the temperature rise, these cores can be used at a lower-temperature class than shown. Similarly, if higher operating temperatures are acceptable, a change in insulation can permit operation at a higher temperature class (see Table 2.7). The need to reduce weight and volume, especially in airborne equipment, continually forces the designer to use materials at their fullest capabilities: thermal, dielectric, and magnetic.

Many small 60-Hz transformers have core loss which is small compared to winding or copper loss. This condition occurs because inductance is limited by exciting current rather than by core loss. As size or frequency increases, this limitation disappears, and core loss is limited only by design considerations. Under such circumstances, the ratio of core to copper loss for maximum rating in a given size may be found as follows. Let

$$W_e = \text{core loss}$$
$$W_s = \text{copper loss}$$
$$K_1, K_2, \text{etc.} = \text{constants}$$
$$E = \text{secondary voltage}$$
$$I = \text{secondary current}$$

For a transformer with a given core, winding, volt-ampere rating, and frequency, $W_e \approx K_1 E^2$. For a given winding, $W_s = K_2 I^2$. Also, for a given size, $W_e + W_s = K_3$, a quantity determined by the permissible temperature rise. Hence the transformer volt-ampere rating is approximately

$$EI = \sqrt{W_e W_s / K_1 K_2}$$
$$= K_4 \sqrt{W_e (K_3 - W_e)}$$

For a maximum, the rating may be differentiated with respect to W_e and the derivative equated to zero:

$$0 = K_3 - 2W_e$$

whence

$$W_e = K_3/2$$

so that $W_s = K_3/2$, or copper and core losses are equal for maximum rating.

Although this equality is not critical, and is subject to many limitations, such as core shape, voltage rating, and method of cooling, it does serve as a guidepost to the designer. If a transformer design is such that a large disparity exists between core and copper losses, size or temperature rise often may be reduced by a redesign in the direction of equal losses.

3.11   400-Hz TRANSFORMER DESIGN

Given a primary of 120 V, 400 Hz, design a transformer to deliver 0.2 A at +450 V dc using a 5U4G tube in a single-phase full-wave circuit with a 0.5-μF capacitor input filter.

Figures 3.5 and 3.6 show whether the product of ωCR_L will produce the necessary dc output without exceeding the rectifier peak inverse voltage rating and peak current rating.

$$\omega CR_L = 6.28 \times 400 \times 0.5 \times 10^{-6} \times (450/0.2) = 2.83$$

For R_S assume a peak current rating of 0.5 A. The average anode characteristics show a 97-V tube drop, or $97/0.5 = 194\,\Omega$ at peak current. $R_S/R_L = 194/2250 = 0.086$. Add 5% for transformer windings; estimated $R_S/R_L = 13.6\%$.

Check on peak current from Fig. 3.6.

$$n\omega CR_L = 5.66$$

$$\hat{I}_P = 5I_P = 5 \times 0.1 = 0.5\,\text{A}$$

the peak value assumed. The rms current in the tube plates and secondary windings is $2 \times 0.1 = 0.2\,\text{A}$. The output voltage, from Fig. 3.5, is 69% of the peak ac volts per side. Hence the secondary rms volts per side = $450 \times 0.707/0.69 = 460\,\text{V}$, and the secondary volt-amperes = $2 \times 460 \times 0.2 = 184$. The anode transformer must deliver $2 \times 460 = 920\,\text{V}$ at 0.2 A rms. The anode volt-amperes = $0.707 \times 184 = 130$.

The inverse peak voltage is the peak value of this voltage plus the dc output, because the tube filament is at the dc value, plus a small amount of ripple, while one anode has a maximum of peak negative voltage during the nonconducting interval. Thus the peak inverse voltage is $460 \times 1.41 + 450 = 1100\,\text{V}$, which is within the tube rating.

The choice of the core for this transformer is governed by size and cost considerations. Assume that the core works at 15,000 G. The loss for 0.005-in. silicon steel and 0.004-in. grain-oriented silicon steel is 12.2 and 7.5 W/lb, respectively (see Fig. 3.14). Laminations have an 80% stacking factor, whereas the Hipersil Z type of C core has 91%. In this thickness the grain-oriented Hipersil compares still better with the M-19 material than Fig. 3.14 would indicate and therefore will be used in this transformer.

Let two type C cores be used, each with the following dimensions:

Strip width	0.75 in.	Window height	0.50 in.
Build	0.27 in.	Window width	1.12 in.
Net core area × 2	0.369 in.²	Core weight	0.196 lb.

Turns are figured from equation 3.4. Primary turns are then

$$N = \frac{3.49 \times 120 \times 10^6}{400 \times 0.369 \times 15,000} = 189$$

Primary: 189 turns, AWG No. 22 polyester enamel wire, DCR = 0.82 Ω
Secondary: 1394 turns, AWG No. 33 polyester enamel wire, DCR = 83 Ω

$$\text{Primary } I^2R \text{ loss at } 100°C \quad = 1.18 \text{ W}$$
$$\text{Secondary } I^2R \text{ loss at } 100°C = 4.32 \text{ W}$$
$$\text{Core loss, } 0.392 \times 9.0 \text{ W/lb} = 3.53 \text{ W}$$
$$\text{Total losses} \qquad\qquad\quad = 9.03 \text{ W}$$

With an open-type mounting and polyamide sheet insulation, this transformer has a temperature rise of 75°C.

3.12 POLYPHASE TRANSFORMERS

A polyphase transformer is one that is energized with two or more electrically displaced phases. The most common systems are those with two phases displaced by 90° and with three phases displaced by 120°. There are literally scores of polyphase rectifier circuits recognized in the literature and in industry standards, most of which are derived from three-phase wye and delta rectifier circuits. For that reason only three-phase circuits will be discussed.

Three-phase transformers are used in large, multikilowatt power rectifiers; however, in much lower-power systems, tens to hundreds of watts, where line currents must be reasonably well balanced, three-phase rectifier circuits are also used. Many electronic circuits fall into the latter category, particularly for airborne applications. Often three single-phase transformers are used, but in applications where weight is important, a core made from E-shaped laminations or an E-shaped cut core will result in significant weight savings when compared with three single-phase transformers.

A three-phase transformer supplying a rectifier load may function in a manner that is different from the same transformer supplying a balanced unrectified load. The neutral of a wye winding supplying an ac load operates essentially at ground. The neutral of the same transformer with a full-wave rectified load has a sawtooth wave at three times the fundamental frequency, varying between 25 and 75% of the output dc voltage. This voltage can be coupled back to the input line and disrupt low-noise circuitry which is also connected to those lines. The same condition can exist with a delta winding because a virtual neutral exists in the delta.

The selection of winding connections for the transformer depends on the requirements of the transformer, and since electronic transformers often supply several different rectified outputs, the windings may be a combination of wyes and deltas. Therefore, the advantages and disadvantages of each type of winding will be considered rather than the merits of delta-wye, wye-delta, and wye-wye connections.

The delta winding provides a circulating path for the third harmonic currents which are a part of the magnetizing current. The delta winding also operates fairly well on some unbalanced loading. It does not allow the neutral to be grounded. The delta is suited to lower-voltage, higher-current windings and in some instances can operate at reduced power in an open delta when three single-phase transformers are being used.

The wye winding can be operated with the neutral grounded to eliminate unbalanced voltages when the load is unbalanced, and when grounded it also provides a path for the third harmonics in the magnetizing current. However, this may cause interference on the power system neutral. The wye is suited to higher voltages and lower currents. Most high-voltage windings are wye connected because each phase must support only 0.577% of the line-to-line voltage and permits some voltage grading of the neutral connection. An example of a three-phase transformer design is included in Chapter 9.

3.13   DESIGN CHART

In preceding sections it has been stated that special conditions require tailored designs. Windings for simple low-voltage 60-Hz transformers may be chosen from Fig. 3.15. This chart is based on the following conditions:

1. Two untapped concentric windings; primary wound first
2. Operating voltage in both windings less than 1000 V
3. Power supply frequency 60 Hz
4. Maximum temperature rise 40°C in 65°C ambient
5. Resistive loads
6. Equal I^2R losses in primary and secondary
7. Solventless resin-impregnated coils
8. Open-type assemblies
9. Grain-oriented silicon-steel cores

It was found that 40°C rise in the four smallest sizes resulted in excessive voltage regulation. For example, a small filament transformer would deliver correct filament voltage at room ambient temperature of 25°C, but at 105°C this voltage dropped to less than the published tube limit. Hence the winding regulation in the two smallest transformers was limited to 15%, and in the next two larger sizes to 10%. In still larger sizes, the 40°C temperature limit held the regulation to less than 10%.

In using the chart, ratings rarely fall exactly on the VA values assigned to each core. Hence a core is generally chosen with somewhat greater than required rating. Lower regulation and temperature rise than maximum then result. Wire size in quadrant I also increases in discrete sizes, and if the chart indication falls between two sizes the smaller size should be used.

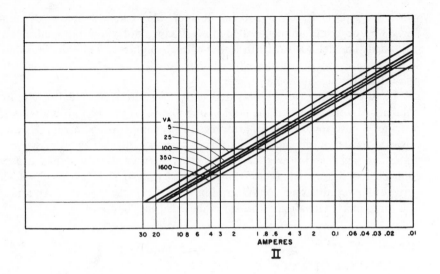

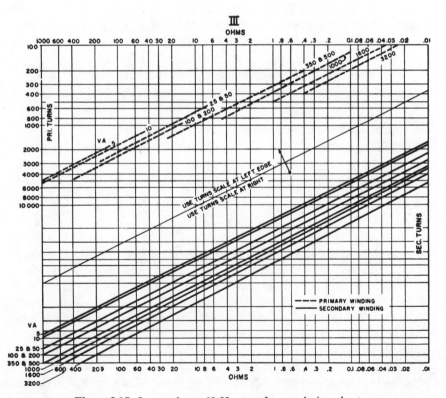

Figure 3.15. Low-voltage 60-Hz transformer design chart.

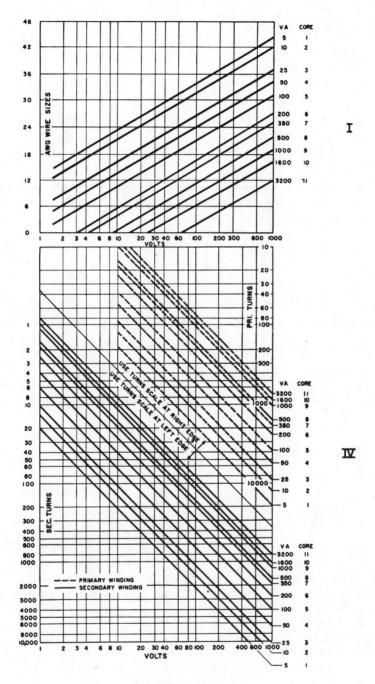

Figure 3.15. (Cont.)

Instructions for Using Fig. 3.15

1. Choose a core from Table 3.3 which has a VA rating equal to or greater than that required.
2. From rated primary and secondary voltages, find number of turns for both windings in quadrant IV.
3. From rated primary and secondary currents, find wire size for both windings in quadrant I.
4. Project turns across to quadrant III to obtain winding resistances.

Departures from the assumed conditions preclude direct application of Fig. 3.15, but the chart is still useful as a starting point in design. For some common modifications, the following notes apply.

1. For each additional secondary winding reduce core maximum rated volt-amperes by 10%. Choose wire size from quadrant II.
2. For 50-Hz transformers, reduce the core maximum rated VA 10%.
3. When the permissible temperature rise is higher than 40°C, the core maximum volt-amperes equal (VA in table) $\times \sqrt{\text{temperature rise}/40°C}$.

Example. A transformer is required for 115/390 V, 60 Hz, to deliver 77 VA. This rating falls between the maxima for cores 4 and 5. Using core 5 at 115 V, we read, from Fig. 3.15, for the primary, 440 turns of No. 22 wire and 3 Ω dc resistance; for the secondary, 1700 turns of No. 27 wire and 40 Ω dc resistance.

TABLE 3.3   Transformer Size, Rating, and Regulation

Core	Maximum VA Rating	Percent Regulation	Total Weight (lb)	Overall Dimensions (in.)
1	5	15	0.38	$1\frac{3}{4} \times 1\frac{3}{4} \times 1\frac{3}{4}$
2	10	15	0.68	$1\frac{7}{8} \times 2\frac{3}{8} \times 1\frac{3}{4}$
3	25	10	1.2	$2\frac{1}{4} \times 2\frac{7}{8} \times 2\frac{1}{4}$
4	50	10	2.2	$2\frac{1}{2} \times 3\frac{1}{8} \times 2\frac{1}{2}$
5	100	8	3.8	$3\frac{1}{8} \times 3\frac{3}{4} \times 3$
6	200	6	6.4	$3\frac{7}{8} \times 4\frac{3}{4} \times 3\frac{5}{8}$
7	350	4	11.0	$4\frac{3}{8} \times 5\frac{3}{8} \times 4$
8	500	3	15	$5\frac{1}{8} \times 6\frac{1}{8} \times 5$
9	1000	2.2	24	$5\frac{7}{8} \times 6\frac{3}{4} \times 6\frac{1}{8}$
10	1600	1.8	36	$7\frac{1}{4} \times 8\frac{1}{4} \times 7\frac{1}{2}$
11	3200	1.2	75	$9\frac{3}{4} \times 12\frac{3}{4} \times 8$

3.14 INDUCTORS

Inductors are used in electronic power equipment to smooth out ripple voltage in dc supplies, so they carry direct current in the coils. It is common practice to build such inductors with air gaps in the core to prevent dc saturation. The air gap, size of the core, and number of turns depend on three interrelated factors: inductance desired, direct current in the winding, and ac volts across the winding.

The number of turns, the direct current, and the air gap determine the dc flux density, whereas the number of turns, the volts, and the core size determine the ac flux density. If the sum of these two flux densities exceeds saturation value, noise, low inductance, and nonlinearity result. Therefore, an inductor must be designed with knowledge of all three of the conditions above.

Magnetic flux through the coil has two component lengths of path: the air gap l_g, and the length of the core l_c. The core length l_c is much greater geometrically than the air gap l_g, as indicated in Fig. 3.11, but the two components do not add directly because their permeabilities are different. In the air gap, the permeability is unity, whereas in the core its value depends on the degree of saturation of the iron. The effective length of the magnetic path is $l_g + l_c/\mu$, where μ is the permeability for the steady or dc component of flux.

Inductor design is, to a large extent, the proportioning of values of air gap and magnetic path length divided by permeability. If the air gap is relatively large, the inductor inductance is not much affected by changes in μ; it is then called a *linear* inductor. If the air gap is small, changes in μ due to current or voltage variations cause inductance to vary; then the inductor is nonlinear.

When direct current flows in an iron core inductor, a fixed magnetizing force H_{dc} is maintained in the core. This is shown in 3.16 as the vertical line H_{dc} to the right of zero H in a typical ac hysteresis loop, the upper half DB_mD' of which corresponds to that in Fig. 2.17. Increment ΔH of ac, superposed on H_{dc}, causes flux density increment ΔB, with permeability μ_Δ equal to the slope of dashed line AB_m. ΔB is twice the peak ac induction B_{ac}. It will be recalled from Fig. 2.5 that the normal induction curve OB_m is the locus of the end points of a series of successively smaller major hysteresis loops. Since the top of the minor loop always follows the left side of a major loop, as H_{dc} is reduced in successive steps the upper ends of corresponding minor loops terminate on the normal induction curve.

Dashed-line slopes of a series of minor loops are shown in Fig. 3.16, the midpoints of which are C, C' C'', and C'''. Increment of induction ΔB is the same for each minor loop. It will be seen that the width of the loop ΔH is smaller, and hence μ_Δ is greater, as H_{dc} is made smaller.

Midpoints C, C', and so on, form the locus of dc induction. The slope of straight line OC is the dc permeability for core magnetization H_{dc}. It is much greater than the slope of AB_m. Hence incremental permeability is much smaller

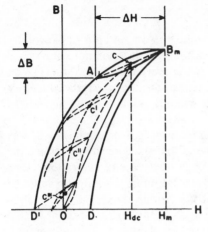

Figure 3.16. Incremental permeability with different amounts of dc magnetization.

than dc permeability. This is true in varying degree for all the minor loops. The smaller ΔB is, the less the slope of a minor loop becomes, and consequently the smaller the value of incremental permeability μ_Δ. The curve in Fig. 3.17 marked μ is the normal permeability of 4% silicon steel for steady values of flux, in other words, for the dc flux in the core. It is 4 to 20 times as great as the incremental permeability μ_Δ for a small alternating flux superposed on the dc flux. The ratio of μ to μ_Δ gradually increases as dc flux density increases.

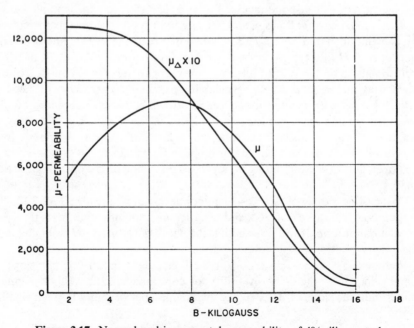

Figure 3.17. Normal and incremental permeability of 4% silicon steel.

Because of the low value of μ_Δ for minute alternating voltages, the effective length of magnetic path $l_g + l_c/\mu_\Delta$ is considerably greater for alternating than for steady flux. But the inductance varies inversely as the length of ac flux path. If, therefore, the incremental permeability is small enough to make l_c/μ_Δ large compared to l_g, it follows that small variations in l_g do not affect the inductance much. For this reason the exact value of the air gap is not important with small alternating voltages.

Inductor size, with a given voltage and ratio of inductance to resistance, is proportional to the stored energy LI^2. For the design of inductors carrying direct current, that is, the selection of the right number of turns, air gap, and so on, a simple method was originated by Hanna (1927). By this method, magnetic data as reduced to curves such as Fig. 3.18, plotted between LI^2/V and NI/l_c from which reactors can be designed directly. The various symbols in the coordinates are:

L = ac inductance, H

I = direct current, A

V = volume of iron core, in.^{3}

    = $A_c l_c$ (see Fig. 3.11 for core dimensions)

A_c = cross section of core, in.^{2}

l_c = length of core, in.

N = number of turns in winding

l_g = air gap, in.

Each curve of Fig. 3.18 is the envelope of a family of fixed air-gap curves such as those shown in Fig. 3.19. These curves are plots of data based on a constant small ac flux (10 G) in the core but a large and variable dc flux. Each curve has a region of optimum usefulness, beyond which saturation sets in and its place is taken by a succeeding curve having a larger air gap. A curve tangent to the series of fixed air-gap curves is plotted as in Fig. 3.18, and the regions of optimum usefulness are indicated by the scale l_g/l_c. Hence Fig. 3.18 is determined mainly by the dc flux conditions in the core and represents the most LI^2 for a given amount of material.

Figure 3.19 illustrates how the exact value of air gap is of little consequence in the final result. The dashed curve connecting B and C is for a 6-mil gap. Point Y' represents the maximum inductance that could be obtained from a given core for $NI/l_c = 19$. Point Y is the inductance obtained if a gap of either 4 or 8 mils is used. The difference in inductance between Y and Y' is 4%, for a difference in air gap of 33%. An example will show how easy it is to make an inductor according to this method.

Example. Assume a stack of silicon-steel laminations having a cross section 0.88 in. by 0.88 in., and with iron filling 92% of the space. The length of the flux path l_c in this core is 7.50 in. It is desired to know how many turns of wire and

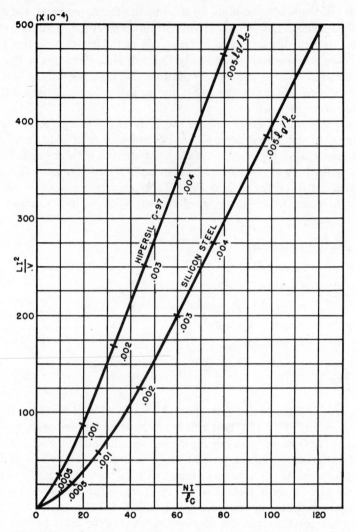

Figure 3.18. Inductor energy per unit volume versus ampere-turns per inch of core.

what air gap are necessary to produce 70 H when 20 mA of direct current is flowing in the winding.

This problem is solved as follows:

$$A_c = (0.875)^2 \times 0.92 = 0.71 \text{ in.}^2$$

$$V = 0.71 \times 7.5 = 5.3$$

$$\frac{LI^2}{V} = \frac{70 \times 4 \times 10^{-4}}{5.3} = 53 \times 10^{-4}$$

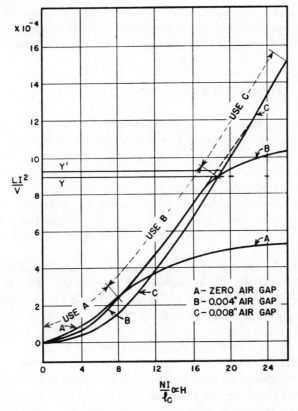

Figure 3.19. Fixed air-gap curves. For $B_{dc} \gg B_{ac}$ air gap is not critical.

In Fig. 3.18 the abscissa corresponding to $LI^2/V = 53 \times 10^{-4}$ is $NI/l_c = 25$ for silicon steel. The ratio of air gap to core length l_g/l_c is between 0.0005 and 0.001.

$$NI/l_c = 25$$

$$N = (25 \times 7.5)/0.020 = 9350 \text{ turns}$$

The total air gap is nearly 0.001×7.5 or 7.5 mils; the gap at each joint is half of this value, or 3.75 mils.

The conditions underlying Hanna's method of design are met in most applications. In receivers and amplifiers working at low audio levels, the alternating voltage is small and hence the alternating flux is small compared to the steady flux. Even if the alternating voltage is of the same order as the direct voltage, the alternating flux may be small, especially if a large number of turns is necessary to produce the required inductance; for a given core the alternating flux is inversely proportional to the number of turns. Dc resistance of the coil is usually fixed by the regulation or size requirements. Heating seldom affects size.

3.15   INDUCTORS WITH LARGE AC FLUX

With the increasing use of higher voltages, it often happens that the ac flux is no longer small compared to the dc flux. This occurs in high-impedance circuits where the direct current has a low value and the alternating voltage has a high value. The inductance increases by an amount depending on the values of ac and dc fluxes. Typical increase of inductance is shown in Fig. 3.20 for an inductor working near the saturation point. Increasing ac flux soon adds to the saturation, which prevents further inductance increase and accounts for the flattening off in Fig. 3.20. Saturation of this sort may be avoided by limiting the value of the dc flux.

To illustrate the effect of these latter conditions, suppose that an inductor has already been designed for negligibly small alternating flux and operates as shown by the minor loop with center at G (Fig. 3.21). Without changing anything else, suppose that the alternating voltage across the inductor is greatly increased, so that the total ac flux change is from zero to B_m. (Assume that the inductor still operates about point G.) The hysteresis loop, however, becomes the unsymmetrical figure $OB_mD'O$. The average permeability during the positive flux swing is represented by the line GB_m, and during the negative flux swing by OG. The slope of GB_m is greater than that of the minor loop; hence the first effect exhibited by the inductor is an increase of inductance.

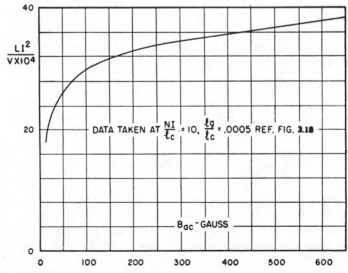

Figure 3.20. Increase of inductance with ac induction.

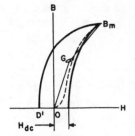

Figure 3.21. Change of permeability with ac induction.

The increase of inductance is nonlinear, and this has a decided effect on the performance of the apparatus. An inductance bridge measuring such an inductor at the higher ac voltage would show an inductance corresponding to the average slope of lines OG and GB_m. That is, the average permeability during a whole cycle is the average of the permeabilities which obtain during the positive and negative increments of induction, and it is represented by the average of the slopes of lines OG and GB_m. But if the inductor were put in the filter of a rectifier, the measured ripple would be higher than a calculated value based on the bridge value of inductance. This occurs because the positive peaks of ripple have less impedance presented to them than do the negative peaks, and hence they create a greater ripple at the load. Suppose, for example, that the ripple output of the rectifier is 500 V and that this would be attenuated to 10 V across the load by a linear inductor having a value of inductance corresponding to the average slope of lines OG and GB_m. With the inductor working between zero and B_m, suppose that the slope of OG is 5 times that of GB_m. The expected average ripple attenuation of 50:1 becomes 16.7:1 for positive flux swings, and 83.3:1 for negative, and the load ripple is

$$\frac{1}{2}\left(\frac{500}{16.7} + \frac{500}{83.3}\right) = 18 \text{ V}$$

or an increase of nearly 2:1 over what would be anticipated from the measured value of inductance.

This nonlinearity could be reduced by increasing the air gap somewhat, thereby reducing H_{dc}. Moreover, the average permeability increases, and so does the inductance. It will be apparent that decreasing H_{dc} further means approaching in value the normal permeability. This can be done only if the maximum flux density is kept low enough to avoid saturation. Conversely, it follows that if saturation is present in an inductor, it is manifested by a decrease in inductance as the direct current through the winding is increased from zero to full-load value.

In an inductor having high ac permeability the equivalent length of core l_c/μ is likely to be small compared to the air gap l_g. Hence it is vitally important to keep the air gap close to its proper value. This is, of course, in marked contrast to inductors not subject to high ac induction.

If an inductor is to be checked to see that no saturation effects are present, access must be had to an inductance bridge. With the proper values of alternating voltage across the inductor, measurements of inductance can be made with various values of direct current through it. If the inductance remains nearly constant up to normal direct current, no saturation is present, and the inductor is suitable for the purpose. If, on the other hand, the inductance drops considerably from zero direct current to normal direct current, the inductor very probably is nonlinear. Increasing the air gap may improve it; otherwise, it should be discarded in favor of a inductor which has been correctly designed for the purpose.

Filter inductors subject to the most alternating voltage for a given direct voltage are those used in inductor input filters of single-phase rectifiers. The inductance of this type of inductor influences the following:

Value of ripple in rectified output

No-load to full-load regulation

Transient voltage dip when load is suddenly applied, as in keyed loads

Peak current through rectifying devices during each cycle

Transient current through rectifying devices when voltage is first applied to rectifier

It is important that the inductance be the right value. Several of these effects can be improved by the use of swinging or tuned inductors. In a swinging inductor, saturation is present at full load; therefore the inductance is lower at full load than at no load. The higher inductance at no load is available for the purpose of decreasing voltage regulation. The same result is obtained by shunt-tuning the inductor, but here the inductance should be constant from no load to full load to preserve the tuned condition.

In swinging inductors, all or part of the core is purposely allowed to saturate at the higher values of direct current to obtain high inductance at low values of direct current. They are characterized by smaller gaps, more turns, and larger size than inductors with constant inductance ratings. Sometimes two parallel gaps are used, the smaller of which saturates at full direct current. When the function of the inductor is to control current by means of large inductance changes, no air gap is used. Design of such inductors is discussed in Chapter 8.

The insulation of an inductor depends on the type of rectifier and how it is used in the circuit. Three-phase rectifiers, with their low ripple voltage, do not require the turn and layer insulation that single-phase rectifiers do. If the inductor is placed in the ground side of the circuit one terminal requires little or no insulation to ground, but the other terminal may operate at a high voltage to ground. In single-phase rectifiers the peak voltage across the inductor is E_{dc}, so the equivalent rms voltage on the insulation is $0.707E_{dc}$. But for figuring B_{max} the rms voltage is $0.707 \times 0.67E_{dc}$. Inductor voltages are discussed in Chapter 4.

3.16   LINEAR INDUCTOR DESIGN

A method of design for linear inductors is based on three assumptions which are justified in the foregoing:

1. The air gap is large compared to l_c/μ, μ being the dc permeability.
2. Ac flux density depends on alternating voltage and frequency.
3. Ac and dc fluxes can be added or subtracted arithmetically.

From assumption 1, the relation $B = \mu H$ becomes $B = H$. Because of fringing of flux around the gap, an average of $0.85B$ crosses over the gap. Hence $B_{dc} = 0.4\pi N I_{dc}/0.85l_g$. With l_g in inches this becomes

$$B_{dc} = 0.6NI_{dc}/l_g \qquad \text{gauss} \tag{3.6}$$

Transposing equation 3.4 yields

$$B_{ac} = (3.49E \times 10^6)/fA_cN \qquad \text{gauss} \tag{3.7}$$

The sum of B_{ac} and B_{dc} is B_{max}, which should not exceed 11,000 G for 4% silicon steel, 16,000 G for grain-oriented steel, or 10,000 G for a 50% nickel alloy. Curves are obtainable from steel manufacturers which give incremental permeability μ_Δ for various combinations of these two fluxes. Figure 3.22 shows values for 4% silicon steel.

By definition, inductance is the flux linkages per ampere or, in cgs units,

$$L = \frac{\phi N}{10^8 I_M} = \frac{B_{ac}A_cN}{10^8 I_M} \tag{3.8}$$

But

$$B_{ac} = \frac{0.4\pi N I_M}{l_g + l_c/\mu_\Delta}$$

If this is substituted in equation 3.8, we have

$$L = \frac{3.19N^2 A_c \times 10^{-8}}{l_g + l_c/\mu_\Delta} \qquad \text{henrys} \tag{3.9}$$

provided that dimensions are in inches. The term A_c in equation 3.9 is greater than in equation 3.7 because of the space factor of the laminations; if the gap is large A_c is greater still because the flux across it fringes. With large gaps, inductance is nearly independent of μ_Δ, at least with moderate values of B_{max}. With small gaps, permeability largely controls. There is always a certain amount

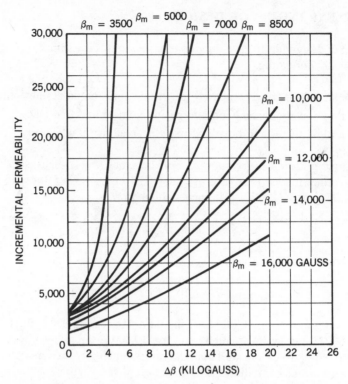

Figure 3.22. Incremental permeability for silicon steel with high ac induction.

of gap even with punchings stacked alternately in groups of 1. Table 3.4 gives the approximate gap equivalent of various degrees of interleaving laminations for magnetic path l_c of 5.5 in.

Example. An input inductor is required for the filter of a 1300-V, 0.25-A single-phase full-wave 60-Hz rectifier. Let $N = 2800$ turns, net $A_c = 2.48$ in.^{2}, gross $A_c = 2.76$ in.^{2}, $lc = 9$ in., and $l_g = 0.050$ in. The 120-Hz voltage for figuring B_{ac} is

TABLE 3.4   Equivalent Gaps with Interleaved Laminations

0.014-in. Laminations Alternately Stacked	Equivalent Air Gap (Total) with Careful Stacking (in.)
In groups of 1	0.0005
In groups of 4	0.001
In groups of 8	0.002
In groups of 12	0.003
In groups of 16	0.004
Butt stacking with zero gap	0.005

$0.707 \times 0.67 \times 1300 = 605$ V.

$$B_{dc} = \frac{0.6 \times 2800 \times 0.25}{0.050} = \underline{8{,}400\ G}$$

$$B_{ac} = \frac{3.49 \times 605 \times 10^6}{120 \times 2.48 \times 2800} = \underline{2{,}534\ G}$$

$$B_{max} = \underline{10{,}934\ G}$$

Figure 3.22 shows

$$\mu_\Delta = 5200$$

$$L = \frac{3.19 \times (2800)^2 \times 2.76 \times 10^{-8}}{0.050 + \frac{9}{5200}} = 13.3\ H$$

3.17   LINEAR INDUCTOR CHART

In Section 3.16 it was assumed that the core air gap is large compared to l_c/μ, where μ is the dc permeability. In grain-oriented steel cores the air gap may be large compared to l_c/μ_Δ, because of the high *incremental* permeability of these cores. When this is true, variations in μ do not affect the total effective magnetic path length or the inductance to a substantial degree. Inductor properties may then be taken from Fig. 3.23. To keep the inductor linear, it is necessary to limit the flux density. For grain-oriented silicon-steel cores, inductance is usually linear within 10% if the dc component of flux B_{dc} is limited to 12,000 G and the ac component B_{ac} to 3000 G.

Dashed lines in quadrant I are plots of turns versus core area for a given wire size and for low-voltage coils, where insulation and margins are governed largely by mechanical considerations. Core numbers in Fig. 3.23 have the same dimensions and weight as in Table 3.3.

If the cores increased in each dimension by exactly the same amount, the lines in quadrant I would be straight. In an actual line of cores, several factors cause the lines to be wavy:

1. Ratios of core window height to window width and core area deviate from constancy.
2. Coil margins increase stepwise.
3. Insulation thickness increases stepwise.

Ac flux density in the core may be calculated by equation 3.7 and B_{dc} by equation 3.6. If B_m materially exceeds 15,000 G, saturation is reached, and the inductor may become nonlinear or noisy.

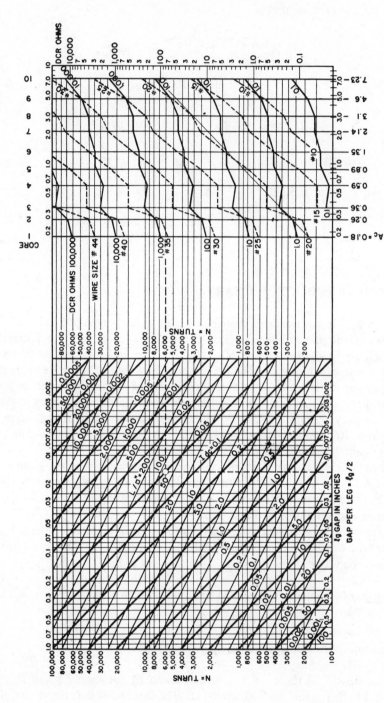

Figure 3.23. Linear inductor design chart.

94

Instructions for Using Fig. 3.23

1. Estimate core to be used.
2. Divide required inductance by area (A_c) of estimated core to obtain a value of $L/$in.^{2}.
3. In second quadrant, locate intersection of $L/$in.^{2} and rated I_{dc}.
4. On this intersection, read total gap length (l_g) and number of turns (N). Gap per leg $= l_g/2$.
5. Project intersection horizontally into first quadrant to intersect vertical line which corresponds to estimated core. This second intersection gives dc resistance and wire size.

Example. Required: 15 H at $I_{dc} = 50$ mA. Estimate core 1. $L/$in.$^2 = 84.3$, $l_g = 0.015$ in., $N = 6000$, DCR $= 800\ \Omega$. Wire size $=$ No. 36. (Example shown starting with dashed circle, Fig. 3.23.)

A similar chart may be drawn for silicon-steel laminations, but to maintain linearity lower values of flux density should be used.

3.18   AIR-GAP FLUX FRINGING

In Section 3.16, equation 3.9 was developed for inductance of a linear reactor with an air gap. It is assumed that 85% of the core flux is confined to the cross section of core face adjoining the gap. The remaining 15% of the core flux "fringes" or leaves the sides of the core, thus shunting the gap. Fringing flux *decreases* the total reluctance of the magnetic path and *increases* the inductance to a value greater than that calculated from equation 3.9. Fringing flux is a larger percentage of the total for larger gaps. Very large gaps are sometimes broken up into several smaller ones to reduce fringing.

If it is again assumed that the air gap is large compared to l_c/μ, the reluctance of the iron can be neglected in comparison with that of the air gap. For a square stack of punchings, the increase of inductance due to fringing is

$$\frac{L'}{L} = \left(1 + \frac{2l_g}{\sqrt{A_c}} \log_\epsilon \frac{2S}{l_g}\right) \tag{3.10}$$

Equation 3.10 is plotted in Fig. 3.24 with core shape $\sqrt{A_c}/S$ as abscissas and gap ratio l_g/S as parameter (see Partridge, 1936).

If the air gap is enclosed by a coil, as at the top of Fig. 3.24, flux fringing is reduced because of the magnetizing force set up near the gap by the ampere-turns of the coil. A coil fitting tightly all around the core would produce no fringing at all. As the distance from inside of the coil to the core increases, so does the fringing. Fringing therefore depends upon the coil form thickness; if it

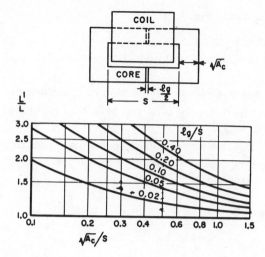

Figure 3.24. Increase of inductance with flux fringing at core gap.

materially exceeds the air gap per leg, fringing is nearly the same as it would be in a core gap which is not enclosed by a coil. Figure 3.24 is based on a thick coil form.

3.19   SIMILITUDE IN DESIGN

Charts such as Fig. 3.15 show that ratings are related to size in an orderly sequence, provided that certain proportions between core dimensions are maintained. Figure 3.15 is for 60 Hz. If a transformer is desired for another frequency, its size may be estimated from Table 3.3, provided that the same core proportions apply, and similar values of induction and temperature rise are used. If the new conditions are widely different, due allowance must be made for them or the estimate will not be accurate.

Table 3.3 and Figs. 3.15 and 3.23 are examples of *similitude*. If all variations between ratings are taken into account, similitude provides a very accurate basis for estimating new sizes; for the transformer designer there is no better basis for starting a new design.

3.20   INDUCTOR CURRENT INTERRUPTION

Sudden interruption of current through an inductor may cause high voltages to develop in the winding. This may be seen by considering the voltage across an inductor with linear inductance L and varying current i in the winding. Let current i be substituted for I_M in equation 3.8; it may be transposed to give

$$\phi = 10^8 Li/N \tag{3.8a}$$

where L is in henrys and i in amperes. If this expression for ϕ be substituted in equation 1.1, we obtain

$$e = -L\frac{di}{dt} \tag{3.11}$$

Equation 3.11 states that the magnitude of voltage across an inductor is equal

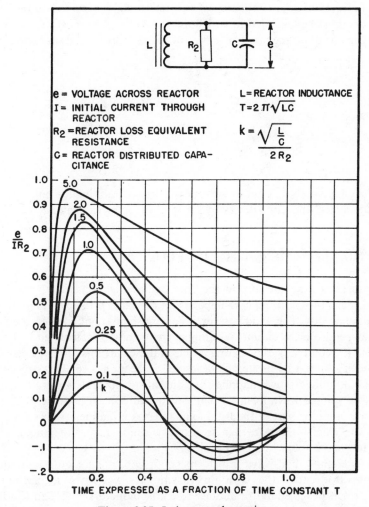

Figure 3.25. Inductor voltage rise.

to the inductance multiplied by the rate of current change with time. The sense or direction of this voltage is always such as to oppose the current change. Therefore, if current interruption takes place instantaneously, inductive voltage is infinitely large. In an actual inductor losses and capacitance are always present; hence interruption of inductor current forces the inductor voltage to discharge into its own capacitance and loss resistance. The curves of Fig. 3.25 show how the inductor voltage e rises when steady current I flowing in the inductor is suddenly interrupted. The maximum value to which voltage e could rise under any condition is IR_2, where R_2 is the equivalent loss resistance. R_2 depends mostly on the inductor iron loss at the resonance frequency determined by inductor inductance L and capacitance C. This frequency is $1/T$, where T is $2\pi\sqrt{LC}$. Conditions for high voltage across the inductor occur with high values of k, the ratio of $\sqrt{L/C}$ to $2R_2$. If subject to sudden current interruptions, inductors must be insulated to withstand this voltage, or must be protected by spark gaps or other means. The curves of Fig. 3.25 are based on equation 3.12:

$$\frac{e}{IR_2} = \frac{k}{\sqrt{k^2 - 1}} (\epsilon^{m_2 t} - \epsilon^{m_1 t}) \tag{3.12}$$

where

$$m_1, m_2 = \frac{-2\pi}{T} (k \pm \sqrt{k^2 - 1})$$

If there is appreciable circuit or wiring capacitance shunting the inductor after it is disconnected, this contributes to the total inductor capacitance C.

3.21   TRANSFORMERS WITH DC FLUX

When there is a net dc flux in the core, as in single-phase half-wave anode transformers, the choice of core depends on the same principles as in inductors with large ac flux. The windings carry nonsinusoidal load current, the form of which depends on the circuit. Winding currents may be calculated with the aid of Table 1.1. Generally, the heating effects of these currents are large. Maximum flux density should be limited as described in Section 3.16. This precaution is essential in limited power supplies like aircraft or portable generators, lest the generator voltage wave form be badly distorted. On large power systems the rectifier is a minor part of the total load and has no influence on voltage wave form. The chief limitation then is primary winding current, and maximum induction may exceed the usual limits.

In single-phase half-wave transformers, air gaps are sometimes provided in the cores to reduce the core flux asymmetry described in Section 1.12. Transformers designed in this manner resemble inductors in that core induction is calculated as in Sections 3.14 to 3.17, depending on the operating conditions.

Even in transformers with no air gap, there is a certain amount of incidental reluctance at the joints in both stacked laminations and type C cores. This small gap reduces the degree of core saturation that would exist in half-wave transformers with unbroken magnetic paths.

3.22   POWER TRANSFORMER TESTS

A power transformer is tested to discover whether the transformer will perform as required, or whether it will give reliable service life. Some tests perform both functions.

1. *DC Resistance.* This test is usually made on transformers at the factory as a check on the correctness of wire size in each winding. Variations are caused by wire tolerances, and by difference in winding tension between two lots of coils or between two coil machine operators. About 10% variation can be expected in the dc resistance of most coils, but this value increases to 20% rather suddenly in sizes smaller than No. 40. The test is made by means of a resistance bridge or specially calibrated meter.

2. *Turns Ratio.* Once the correct number of turns in each winding is established, correct output voltage can be assured for a coil of given design by measuring the turns. A simple way of doing this is by use of the turns ratio bridge in Fig. 3.26. If the turns are correct, the null indicated by the meter occurs at a ratio of resistances

$$R_1/R_2 = N_1/N_2 \qquad\qquad (3.13)$$

If there is an error in the number of turns of one winding, the null occurs at the wrong value R_1/R_2. A source of 1000 Hz is preferable to one of 60 Hz for this

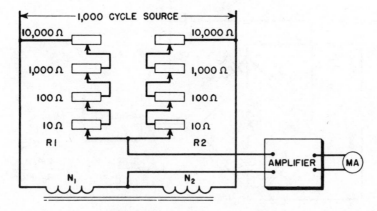

Figure 3.26. Turns ratio bridge.

test. The smaller current drawn by the transformer reduces IR and IX errors. Harmonics in the source obscure the null, and so the source should be filtered. The null is often made sharper by switching a small variable resistor in series with R_1 or R_2 to offset any lack of proportion in resistances of windings N_1 or N_2.

An accuracy of 0.1% can usually be attained with four-decade resistances. Polarity of winding is also checked by this test, because the bridge will not balance if one winding is reversed.

3. *Open-Circuit Inductance (OCL)*. There are several ways of measuring inductance. If the Q (or ratio of coil reactance to ac resistance) is high, the check may be made by measuring the current drawn by an appropriate winding connected across a source of known voltage and frequency. This method is limited to those cases where the amount of current drawn can be measured. A more general method makes use of an inductance bridge, of which one form is shown in Fig. 3.27.

If direct current normally flows in the winding, it can be applied through a large choke as shown. Inductance is then measured under the conditions of use. Source voltage should be adjustable for the same reason and should be filtered to produce a sharp null. R_c is provided to compensate for coil ac resistance. Without it an accurate measurement is rarely attained. Enough data are provided by the test to calculate ac resistance as well as inductance.

When Q is low, as it is in coils with high resistance, better accuracy is obtained with the Maxwell bridge, which is like the Hay bridge except that X_c and R_c are paralleled. Then the equations for bridge balance become

$$L_x = R_1 R_2 C \qquad R_x = R_1 R_2 / R_c \qquad (3.14)$$

The Maxwell bridge has the further advantage that the null is independent of the source frequency.

4. *Temperature Rise*. Tests to determine whether a transformer overheats are

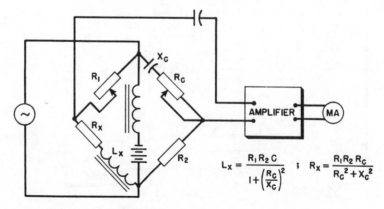

$$L_x = \frac{R_1 R_2 C}{1 + \left(\frac{R_c}{X_c}\right)^2} \quad ; \quad R_x = \frac{R_1 R_2 R_c}{R_c^2 + X_c^2}$$

Figure 3.27. Modified Hay bridge for measuring inductance.

made by measuring the winding resistances before and after a heat run, during which the transformer is loaded up to its rating. Where several secondaries are involved, each should deliver rated voltage and current. Power is applied long enough to allow the transformer temperature to become stable; this is indicated by thermometer readings of core or case temperature taken every half hour until successive readings are the same. Ambient temperature at a nearby location should also be measured throughout the test. The average increase in winding resistance furnishes an indication of the average winding temperature. Figure 3.28 furnishes a convenient means for finding this temperature.

5. *Regulation.* It is possible to measure voltage regulation by connecting a voltmeter across the output winding and reading the voltage with load off and on. This method is not accurate because the regulation is usually the difference between two relatively large quantities. Better accuracy can be obtained by multiplying the rated winding currents by the measured winding resistances and using equation 1.13. If the winding reactance drop is small this equation works well for resistive loads. To measure winding reactance drop, a short-circuit test is used. With the secondary short-circuited, sufficient voltage is applied to the primary to cause rated primary current to flow. The quotient E/I is the vector sum of winding resistances and reactances. Reactance is found from

$$X = \sqrt{Z^2 - R^2} \qquad (3.15)$$

where R includes the resistance of both windings and the meter.

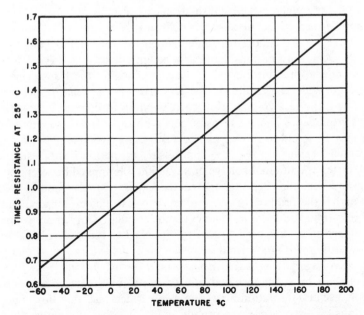

Figure 3.28. Copper resistance versus temperature in terms of resistance at 25°C.

Sometimes it is more convenient to measure the leakage inductance with secondary short-circuited on a bridge and multiply by $2\pi f$.

6. *Output Voltage.* Although the method described under test 5 is accurate for two-winding transformers, it is not applicable to multi-secondary transformers unless they are tested first with newly calibrated meters to see that all windings deliver proper voltage at full load. Once this is established, values of winding resistance and reactance thereafter can be checked to control the voltage. The interdependence of secondary voltages when there is a common primary winding makes such an initial test desirable. This is particularly true in combined filament and plate transformers, for which the best test is the actual rectifier circuit.

7. *Losses.* Often it is possible to reduce the number of time-consuming heat runs by measuring losses. The copper loss is readily calculated by multiplying the measured values of winding resistance (corrected for operating temperature) by the squares of the respective rated currents. Core loss is measured with open secondary by means of a low-reading wattmeter at rated voltage in the primary circuit. If these losses correspond to the allowable temperature rise, the transformer is safely rated.

8. *Insulation.* There is no test to which a transformer is subjected which has such a shaky theoretical basis as the insulation test. Yet it is the one test it must pass to be any good. Large quantities of transformers can be built with little or no insulation trouble, but the empirical nature of standard test voltages does not assure insulation adequacy. It has been found over a period of years that, if insulation withstands the standard rule of twice normal voltage plus 1000 V rms at 60 Hz for 1 minute, reasonable insulation life is usually obtained. It is possible for a transformer to be extremely under-insulated and still pass this test; conversely, there are conditions under which the rule would be a handicap. Therefore, it can only be considered as a rough guide.

The manner of making insulation tests depends on the transformer. Low-voltage windings categorically can be tested by short-circuiting the terminals and applying the test voltage from each winding to core or case with other windings grounded. Filament transformers with secondaries insulated for high voltage may be tested in similar manner. But a high-voltage plate transformer with grounded center tap requires unnecessary insulation if it is tested by this method. Instead, a nominal voltage of, say, 1500 V is applied between the whole winding and ground; after that the center tap is grounded and a voltage is applied across the primary of such value as to test the end terminals at twice normal plus 1000 V. Similar test values can be calculated for windings operating at dc voltages other than zero. Such a test is called an induced voltage test. It is performed at higher than normal frequency to avoid saturation. An advantage of induced voltage testing is that it tests the layer insulation.

If insulation tests are repeated one or more times they may destroy the insulation, because insulation breakdown values decrease with time. Successive applications of test voltage are usually made at either decreased voltage or decreased time. In view of their dubious value, repeated insulation tests are best omitted.

Corona or partial discharge tests are not open to this objection. These tests are normally run at from 105 to 130% of normal voltage. Partial discharge detection is discussed in Section 9.11.

Transformers which are subjected to voltage surges may be given impulse tests to determine whether the insulation will withstand the surges. Power line surges are the most difficult to insulate for. The electric power industry has standardized on certain impulse voltage magnitudes and wave shapes for this testing (see ASA Standard C57.22-1948, paragraph 22.116). The ratio of impulse voltage magnitude to 60-Hz, 1-minute insulation test voltage is called the *impulse ratio*. This ratio is much greater for oil-insulated transformers than for dry-type transformers, and is discussed further in Chapter 4.

4 RECTIFIER PERFORMANCE

4.1 RIPPLE

Filters used with rectifiers allow the rectified direct current to pass through to the load without appreciable loss, but ripple in the rectified output is attenuated to the point where it is not objectionable. Filtering must sometimes be carried out to a high degree. From the microphone to the antenna of a high-power broadcast station, there may be a power amplification of 2×10^{15}. The introduction of a ripple voltage as great as 0.005% of output voltage at the microphone would produce a noise in the received wave loud enough to spoil the transmitted program. A rectifier used at the low power levels must be unusually well filtered to prevent noticeable hum from being transmitted.

Different types of rectifiers have differing output voltage waveforms which affect the filter design to a large extent. Certain assumptions, generally permissible from the standpoint of the filter, will be made in order to simplify the discussion. These assumptions are:

1. The alternating voltage to be rectified is a sine wave.
2. The rectifying device passes current in one direction but prevents any current flow in the other direction.
3. Transformer and rectifier voltage drops are negligibly small.
4. Filter capacitor and inductor losses are negligible.

4.2 SINGLE-PHASE RECTIFIERS

Single-phase half-wave-rectified voltage across a resistive load R is shown in Fig. 4.1. It may be resolved by Fourier analysis into the direct component whose value is $0.318E_{pk}$ or $0.45E_{ac}$ and a series of alternating components. The fundamental alternating component has the same frequency as that of the supply.

Single-phase half-wave rectifiers are used only when the low-average value of load voltage and the presence of large variations in this voltage are permissible. The chief advantage of this type of rectifier is its simplicity. A method of overcoming both of its disadvantages is illustrated in Fig. 4.2, where a capacitor C shunts the load. By using the proper capacitor it is often possible to increase the value of E_{dc} to within a few percent of the peak voltage E_{pk}. The principal disadvantage of this method of filtering is the large current drawn by the capacitor during the charging interval, as shown in Fig. 3.3(b). This current is limited only by transformer and rectifier regulation, yet it must not be so large as to cause damage to the rectifier. The higher the value of E_{dc} with respect to E_{ac}, the larger is the charging current taken by C (see Figs. 3.4 and 3.5). Therefore, if a smooth current is desired, some other method of filtering must be used.

To obtain less voltage variation or ripple amplitude, after the limiting capacitor size has been reached, an inductive reactor may be employed. It may be placed on either the rectifier or the load side of the capacitor, depending on whether the load resistance R is high or low, respectively [see Fig. 4.3(a) and (b)]. In the former the voltage E_{dc} has less than the average value of $0.45E_{ac}$, because the inductor delays the buildup of current during the positive half-cycle of voltage, yet the inductor in this case should have a high value of reactance X_L, compared to the capacitive reactance X_C in order to filter effectively. When R is

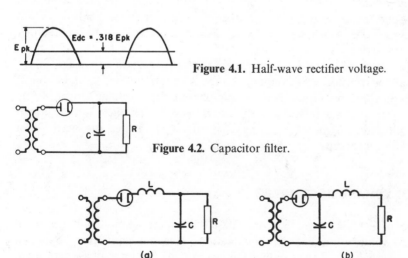

Figure 4.1. Half-wave rectifier voltage.

Figure 4.2. Capacitor filter.

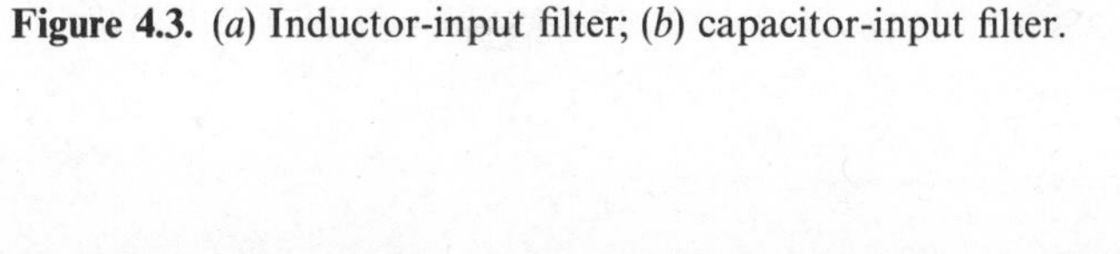

(a)   (b)

Figure 4.3. (a) Inductor-input filter; (b) capacitor-input filter.

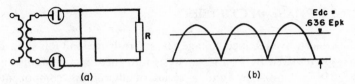

Figure 4.4. (*a*) Single-phase full-wave rectifier; (*b*) rectified full-wave voltage.

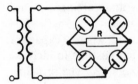

Figure 4.5. Bridge rectifier.

low, the reactance X_L should be high compared to R. In Fig. 4.3(*a*) the ripple amplitude across R is $-X_C/(X_L - X_C)$ times the amplitude generated by the rectifier if R is high compared to X_C. Also in Fig. 4.3(*b*), the ripple amplitude across R is R/X_L times the ripple obtained with a capacitor only. R here is small compared to X_L.

Large values of inductance are required to cause continuous current flow when the inductor is on the rectifier side of the capacitor in a half-wave rectifier circuit. Since current tends to flow only half the time, the rectifier output is reduced accordingly. This difficulty is eliminated by the use of the full-wave rectifier of Fig. 4.4. The alternating components of the output voltage have a fundamental frequency double that of the supply, and the amplitudes of these components are much less than for the half-wave rectifier. The higher ripple frequency causes L and C to be doubly effective; the smaller amplitude results in a smaller percentage of ripple input to the filter. Current flow is continuous and E_{dc} has double the value it had in Fig. 4.1. For these reasons this type of rectifier is widely used.

A full-wave rectifier uses only one-half of the transformer at a time: that is, E_{ac} is only half the transformer secondary voltage. A circuit that utilizes the whole of this voltage in producing E_{dc} is the single-phase bridge rectifier shown in Fig. 4.5. The output voltage relations are the same as those for Fig. 4.4(*b*). Although this circuit requires more rectifying elements it eliminates the need for a transformer center tap.

4.3 POLYPHASE RECTIFIERS

The effect of rectifying more than one phase is to superpose more voltages of the same peak value but in different time relations to each other. Figure 4.6 gives a comparative picture of the rectified output voltage for three-phase half-wave and full-wave rectifiers. Increasing the number of phases increases the value of

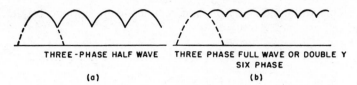

THREE-PHASE HALF WAVE   THREE PHASE FULL WAVE OR DOUBLE Y
                                        SIX PHASE

(a)   (b)

Figure 4.6. Polyphase rectifier output waves.

E_{dc}, increases the frequency of the alternating components, and decreases the amplitude of these components. Ripple frequency is p times that of the unrectified alternating voltage, p being 1, 2, 3, and 6 for the respective waves. Roughly speaking, p may be taken to represent the number of phases, provided that due allowance is made for the type of circuit, as in Fig. 4.7. Rectifiers with $p = 3$ or 6 are derived from three-phase supply lines and, by special connections, rectifiers with $p = 9$, 12, or more are obtained. The frequency of any ripple harmonic is mp, where m is the order of the harmonic.

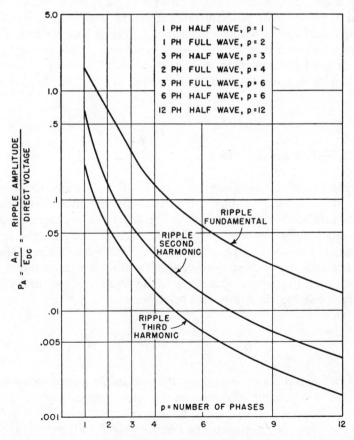

Figure 4.7. Rectifier ripple voltage.

The ripple voltage for any of these rectifiers can be found by the Fourier relation:

$$A_n = \frac{2}{T} \int_{-T/2}^{T/2} f(t) \cos n\omega t \, dt \qquad (4.1)$$

where A_n = amplitude of nth ripple harmonic
$\quad\quad T$ = ripple fundamental period
$\quad\quad t$ = time (with peak of rectified wave as $t = 0$)
$\quad\quad \omega = 2\pi/Tp = 2\pi \times$ supply line frequency
$\quad\quad f(t)$ = ripple as a function of time
$\quad\quad\quad = E_{pk} \cos \omega t, \; T/2 > \omega t > -T/2$

The voltage peak is chosen as $t = 0$ to obtain a symmetrical function $f(t)$ and eliminate a second set of harmonic terms like those in equation 4.1, but with $\sin n\omega t$ under the integral.

Ripple amplitude is given in Fig. 4.7 for the ripple fundamental and second and third harmonics with inductor input filters. In this curve the ratio P_A of ripple amplitude to direct output voltage is plotted against the number of phases p. It should be noted that P_A diminishes by a considerable amount for the second and third harmonics. In general, if a filter reduces the percentage of fundamental ripple across the load, the harmonics may be considered negligibly small.

4.4 MULTISTAGE FILTERS

In the inductor input filter shown in Fig. 4.3(a), the rectifier is a source of nonsinusoidal alternating voltage connected across the filter. It is possible to replace the usual circuit representation by Fig. 4.8(a). For any harmonic, say the nth, the voltage across the whole circuit is the harmonic amplitude A_n, and the voltage across the load is $P_R E_{dc}$, P_R being ripple allowable across the load, expressed as a fraction of the average voltage. Since the load resistance R is high compared to X_C, the two voltages are nearly in phase, and they bear the same ratio to each other as their respective reactances, or

$$\frac{P_A}{P_R} = \frac{X_L - X_C}{X_C} = \frac{X_L}{X_C} - 1 \qquad (4.2)$$

From the type of rectifier to be used and the permissible amount of ripple in the load voltage, it is possible to determine the ratio of inductance to capacitive reactance.

When the magnitude of P_R must be kept very small, the single-stage filter of Fig. 4.8(a) may require the inductor and the capacitor to be abnormally large. It

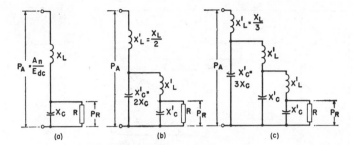

Figure 4.8. Inductor-input filter circuits.

is preferable under this condition to split both the inductor and the capacitor into two separate equal units and connect them like the two-stage filter of Fig. 4.8(b). A much smaller total amount of inductance and of capacitance will then be necessary. For this filter

$$\frac{P_A}{P_R} = \left(\frac{X'_L - X'_C}{X'_C}\right)^2 \tag{4.3}$$

X'_L and X'_C being the reactances of each inductor and capacitor in the circuit. Similarly, the three-stage filter of Fig. 4.8(c) may be more practicable for still smaller values of P_R. In the latter filter,

$$\frac{P_A}{P_R} = \left(\frac{X'_L - X'_C}{X'_C}\right)^3 \tag{4.4}$$

and in general, for an n-stage filter,

$$\frac{P_A}{P_R} = \left(\frac{X'_L - X'_C}{X'_C}\right)^n \tag{4.5}$$

It is advantageous to use more than one stage only if the ratio of P_A/P_R is high. That the gain from multistage filters is realized only for certain values of P_A/P_R is shown by Fig. 4.9. The lower curve shows the relation between P_A/P_R and X_L/X_C for the single-stage filter. The second curve shows the increase in P_A/P_R gained by splitting the same amount of L and C into a two-stage filter; as indicated in Fig. 4.8(b), the inductor and capacitor have one-half their "lumped" value. The upper curve indicates the same increase for a three-stage filter, each inductor and capacitor of which have one-third of their "lumped" or single-stage filter value. The attenuation in multistaging is enormous for high X_L/X_C. For lower ratios there may be a loss instead of a gain, as shown by the intersection of the two upper curves. These curves intersect the lower curve if all are extended farther to the left. This may be a puzzling condition, but consider that for

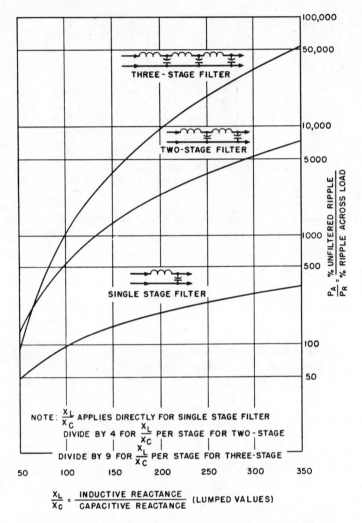

Figure 4.9. Attenuation in one-, two-, and three-stage filters.

$X_L/X_C = 50$ in the single-stage filter, the ratio is $1/3X_L/3X_C$ or $50/9$ in the three-stage filter; accounting for the rather small advantage in the latter is not difficult.

Other factors may influence the number of filter stages. In some applications, modulation or keying may require that a definite size of filter capacitor be used across the load. Usually, these conditions result in a single-stage filter, where otherwise more stages might be most economical.

Table 3.1 shows filter inductors in the negative lead, which may be either at ground or high potential. If low ripple is required in the filtered output, it is usually preferable to locate the filter inductors in the positive voltage lead. Otherwise, there is a ripple current path through the anode transformer winding capacitance to ground which bypasses the filter inductor. Ripple, then, has a residual value which cannot be reduced by additional filtering. In the three-phase, zigzag, full-wave circuit, with center tap used for half voltage output, separate inductors should be used in the positive leads; placing a common inductor in the negative lead introduces a high-amplitude ripple into the high-voltage output.

In rectifiers with low ripple requirements, the windings should be accurately center-tapped to avoid low-frequency ripple, which is difficult to filter. Three-phase leg voltages should be balanced for the same reason.

4.5 CAPACITOR INPUT FILTERS

One of the assumptions implied at the beginning of this chapter, namely, that the transformer and rectifier voltage drops are negligibly small, cannot usually be made when capacitor input filters are used, because of the large peak currents drawn by the capacitor during the charging interval. Such charging currents drawn through finite resistances affect both the dc output voltage and the ripple in a complicated manner, and simple analysis such as that given for inductor input filters is no longer possible. Figure 4.10 is a plot of the ripple in the load of capacitor input filters with various ratios of source to load resistance, and for three types of single-phase rectifiers. These curves are useful also when resistance

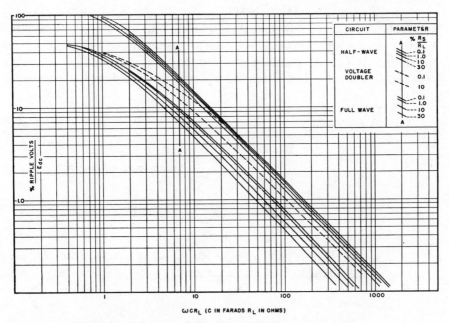

Figure 4.10. Rms ripple voltage of capacitor-input filters.

is used in place of inductance at the input of a filter. ω is 2π times the ac supply frequency, C is the capacitance, R_L the load resistance, and R_S the source resistance.

When LC filter stages follow a capacitor input filter, the ripple of the latter is reduced as in Fig. 4.9 except that the value of P_A must be taken from Fig. 4.10. When an RC filter stage follows any type of filter, the ripple is reduced in the ratio R/X_C represented by the RC stage.

4.6   RECTIFIER REGULATION

The regulation of a rectifier comprises three distinct components:

1. Dc resistance or IR drop
2. Commutation reactance or IX drop
3. Capacitor charging effect

The first component can be reduced to a small value by the use of rectifiers, transformers, and inductors having low resistance. Mercury-vapor tubes are of noteworthy usefulness in this respect, as the internal voltage drop is low and almost independent of load current variations.

Commutation reactance can be kept to a low value by proper transformer design, particularly where the ratio of short-circuit current to normal load current is high.

During part of each cycle, both elements of a single-phase full-wave rectifier are conducting. During this interval one element loses its current and the other builds up to normal current. Because of the inevitable reactance of the transformer, this change does not take place immediately but during an angle θ as in Fig. 4.11. Short-circuit current is initiated which would rise as shown by the

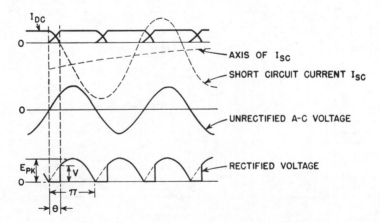

Figure 4.11. Commutation current effect on rectifier voltage.

dashed lines in Fig. 4.11 if it could pass through the rectifying elements; it prevents the rectified voltage wave from returning to its normal shape, so that for a portion of each cycle the rectified output is zero.

Let the transformer winding resistance be temporarily neglected; if the current could rise to maximum, the short-circuit value would be $2E_{pk}/X$, where X is the leakage reactance of the whole secondary, but it is limited by the rectifier to I_{dc}. The short-circuit current rises to $(1 - \cos \theta)$ times maximum in the commutation period, or

$$[2E_{pk}(1 - \cos \theta)]/X = I_{dc}$$

The average voltage from zero to the reignition point V is

$$(E_{pk}/\pi)(1 - \cos \theta)$$

Combining these relations gives, for the average voltage cut out of the rectified voltage wave by commutation,

$$V_{av} = I_{dc}X/2\pi \tag{4.6}$$

By similar reasoning, the commutation reactance drop for polyphase rectifiers is

$$V_{av} = pI_{dc}X'/2\pi \tag{4.7}$$

where X' is the transformer leakage reactance from line to neutral on the secondary side, and p is the number of phases in Fig. 4.7.

In this equation the leakage reactance per winding is associated with the voltage across that winding. This is accurate when each phase is supplied by a separate transformer. But it fails for $p = 2$ in the single-phase full-wave rectifier, using a single transformer, where half of the secondary voltage is rectified each half-cycle. In such a rectifier, during commutation, the whole secondary is effective and so is the leakage reactance in the whole secondary. This reactance has four times the leakage reactance of each secondary half-winding, but only twice the half-winding voltage acts across it. Hence equation 4.6 must be used for the single-phase rectifier; here X is the reactance of the entire secondary.

When high-winding resistance limits short-circuit current, commutation has less effect than equation 4.6 would indicate. This condition prevails in small rectifiers; the IX drop is negligibly small because of the small transformer dimensions. For example, in the transformer designed in Fig. 3.12, the leakage inductance is 0.166 H. The commutation reactance drop is, from equation 4.6,

$$0.115 \times 0.166 \times 2\pi \times 60/2\pi = 1.15 \text{ V}$$

or 0.1%. This is negligible compared to the 3.7% regulation calculated in Fig.

3.12. In this case the short-circuit current would be limited by winding resistance rather than by leakage inductance.

In large rectifiers, both the IX drop in the transformer and the IR drop in all components must be kept small to prevent excessive voltage drop, overheating, and reduced efficiency. In large rectifiers the IX drop is the dominant cause of regulation. An example with 60-kVA rating has 0.7% IR drop and 6% IX drop.

In medium-sized rectifiers the IR and IX drops may be equal, or at least comparable in value. In such rectifiers these two components of regulation do not add arithmetically. Commutation interval θ (Fig. 4.11) depends on the short-circuit reactance when resistance is negligible, but if resistance is appreciable, θ is related to the ratio X/R exponentially (see Prince and Vodges, 1927). The increase in regulation caused by commutation reactance may be found from Fig. 4.12 in terms of dc output voltage E_{dc}. In this figure the regulation of three widely used rectifiers (single-phase full-wave, three-phase half-wave, and three-phase full-wave) is given in a manner that enables one to proceed directly from the IR component of regulation to total regulation.

X and R are ohms per phase except that the X/R ratio is for the whole secondary in single-phase full-wave rectifiers. R in the X/R ratio includes primary R in all cases. R in $I_{dc}R/E_{dc}$ is for two windings in three-phase full-wave rectifiers. To obtain the total regulation, project $I_{dc}R/E_{dc}$ vertically to the one-phase or three-phase line in Fig. 4.12. Project this point to the left to the proper X/R line. The abscissa at the left gives the total regulation. An example is indicated by the dashed line. In this example, the rectifier is three-phase full-wave.

$$E_{dc} = 2000 \text{ V} \qquad \frac{X}{R} = 2$$

$$I_{dc} = 1 \text{ A} \qquad I_{dc}R/E_{dc} = 60/2000 = 3\%$$

$$R = 60 \,\Omega$$

$$X = 120 \,\Omega$$

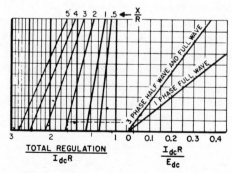

Figure 4.12. Increase in rectifier regulation due to transformer reactance.

Total regulation = $1.68 \times 3 = 5.04\%$. If the IX regulation had been added directly to IR, it would be $6\% + 3\% = 9\%$, and the calculated regulation would be nearly 4% higher than actual.

4.7 CAPACITOR EFFECT

If the rectifier had no filter capacitor, the rectifier would deliver the average value of the rectified voltage wave, less regulation components 1 and 2 of Section 4.6. But with a filter capacitor, there is a tendency at light loads for the capacitor to charge up to the peak value of the rectified wave. At zero load, this amounts to 1.57 times the average value, or possible regulation of 57% in addition to the IR and IX components, for single-phase full-wave rectifiers. This effect is smaller in magnitude for polyphase rectifiers, although it is present in all rectifiers to some extent.

Suppose that the rectifier circuit shown in Fig. 4.4(a) delivers single-phase full-wave rectifier output as shown in Fig. 4.4(b) to an inductor input filter and thence to a variable load. In such a circuit, the filter inductor keeps the capacitor from charging to a value greater than the average E_{dc} of the rectified voltage wave at heavy loads. At light loads the dc output voltage rises above the average of the rectified wave, as shown by the typical regulation curve of Fig. 4.13.

Starting at zero load, the dc output voltage E_0 is 1.57 times the average of the rectified wave. As the load increases, the output voltage falls rapidly to E_1 as the current I_1 is reached. For any load greater than I_1, the regulation is composed only of the two components IR and IX. It is good practice to use a bleeder load I_1, so that the rectifier operates between I_1 and I_2.

Filter elements X_L and X_C determine the load I_1 below which voltage rises rapidly. The filter, if it is effective, attenuates the ac ripple voltage so that there exists a dc voltage with a small ripple voltage superposed. An inductor input filter attenuates the harmonic voltages much more than the fundamental, and

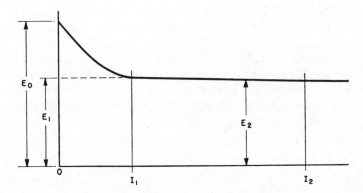

Figure 4.13. Rectifier regulation curve.

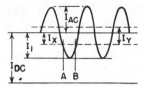

Figure 4.14. Ac and dc components of filter current.

since the harmonics are smaller to begin with, the main function of the filter is to take out the fundamental ripple voltage. This has a peak value, according to Fig. 4.7, of 66.7% of the average rectified dc voltage for a single-phase full-wave rectifier. Since this ripple is purely ac, it encounters ac impedances in its circuit. If we designate the inductor impedance as X_L and the capacitor impedance as X_C, both at the fundamental ripple frequency, the impedance to the fundamental component is $X_L - X_C$, the load resistance being negligibly high compared to X_C in an effective filter. The dc voltage, on the other hand, produces a current limited mainly by the load resistance, provided that the inductor IR drop is small.

Ac and dc components are shown in Fig. 4.14 with the ripple current I_{ac} superposed on the load direct current I_{dc}. If the direct current is made smaller by increased load resistance, the ac component is not affected because load resistance has practically no influence in determining its value. Hence a point will be reached, as the dc load current is diminished, where the peak value of ripple current just equals the load direct current. Such a condition is given by dc load I_1 which is equal to I_{ac}. If the dc load is further reduced, say to the value I_x, no current flows from the rectifier in the interval A–B of each ripple cycle. The ripple current is not a sine wave but is cut off on the lower halves, as in the heavy line of Fig. 4.15. Now the average value of this current is not I_x but a somewhat higher current I_y. That is, the load direct current is higher than the average value of the rectified sine wave divided by the load resistance. This increased current is caused by the tendency of the capacitor to charge up to the peak of the voltage wave between such intervals as A–B; hence the term "capacitor effect," which is applied to the voltage increase. The limiting value of voltage is the peak value of the rectified voltage, which is 1.57 times the sine-wave average, at zero load current.

To prevent capacitor effect the inductor must be large enough so that I_{ac} is equal to or less than the bleeder current I_1. This consideration leads directly to the value of inductance. The bleeder current I_1 is E_1/R_1, where R_1 is the value of the bleeder resistance. The ripple current is the fundamental ripple voltage

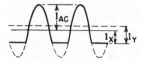

Figure 4.15. Capacitor effect at light load.

divided by the ripple circuit impedance, or

$$I_{ac} = \frac{0.667E_1}{X_L - X_C}$$

Equating I_1 and I_{ac} we have, for a single-phase full-wave rectifier,

$$R_1 = \frac{X_L - X_C}{0.667} \tag{4.8}$$

Here we see that the value of capacitance also has an effect, but it is minor relative to that of the inductor. In a well-designed filter the inductor reactance X_L is high compared to X_C. Therefore, the predominant element in fixing the value of R_1 (and of I_1) is the filter inductor.

Polyphase rectifiers have similar effects, but the rise in voltage is not so great because of the smaller difference between peak and average dc output. The bleeder resistance for eliminating capacitor effect can be found in general from

$$R_1 = \frac{X_L - X_C}{P_1} \tag{4.9}$$

where P_1 is the fundamental ripple peak amplitude from Fig. 4.7 and X_L and X_C are the filter reactances at the fundamental ripple frequency.

Between load I_1 and zero load, the rate of voltage rise depends on the filter. Figure 4.16 shows the voltage rise as a function of the ratio $(X_L - X_C)/R_L$ for a single-phase full-wave rectifier. A curve of ripple in terms of full load is given. Figure 4.16 is a plot of experimental data taken on a rectifier with $IR + IX$ regulation of 5%. Reactances X_L and X_C are computed for the fundamental ripple frequency.

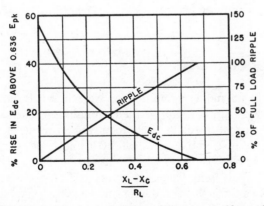

Figure 4.16. Voltage rise in single-phase, full-wave rectifier at light loads.

Capacitor input filters have the voltage regulation curves shown in Figs. 3.4, 3.5, and 3.6 for their respective circuits. At light loads these filters may give reasonably good regulation, but it is possible to get very poor regulation at heavier loads, as can be seen from the curves. Rectifier series resistance plays an important part in the voltage regulation of this type of filter. The effect of transformer leakage reactance can be found from Fig. 4.16.

4.8   TUNED RECTIFIER FILTERS

Sometimes an inductor input filter is tuned as in Fig. 4.17. The addition of capacitor C_1 increases the effective reactance of the inductor to the fundamental ripple frequency. Both regulation and ripple of this type of filter are improved. The filter is not tuned for the ripple harmonics, so the use of high-Q filter inductors is not necessary. An increase in effectiveness of the filter inductor of about $3:1$ can be realized in a single-phase full-wave rectifier circuit. Tuned filters are less effective with three-phase rectifiers because slight phase unbalance introduces low-frequency ripple which the filter does not attenuate.

Filters may be tuned as in Fig. 4.18, where the filter capacitor C_1 is connected to a tap near the right end of inductor L, and the other filter capacitor C_2 is chosen to give series resonance and hence zero reactance across the load at the fundamental ripple frequency. Because of inductor losses, the impedance across R_L is not zero, but the resulting ripple across load resistor R_L can be made lower than without the use of capacitor C_1. Ripple is attenuated more than in the usual inductor input filter, but regulation is not substantially different.

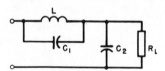

Figure 4.17. Shunt-tuned filter.

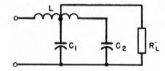

Figure 4.18. Series-tuned filter.

4.9   RECTIFIER CURRENTS

If the inductor in an inductor input filter were infinitely large, the current through it would remain constant. If the commutation reactance effect is not considered, the current through each rectifying element of a single-phase rectifier would be a square wave, as shown by I_1 and I_2 of Fig. 3.3(a). The peak value of this current wave is the same as the dc output of the rectifier and the rms value is $0.707I_{dc}$. With finite values of inductance an appreciable amount of ripple current flows through the inductor and effectively modulates I_1 and I_2, thus producing a larger rms inductor current like the waveform of Table 1.1.

Capacitor input filters draw current from the rectifier only during certain portions of the cycle, as shown in Fig. 3.3(*b*). For a given average direct current, the peak and rms values of these current waves are much higher than for inductor input filters. Values for the single-phase rectifiers are given in Fig. 3.6. If an *LC* filter stage follows the input capacitor, the inductor rms current is the output direct current plus the ripple current in quadrature.

Polyphase rectifiers are ordinarily of the inductor input type, because they are used mostly for large power, and therefore any appreciable amount of series resistance cannot be tolerated. For this reason, the low *IR* drop rectifiers, such as mercury-vapor rectifiers, are commonly used. Such rectifiers do not possess sufficient internal drop to limit the peak currents drawn by capacitor input filters to the proper values.

In a shunt-tuned power supply filter such as shown in Fig. 4.17, the current drawn from the rectifier is likely to be peaked because two capacitors C_1 and C_2 are in series, without intervening resistance or inductance. This peak subsides quickly because of the influence of inductor *L*, but an oscillation may take place on top of the rectifier current wave as shown in Fig. 4.19. The rectifier must be rated to withstand this peak current. At the end of commutation the voltage jumps suddenly from zero to *V* (Fig. 4.11). Peak rectifier current may be as much as

$$I_{\text{pk}} = V/\omega L_s \qquad (4.10)$$

L_s is half the transformer leakage inductance and $\omega = 2\pi \times$ frequency of oscillation determined by L_s in series with C_1 and C_2. This peak current is superposed on I_{dc}. It flows through the anode transformer and the rectifier, but the current in inductance *L* (Fig. 4.17) is determined by the ripple voltage amplitude and the inductor reactance. Series resistance R_s reduces this peak current to the value

$$I_{\text{pk}} = (V/\omega L_s)\epsilon^{-\pi R_s/4\omega L_s} \qquad (4.11)$$

It is obtained by applying a step function voltage to the series $R_s L_s C$ circuit. The criterion for oscillations is

$$R_s < 2\sqrt{L_s/C} \qquad (4.12)$$

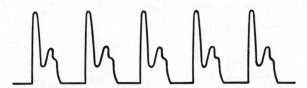

Figure 4.19. Anode current with shunt-tuned filter.

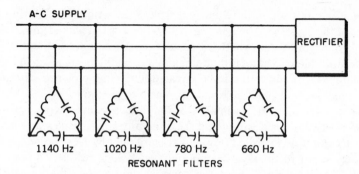

A-C SUPPLY

RECTIFIER

1140 Hz   1020 Hz   780 Hz   660 Hz

RESONANT FILTERS

Figure 4.20. Ac line filter for large power rectifier.

where C is the capacitance of C_1 and C_2 in series. Many rectifiers have peak current ratings which must not be exceeded by such currents.

Currents shown in Table 3.1 and Figs. 3.3 and 4.19 are reflected back into the ac power, except that alternate current waves are of reverse polarity. Small rectifiers have little effect on the power system, but large rectifiers may produce excessive interference in nearby telephone lines because of the large harmonic currents inherent in rectifier loads. High values of commutation reactance reduce these line current harmonics, but since good regulation requires low commutation reactance, there is a limit to the control possible by this means. Ac line filters are used to attenuate the line current harmonics. A large rectifier, with three-phase series-resonant circuits designed to eliminate the eleventh, thirteenth, seventeenth, and nineteenth harmonics of a 60-Hz system, is shown in Fig. 4.20. Smaller rectifiers sometimes have filter sections such as those in Fig. 4.21; these are rarely used in large installations because of the excessive voltage regulation introduced by the line inductors.

Filters designed to keep RF currents out of the ac lines are often used with high-voltage power supplies. Even if the high-voltage transformer has low radio influence, commutation may cause RF currents to flow in the supply lines unless there is an input filter.

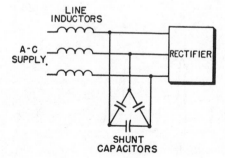

LINE
INDUCTORS

A-C
SUPPLY

RECTIFIER

SHUNT
CAPACITORS

Figure 4.21. Ac line filter for medium-sized power rectifier.

4.10   RECTIFIER TRANSIENTS

The shunt-tuned filter currents mentioned in the preceding section are transient. Since the rectifier current is cut off during each cycle, a transient current may occur in each cycle. When power is first applied to the rectifier, another transient occurs, which may be smaller or larger than the cyclic transient, depending on the filter elements. In inductor input filters the transient current can be approximated by equation 4.11 for a step function applied to a series circuit comprising filter L and C plus R_S. This circuit is valid because the shunting effect of the load is slight in a well-proportioned filter. In capacitor input filters, the same method can be used, but here the inductance is the leakage inductance of the anode transformer. Therefore, equation 4.11 applies, except that the maximum step function voltage is E_{pk}.

Transients that occur when power is first applied differ from cyclic transients in that they are sporadic. Power may be applied at any instant of the alternating voltage cycle, and the suddenly impressed rectifier voltage ranges from zero to E_{pk}. Turn-on transients are difficult to detect because of their random character; they depend on the magnetic history of the transformer core and the time in the alternating cycle at which the voltage is impressed. It may be necessary to go through the turn-on cycle several times to observe the maximum rectifier transient.

Inrush current, which occurs when power is applied to a power transformer, may have an adverse effect on the rectifiers. The current inrush phenomenon is associated with core saturation. For example, suppose that the core induction is at the top of the hysteresis loop in Fig. 2.5 at the instant when power is removed from the rectifier, and that it decreases to a residual value B_r for $H = 0$. Suppose that the next application of power is at such a point in the voltage cycle that the normal induction would be B_m. This added to B_r requires a total induction far above saturation value; therefore, heavy initial magnetizing current is drawn from the line, limited only by primary winding resistance and leakage inductance. This heavy current has a peaked waveform which may induce momentary high voltages by internal resonance in the secondary coils and damage the rectifiers. Or it may trip ac overload relays. This problem is especially acute in large transformers with low regulation. A common remedy is to start the rectifier with external resistors in the primary circuit which are short circuited a few cycles later. Some rectifiers are equipped with voltage regulators which reduce the primary voltage to a low value before restarting.

Ac line transients may cause trouble in three-phase rectifiers by shifting the floating neutral voltage, especially in those having balance coils. Filters such as that in Fig. 4.21 prevent such transients from appearing in the rectified output.

In some applications the load is varied or removed periodically. Examples of this are keyed or modulated amplifiers. Transients occur when the load is applied (key down) or removed (key up), causing, respectively, a momentary drop or rise in anode voltage. If the load is a device that transmits intelligence, the variation in filter output voltage produced by these transients results in the following undesirable effects:

1. Modulation of the transmitted signal
2. Frequency variation in oscillators, if they are connected to the same anode supply
3. Greater tendency for key clicks, especially if the transient initial dip is sharp
4. Loss of signal power

A filter that attenuates ripple effectively is normally oscillatory; hence damping out the oscillations is not practicable. Nor would it remedy the transient dip in voltage, which may increase with nonoscillatory circuits. The filter capacitor next to the load should be large enough to keep the voltage dip reasonably small. An approximation for transient dip in load voltage which neglects the damping effect of load and series resistance is

$$\Delta E_D = \frac{1}{R} \sqrt{\frac{L}{C}} \tag{4.13}$$

where E_D is the transient dip expressed as a fraction of the steady-state voltage across R, and L, C, and R as shown in Fig. 4.3(a). The accuracy of this approximation is poor for dips in transient voltage greater than 20%.

Although the tendency for key clicks in the signal may be reduced by attention to the dc supply filter elements, the clicks may not be entirely eliminated. Where key click elimination is necessary, some sort of key click filter is used, of which Fig. 4.22 is an example. The filter has inductance and capacitance enough to round off the top and back of the wave and eliminate the sharp click-producing corners. Figure 4.23 is an oscillogram showing a keyed wave shape with and without such a filter.

In an inductor input filter, voltage surges are developed across the inductor under the following conditions:

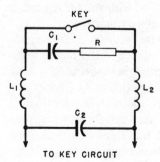

TO KEY CIRCUIT   **Figure 4.22.** Key-click filter.

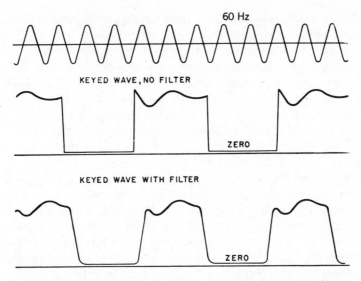

Figure 4.23. Keyed wave shape with and without key-click filter.

1. *Ripple Voltage.* With large rectifier commutation angles, or with grid-controlled rectifiers, a surge occurs once each ripple cycle. In the limit, this surge equals the rectifier peak voltage.

2. *Initial Starting Surge.* This surge adds output dc voltage. Under the worst conditions it raises the voltage at this point to twice normal and occurs every time rectifier plate voltage is applied.

3. *Keying or Modulation Transient.* Surge value depends on constants L, C, and R_L, and is limited by considerations of wave shape. This occurs each time the key is opened or closed, or as load is changed.

4. *Short-Circuit Surge.* If load R_L is suddenly short circuited, it causes full dc voltage to appear across the filter inductor until the input to the rectifier, often a circuit breaker, is opened. Circuits are sometimes arranged so that if the short persists, the circuit breaker recloses three times and then remains open.

5. *Interruption of Inductor Current.* This surge voltage is limited only by losses and capacitance of the circuit; it can be large, as shown by Fig. 3.25. Unless the inductor is designed to produce this voltage, it occurs only accidentally.

Conceivably, surges 1, 2, and 3 may occur simultaneously and add arithmetically. An inductor insulated to withstand surges 1 plus 2 plus 3 would also withstand surge 4. A reasonable value of peak surge voltage comprising these factors is 2.5 times the full dc working voltage. If surges 1 and 5 are too great for reasonable insulation, the inductor should be protected by a spark gap or other means.

If a power supply is disconnected from the supply line while the load is off, interruption of the anode transformer peak magnetizing current may cause high voltages to appear at random in the windings in much the same way as inductor current interruption causes high voltages. This is especially true if the transformer is operating at high core induction. The effect is partly mitigated by the arc energy incident to the opening of the disconnecting switch. But unless the transformer is insulated specifically to prevent dangerously high voltages, it may be necessary to add protective elements in a power supply subject to switching at light loads. The need for such protection may be determined from exciting volt-ampere data and the curves of Fig. 3.25.

Sometimes, insufficient attention is given to the manner in which power supply lines are brought into buildings. This is particularly important where a power supply is supplied by overhead high-voltage lines. Because of their relatively high surge impedance, lightning and switching surges occurring on such lines may cause abnormally high voltages to appear in a power supply and break down the insulation of transformers or other component parts. The likelihood of such surges occurring should be considered before transformers for such applications are designed.

Underground cable power lines impose much less severe hazards: first because they are protected from lightning strokes, and second, because they have much lower impedance (about one-tenth that of overhead lines). Surges on these cables have much lower values compared to those on overhead lines carrying the same rated voltage. Protection against these surges varies with the type of installation.

The best protection of all is provided by an indoor power system with an underground cable connecting it to the rectifier. Good protection is afforded by an oil-insulated, outdoor, surge-proof power distribution transformer and rectifier. No protection at all is provided when overhead lines are connected directly to the power supply input.

With the increased use of dry-type insulated transformers it is desirable to use lightning arresters on overhead lines where they enter the building. Because of their low impulse ratio, dry-type transformers require additional arresters inside the building. When a line surge is discharged by a lightning arrester, there is no power interruption.

4.11 RECTIFIER FILTER CHARTS

There are available to electronic circuit designers a wide range of programs that can be used to design filters for use in power supply circuits and to analyze their performance. Many of these programs can be used with hand-held scientific calculators and other, more sophisticated programs utilize digital computers. Even with these powerful tools it is important for the designer to have an understanding of what factors are involved in the design of the filter.

From the preceding sections it can be seen that various properties of rectifier filters, such as ripple, regulation, and transients, may impose conflicting conditions on the rectifier design. To save time in what otherwise would be a laborious cut-and-try process, charts are used. In Fig. 4.24 the more usual filter properties are presented on a single chart to assist in arriving at the best filter directly. This chart primarily satisfies ripple and regulation equations 4.2 and 4.9 for an inductor input filter.

The abscissa values of the right-hand scale are bleeder conductance in milliamperes per volt, and the left-hand scale, filter capacitance in microfarads. Ordinates of the lower vertical scale are inductance in henrys. Lines representing various amounts of ripple in the load are plotted in quadrant I, labeled both in dB and rms percent ripple. In quadrant II, lines are drawn representing different types of rectifiers and supply line frequencies. A similar set of lines is shown in quadrant IV.

Two orthogonal sets of lines are drawn in quadrant III. Those sloping downward to the right represent resonant frequency of the filter L and C, and also load resistance R_L. The other set of lines is labeled $\sqrt{L/C}$, which may be regarded as the filter impedance. It can be shown that the transient properties of the filter are dependent on the ratio of $\sqrt{L/C}$ to R_L.

The L scale requires a correction to compensate for the fact that ripple is not exactly a linear function of L but rather of $X_L - X_C$. The curves in the lower part of quadrant IV give the amount of correction to be added when the correction factor is greater than 1%.

Instructions for Using Chart

1. Assume a suitable value of bleeder resistance or bleeder current I_1 in milliamperes per volt of E_{dc}. This is also the steady-state peak ripple current in milliamperes.

2. Trace upward on the assumed bleeder ordinate to intersect the desired value of load ripple, and from here trace horizontally to the left to a diagonal line for the rectifier and supply frequency used. Directly under, read the value of C.

3. Trace downward on the assumed bleeder ordinate to intersect the diagonal line below for rectifier and supply frequency, and read value of L.

4. From the desired ripple value, determine the correction for L on the graph at the lower right and add the indicated correction to the value of L.

5. Using the corrected value of L and the next standard value of C, find the intersection in quadrant III, and read the maximum resonant frequency f_r.

6. Using the same values of L and C as in step 5, read the value of $\sqrt{L/C}$.

7. Under the intersection of $\sqrt{L/C}$ with load resistance R_L, read the values of the four transients in Fig. 4.25 (in percent).

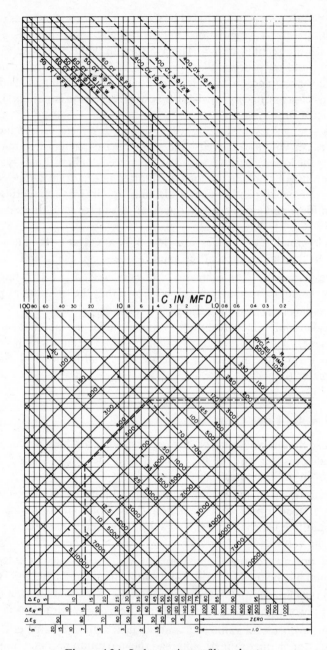

Figure 4.24. Inductor-input filter chart.

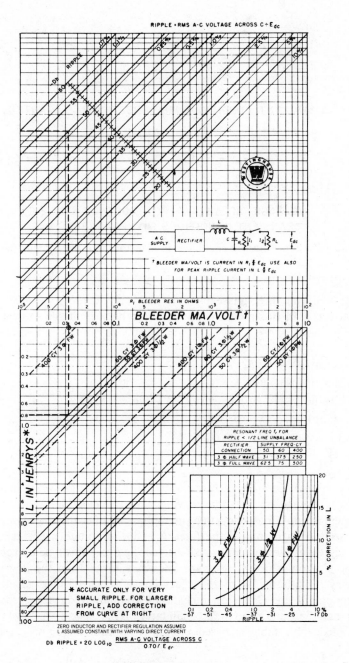

Figure 4.24 (Cont.)

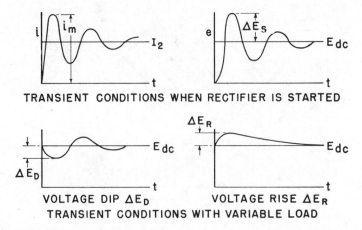

TRANSIENT CONDITIONS WHEN RECTIFIER IS STARTED

VOLTAGE DIP ΔE_D VOLTAGE RISE ΔE_R
TRANSIENT CONDITIONS WITH VARIABLE LOAD

Figure 4.25. Four transient conditions in inductor-input filter circuit and curves.

Example. Given a three-phase full-wave 60-Hz rectifier; $E_{dc} = 3000$ V; $I_2 = 1$ A; $I_1 = 96$ mA; load ripple $= -50$ dB; balanced line. (Example shown as dashed lines, Fig. 4.24.)

Bleeder mA/V $= 0.032$
$C = 4.5 \, \mu F$ (use $5 \, \mu F$)
Scale value of $L = 0.78$ H; corrected value $= 0.82$ H
Resonant frequency $= 75$ Hz
Load resistance $R_L = 3000 \, \Omega$
$i_m = 7I_2 = 7$ A; $\Delta E_D = 12\%$; $\Delta E_R = 15\%$; $\Delta E_S = 80\%$

In polyphase rectifiers the possibility exists of enough phase unbalance to impress a voltage on the filter having a frequency lower than the normal fundamental ripple frequency. In a three-phase rectifier this unbalanced voltage is the second harmonic. If the filter L and C resonate near the unbalance frequency, excessive ripple may be expected. To avoid this type of trouble, the L and the C should have a resonant frequency lower than the line unbalance frequency. Quadrant III of the chart has a series of lines labeled f_r; the intersection of L and C thereon indicates the resonant frequency. f_r should be no higher than the value given in the small table on the chart if excessive ripple is to be avoided. This table is based on 2% maximum unbalance in the phase voltages.

For most practical rectifiers, transient conditions fall within the left-hand portion of quadrant III. The other conditions sometimes help in the solution of problems in which L and C are incidental, for example, the leakage inductance and distributed capacitance of an anode transformer.

Although the chart applies directly to single-stage untuned filters with constant inductance, it can be used with others with modifications:

1. *Shunt-Tuned Inductor per Fig. 4.17.* Figure 4.24 can be used directly for capacitance C, but for a given amount of ripple divide the chart values of inductance by 3 to obtain the actual inductance needed.

2. *Swinging Inductor.* If at light load the filter inductor swings to S times the full-load value of inductance, multiply the capacitance obtained from the chart by the ratio $1/S$ to find the capacitance needed (C_n). The value of L obtained by projecting the bleeder current downward is the maximum or swinging value. It must be divided by S to obtain the full-load value. Transient conditions may then be approximated by using capacitance C_n and the full-load value of inductance.

3. *Two-Stage Filters.* In a filter with two identical stages [Fig. 4.8(b)], the chart can be used if it is recognized that the ripple is that on the load side of the first inductor. For example, if the filter consists of two stages both equal to the example given for the single-stage filter, the ripple would not be -100 dB but -75 dB because of the fact that the output has (per Table 3.1) only 4% ripple, which is -25 dB.

The regulation in a two-stage filter, as far as capacitor effect is concerned, depends on the inductance of the first inductor as in the single-stage filter. Therefore, the chart applies directly to the inductance and capacitance of one stage. The peak ripple current similarly depends on the inductance of the first inductor, regardless of the location of the bleeder resistor. Transients, however, are more complicated, owing to the fact that the two stages interact under transient conditions [see *Proc. IRE.*, *22*, 213 (February 1934)].

4.12   RECTIFIER EFFICIENCY

Losses in a rectifier circuit consist of transformer, rectifier element, and filter losses. If rectification is accomplished with an electron tube, filament power must also be counted as a loss, particularly when the tube rectifier is being compared with a rotating machine or a semiconductor rectifier. Even with this loss, a high-voltage polyphase rectifier of the mercury-vapor or pool type may have 95% efficiency at full load.

4.13   RECTIFIER TESTS

Even though the transformers, inductors, rectifying elements, and capacitors have been tested before assembly of the power supply, performance tests of the power supply are desirable. These generally include tests of output, regulation, efficiency, ripple, and input kilovolt-amperes or power factor. Accurate meters should be used and polyphase circuits should have balanced input voltages.

Wiring is tested at some voltage higher than normal voltage, preferably with the electrical components disconnected to avoid damage to them during the test. Ordinary care in testing is sufficient except for regulation tests. If the regulation is low, the difference in meter readings at no load and at full load may be inaccurate. Differential measurements are sometimes used, such as a voltmeter connected between the power supply output and a fixed source of the same polarity and voltage. Artificial loading of a high-voltage power supply is often a problem. Variable water loads have been used for this purpose. Load tests, preferably with the power supply as a part of the system in which it is to be used, are desired as safeguards against troubles in field operation. Operating tests are essential when the load is keyed or modulated, so that overheating or inadequate transformer operation may be detected.

Ripple is measured either with a special hum-measuring instrument or with a capacitance-resistance network arranged to block direct current from the measuring circuit. Capacitance and resistance values in the measuring circuit should be chosen so as to avoid influencing the ripple or loading the transformer; sometimes capacitor dividers are used for this purpose. The problem of proper values becomes particularly critical with high-voltage low-current power supplies. The effect of stray capacitance is especially important.

5 AMPLIFIER TRANSFORMERS

An amplifier is a device for increasing voltage, current, or power in a circuit. The original waveform may or may not be maintained; the frequency usually is. An amplifier may be mechanical, electromechanical, electromagnetic, or electronic in form, or it may be a combination of these. In this chapter the transformer-coupled electronic amplifier is considered. The amplifier consists of an amplifying device and the transformers, capacitors, and resistors necessary for the desired operation. The amplifying device may be a vacuum tube, bipolar transistor, FET, or any other device that has an input element controlling an output. A voltage or current is impressed the input element of the amplifying device. A higher voltage or current appears in the output circuit.

In the previous chapters, transformers were considered to be single-frequency devices. A transformer can be operated at other frequencies than the one for which it was designed. Unless a transformer is specifically designed for operation over a frequency range, the transformer may not be suitable for the application. Transformers designed to operate at many frequencies at the same time are called wide-band transformers. A wide-band transformer may be operated in either the frequency or the time domain. A wide-band transformer operated in the time domain is called a pulse transformer. Inverter transformers are a special class of pulse transformers.

In the discussion of power transformers, leakage inductance and distributed capacitance were both mentioned. These parameters are not important in the design of low-voltage power transformers. In the design of high-voltage transformers both leakage inductance and distributed capacitance must be controlled. Leakage inductance, distributed capacitance, and open-circuit inductance determine the performance of wide-band transformers. The relationships of these parameters to the operation of transformer-coupled amplifiers and the transformer design techniques that ensure proper operation of the amplifier are discussed in this chapter.

5.1   AMPLIFIERS

Electronic amplifiers are characterized by the use of amplifying devices having three or more elements. Some types of amplifying devices, such as vacuum tubes and field-effect transistors (FETs), are voltage-driven devices. The input elements are dependent on the voltage for control of the input. The input elements draw little or no current except in specific cases. Bipolar transistors are current-fed devices; that is, the input current controls the output current. Voltage is necessary to cause the current to flow through the input impedance. The difference in the design of transformers for voltage- or current-fed devices is primarily the magnitude of the impedances involved.

The addition of the third element to a two-element device alters the way current flows between the negative element (cathode) and positive element (anode). The various elements are called by different names in different devices, but, to generalize the discussion they are called negative element, positive element, and input element to differentiate the functions. A comparison of the element names for different types of devices is given in Table 5.1. The negative element is shown at zero voltage in Fig. 5.1. Depending on whether the control voltage is positive or negative compared to the negative element, the control element either aids or opposes the flow of electrons from the negative to positive elements. The voltage gradients for a vacuum tube are shown in Fig. 5.1. The action of a p-n junction with voltage applied is shown in Fig. 5.2. A combination of p-n junctions is a junction transistor. By applying a control current to one p-n junction of a transistor, the current in the other p-n junction can be changed in proportion to the relative impedances. FETs are a special construction of p-n junctions, in which the output is controlled by the electric field at the control element. A comparison of the two types of transistors is shown in Fig. 5.3. As the

TABLE 5.1   Comparison of Electrode Names in Amplifying Devices

| Generic Name | Vacuum Tube | Bipolar Transistor | | FET, MOSFET, etc. |
		NPN	PNP	
Positive electrode	Plate (anode)	Collector	Emitter	Drain
Negative electrode	Cathode	Emitter	Collector	Source
Control electrode	Grid	Base	Base	Gate

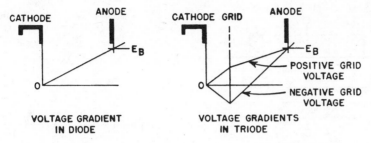

Figure 5.1. Vacuum-tube diode and triode voltage gradients.

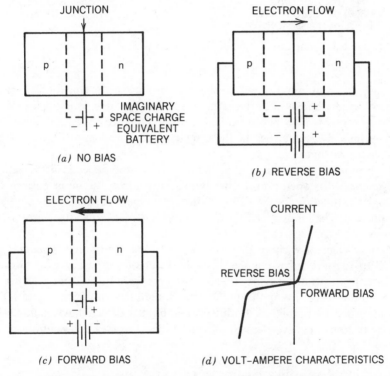

Figure 5.2. Action of a p-n junction with applied voltages.

control voltage is made more and more negative, electron flow is diminished and finally stops. At this point the current from the positive element is zero. This condition is called current cutoff.

As the control voltage is made more positive, the current from the positive element increases. Eventually, a further increase in control voltage causes no additional output current increase. This condition is called saturation.

The operation of NPN bipolar transistors and NFETs can be described

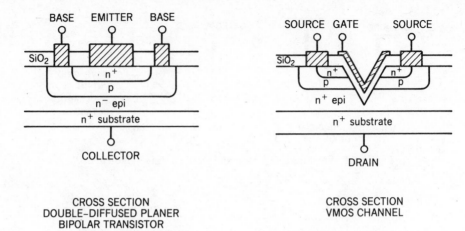

Figure 5.3. Comparison of bipolar and MOSFET transistor construction.

similarly. Current cutoff occurs in these devices when the control element is at zero voltage with respect to the cathode and not negative, as in a vacuum tube. The operation of an NPN bipolar transistor is similar to a vacuum tube drawing grid current. The polarities of the circuit are reversed for PFETs and PNP bipolar transistors.

Both bipolar transistors and field-effect transistors depend on the action of p-n junctions for operation. In the bipolar transistor, the input current flows through the junction. In the FET, the input voltage is capacitatively coupled to the junction. The input current is limited to very small capacitive currents. This is shown in Fig. 5.3.

A three-element vacuum tube is called a triode. Additional elements (grids) added to vacuum tubes change their characteristics. The curves showing the performance of transformer-coupled amplifiers were derived for vacuum-tube applications. They are applicable for the design of transformers for FET and bipolar transistor amplifiers. Both FET and bipolar devices have characteristics similar to four- and five-element vacuum tubes (tetrodes and pentodes).

5.2 TRANSFORMER-COUPLED AMPLIFIERS

Amplifier circuits in which transformers are used can be represented by circuits similar to Fig. 5.4. Here amplifying devices are shown with a voltage impressed on the input of the amplifier. The input circuit consists of a biasing circuit (a constant voltage) and a superimposed alternating voltage e_i. A positive dc voltage E_B is applied from some source through the transformer primary. The output is an alternating voltage e_0 across the primary of the transformer. The secondary of the transformer is connected to a load Z_L. Under certain conditions the circuits shown in Fig. 5.4(a) to (c) can be simplified to that of Fig.

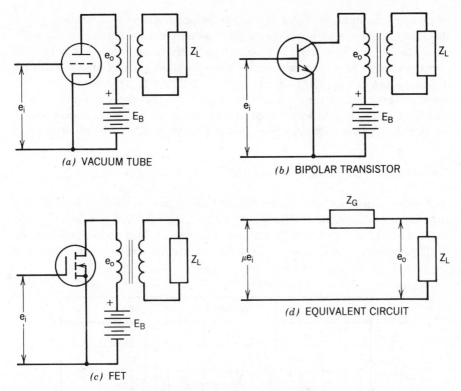

Figure 5.4. Transformer-coupled amplifiers.

5.4(*d*). A fictitious alternating voltage μe_i is impressed on the circuit, where μ is the small-signal gain of the amplifying device. A different symbol is used for the gain of different types of amplifying devices. The symbol μ is used here for convenience in expressing the gain of any type of amplifying device. The internal impedance Z_G is in series with the load Z_L, which is reflected by the transformer to the proper value in the primary circuit for amplifying element operation. That is, Z_L in Fig. 5.4(*d*) is equal to that in Fig. 5.4(*a*) to (*c*) only if the transformer has a 1:1 ratio. For any ratio *a*, the quotient of two *Z*'s is equal to a^2. The winding resistances are regarded as zero. In the absence of an input signal, full voltage E_b appears on the positive element of the amplifying device.

Alternating voltage μe_i causes voltage e_o to appear across the load Z_L. The voltage e_o is not μ times e_i but is related by the following equation:

$$e_o = \mu e_i \frac{Z_L}{Z_G + Z_L} \tag{5.1}$$

This is a general equation for voltage gain. How close the amplifier actually comes to the theoretical gain of the amplifying device can easily be seen to be a function of the internal impedance of the amplifying device. In vacuum tubes this internal impedance is quite high and has a significant impact on the stage gain. The internal impedance of bipolar transistors or FETs is less than it is for vacuum tubes. The device gain is also less. The stage gains of amplifiers using different types of amplifying devices may be comparable. Transformer-coupled amplifiers may be used for voltage amplification, but they are used mostly where matching of the output impedance of one device to the input impedance of another device to achieve optimum power transfer is extremely important.

5.3   AMPLIFIER CLASSIFICATION

Amplifiers using any type of amplifying device can be divided into classes, depending on the mode of operation. A class A amplifier is one in which the input bias and ac input voltage are such that current flows continuously in the output electrode. In a class B amplifier the input bias is almost equal or equal to

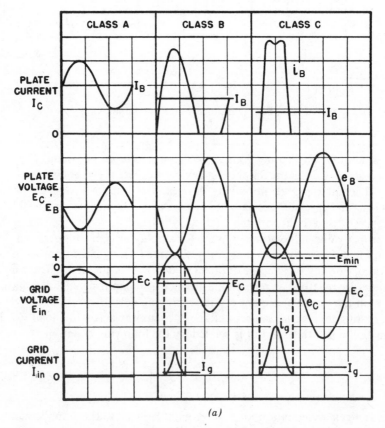

(a)

Figure 5.5. (*a*) Vacuum-tube and (*b*) bipolar transistor amplifier voltages and currents.

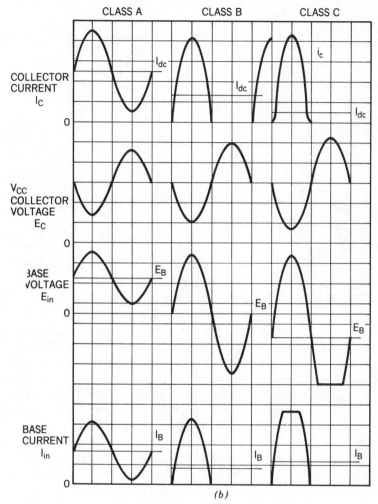

Figure 5.5. (Cont.)

the cutoff value for the amplifying element, so that the output current is nearly zero with no ac input signal. When full ac input voltage is applied, output current flows for approximately one-half of each cycle. A class C amplifier is biased well into the cutoff region, so that output current is zero when no input ac voltage is applied and it flows for appreciably less than one-half of each cycle when an ac input voltage is applied. These classes are illustrated in Fig. 5.5. The output current, output voltage, input ac voltage, and input ac current are shown with the steady or average values, which are, respectively, I_{in}, E_{in}, E_c, and I_c. Relative output and input voltage amplitudes for these three types of amplifiers are shown in Fig. 5.5, and other properties are summarized in Table 5.2.

TABLE 5.2   Amplifier Classes

Amplifier Class:	A	B	C
Anode efficiency			
Theoretical maximum	50%	78.5%[a]	100%
Practical value for low distortion	Up to 30%	40–67%[a]	70–85%[b]
Output proportional to	e_g^2	e_g^2	E_B^2 (grid saturated)
Grid current I_g	None	Small	Large (may $\approx I_B$)
Anode current I_B	Fairly constant	$e_g=0$, I_B low $e_g=$max., I_B high	$e_g=0$, $I_B=0$ $e_g=$max., I_B high

[a]These values are for push-pull amplifiers.
[b]With a high-Q tank circuit, the efficiency depends on excitation power.

Class A amplifiers are characterized by comparatively high no-signal output current. The input ac does not drive the grid of a vacuum tube into the region where it draws current or the base of a bipolar transistor into cutoff. Output current remains comparatively constant when averaged over a complete ac cycle. In class B amplifiers the input is biased at or near the cutoff voltage in the absence of an ac input signal. Positive swings of the ac voltage result in the amplifying device conducting current. The voltage at the positive electrode will dip under these conditions. Negative swings of the ac voltage at the input cause no current flow, but the voltage at the positive electrode will increase. In class C amplifiers the dc bias on the input is such that the amplifying device is well into the cutoff region. Current through the device flows for much less than one-half cycle, and mostly when the voltage on the positive electrode is at a fairly low value. Peak input currents may reach values nearly as great as the output currents of the amplifying device. A sine wave of voltage is maintained by using a resonant circuit on the output of the amplifying device.

Operation is often improved by using two amplifying devices in push-pull, as shown in Fig. 5.6. This is the most common connection for class B amplifiers and it is often used for class A amplifiers. Intermediate between class A and class B amplifiers are those known as class AB, with input bias and efficiency between class A and B amplifiers. These amplifiers are biased closer to cutoff than are class A amplifiers. Amplification of the positive and negative halves of the input sine wave are unsymmetrical. Conduction of the amplifying device does occur in each half-cycle. Class AB amplifiers are usually used in a push-pull connection.

The properties of the various amplifying devices, such as input impedance, output impedance, and amplification factor, can be found in published data. Operating conditions for the device, such as maximum voltage, voltage swings, power output, and device power dissipation, can also be obtained from the manufacturer's data for the amplifying device.

5.4   DECIBELS; IMPEDANCE MATCHING

In working with amplifiers, the ratio of two voltages E_1 and E_2 at the same impedance level is often stated in decibels (dB) according to the definition

$$dB = 20 \log_{10}(E_1/E_2) \tag{5.2}$$

Amplifier voltage gain, transformer ratio, frequency response, and noise levels may all be expressed in decibels. Voltage, current, or power in decibels must be compared to a reference level; otherwise, the term is meaningless. A standard reference level is 1 mW. This is expressed as 0 dBm. Across 600 Ω, the voltage for 0 dBm is $\sqrt{0.001 \times 600} = 0.775$ V; for 20 dBm the voltage is 7.75 V.

Transmission lines at 400 Hz and higher frequencies exhibit properties often ignored at 60 Hz. Line wavelength, characteristic impedance, attenuation, and impedance matching are important at higher frequencies. If a long transmission line has no resistance, its characteristic impedance is given by

$$Z_0 = \sqrt{L/C} \tag{5.3}$$

where L and C are the inductance and capacitance per unit length. If such a line terminates in a pure resistance load equal in ohmic value to Z_0, all the power fed into the line appears in the load without attenuation or reflection. This is called matching the impedance of the line. To save power in amplifiers and avoid reflections, impedance matching of transmission lines is the usual practice. The notion has been extended to include the loading of amplifying devices, but this is stretching the meaning of the term "matching." An amplifying device has an optimum load impedance, but the value depends on the operation of the device and is not necessarily the same as the device internal impedance.

Power transmission lines operating at 60 or 400 Hz are rarely long enough to act as appreciable source impedances. When a short circuit or low impedance fault occurs on the load side of a power transformer, the load current is limited mainly by the transformer short-circuit impedance. In an amplifier, the load current delivered into a short-circuit load is limited mainly by the internal impedance of the amplifying device rather than by the transformer. At certain frequencies the transformer itself may contribute to low load impedance. The greatest difference between power and amplifier transformers is the difference in source impedance. Even the use of the word "impedance" in the two fields of application reflects this difference. In power work, transformer impedance denotes the short-circuit impedance. In amplifier work, the same term refers to the load or source impedance.

5.5   WIDE-BAND TRANSFORMERS

In designing a wide-band transformer, the influence of the circuit on transformer operation must be determined. There are several factors external to the

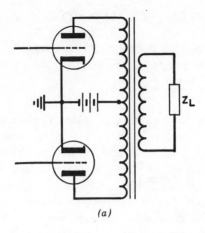

(a)

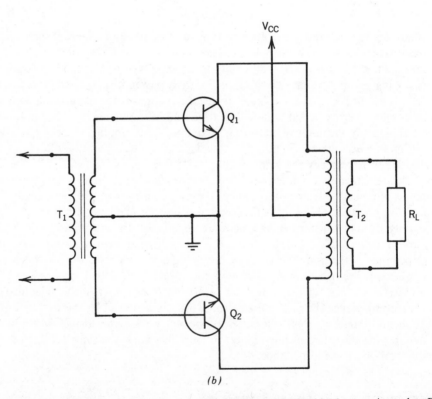

(b)

Figure 5.6.(*a*) Vacuum tube class B push-pull amplifier; (*b*) NPN bipolar transistor class B push-pull amplifier; (*c*) FET class B push-pull amplifier; (*d*) NPN transistor class AB push-pull amplifier.

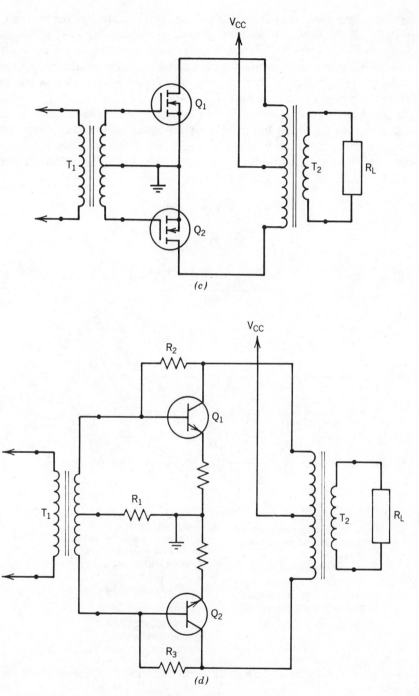

Figure 5.6. (Cont.)

transformer that affect its operation. These are: (1) impedance of the source, (2) impedance of the load, (3) frequency, and (4) relationship of the various impedances to the frequency and amplitude of the signal. Since the input and output impedances usually contain some reactive component, the impedances will change with frequency. The varying impedances have a very important effect on the design of the transformer.

An equivalent circuit that is similar to the one used for power transformers may also be used for wide-band transformers. Any transformer that connects a source to a load may be represented by Fig. 5.7(a). There are other equivalent circuits derived from network theory which could be used. The one shown here is satisfactory for the large majority of practical designs.

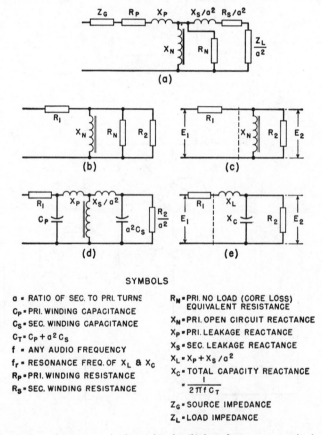

SYMBOLS

a = RATIO OF SEC. TO PRI. TURNS
C_P = PRI. WINDING CAPACITANCE
C_S = SEC. WINDING CAPACITANCE
C_T = $C_P + a^2 C_S$
f = ANY AUDIO FREQUENCY
f_r = RESONANCE FREQ. OF X_L & X_C
R_P = PRI. WINDING RESISTANCE
R_S = SEC. WINDING RESISTANCE

R_N = PRI. NO LOAD (CORE LOSS) EQUIVALENT RESISTANCE
X_N = PRI. OPEN CIRCUIT REACTANCE
X_P = PRI. LEAKAGE REACTANCE
X_S = SEC. LEAKAGE REACTANCE
X_L = $X_P + X_S / a^2$
X_C = TOTAL CAPACITY REACTANCE
        $= \dfrac{1}{2 \pi f C_T}$
Z_G = SOURCE IMPEDANCE
Z_L = LOAD IMPEDANCE

Figure 5.7. (a) Transformer equivalent circuit; (b) low-frequency equivalent circuit; (c) simplified low-frequency circuit; (d) high-frequency equivalent circuit; (e) simplified high-frequency circuit.

5.5.1 Low-Frequency Response

At low frequencies, the leakage reactances are negligibly small. Resistance R_p may be combined with Z_g to form R_1 for a pure resistance source, and R_s with Z_L to form R_2 for a resistance load. At low frequencies both source and load are pure resistances, and the circuit may be simplified to that of Fig. 5.7(b). The a^2 has been dropped, indicating a transformer with a $1:1$ turns ratio referred to the primary side. X_N is the primary open-circuit reactance, or $2\pi f$ times the primary open-circuit inductance (OCL) as measured at low frequencies.

If the shunt resistance R_N is included in the load resistance R_2, the circuit becomes like that of Fig. 5.7(c). Winding resistances are small compared with source and load resistances in well-designed transformers. Similarly, R_N is high compared with the load resistance, especially if core material of good quality is used.

To a good approximation in Fig. 5.7(c), R_1 may represent the source impedance and R_2 the load impedance. On a $1:1$ turns ratio basis, the voltages E_2 and E_1 are proportional to the impedances across which they appear, or

$$\frac{E_2}{E_1} = \frac{jX_N R_2/(jX_N + R_2)}{R_1 + jX_N R_2/(jX_N + R_2)} \tag{5.4}$$

The scalar value of this ratio is found by taking the square root of the sum of the quadrature terms:

$$\frac{E_2}{E_1} = \frac{1}{\sqrt{(1 + R_1/R_2)^2 + (R_1/X_N)^2}} \tag{5.5}$$

Equation 5.5 holds for any values of R_1, R_2, and X_N, but there are three cases that deserve particular attention: (a) $R_2 = R_1$, (b) $R_2 = 2R_1$, and (c) $R_2 = \infty$. Of these, case (a) corresponds to the usual matching transformer with the source and load impedances equal; case (b) is sometimes used; and case (c) is realized practically when the load is the input of a class A amplifier or transistor circuits with low source impedance. For these cases, equation 5.5 becomes

$$\frac{E_2}{E_1} = \frac{1}{\sqrt{4 + (R_1/X_N)^2}} \tag{5.5a}$$

$$\frac{E_2}{E_1} = \frac{1}{\sqrt{2.25 + (R_1/X_N)^2}} \tag{5.5b}$$

$$\frac{E_2}{E_1} = \frac{1}{\sqrt{1 + (R_1/X_N)^2}} \tag{5.5c}$$

These three equations are plotted in Fig. 5.8 to show low-frequency response as "dB down" from median. The median frequency in an audio transformer is the geometric mean of the audio range. For other transformers it is a frequency above the low-frequency cutoff and below the high-frequency cutoff at which the ratio X_N/R_1 is very large. At median frequency the circuit is properly represented by Fig. 5.7(c). The equivalent voltage ratio E_2/E_1 has maxima of 0.5, 0.667, and 1.0 for cases (a), (b), and (c), respectively, at the median frequency, or for $X_N/R_1 = \infty$ in Fig. 5.8. The higher the OCL, the nearer the transformer voltage ratio approaches the median frequency value. Lowering the value of loading resistance R_2 lowers the equivalent voltage ratio. The factors 0.5, 0.667, and 1.0, multiplied by the turns ratio a, give the actual voltage ratio at median frequency. At lower frequencies, these factors are lower.

The transformer loaded by the lowest resistance has the best low-frequency characteristic. A transformer having an open-circuit secondary has twice the voltage ratio and gives the same response at twice the "low-end" frequency as a line-matching transformer of the same turns ratio.

Figure 5.8 is of direct use in determining the proper value of primary OCL. Permissible response deviation at the lowest operating frequency fixes X_N/R_1 and therefore X_N. At the corresponding frequency, this represents a certain value of primary OCL. This inductance determines the size and weight of the transformer. The importance of Fig. 5.8 is evident. Figure 5.8 can also be used to

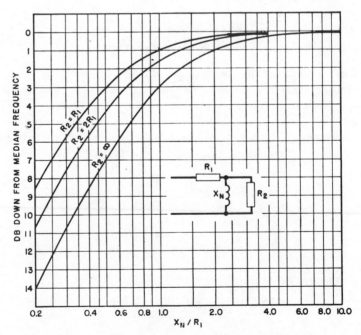

Figure 5.8. Transformer characteristics at low frequencies.

scale the low-frequency response of a transformer for an impedance other than the design impedance.

If the primary and equivalent (1 : 1) secondary winding resistance each are 5% of R_1, the total effect will be a decrease of 10% in the median-frequency voltage ratio of a transformer with matched source and load impedances. There will be corresponding decreases at lower frequencies. The primary resistance of an open secondary transformer has no effect on the median-frequency voltage ratio but has some effect at lower frequencies. The secondary resistance has no effect either at median or at lower frequencies. It is important in the design of a transformer operating with an open secondary to keep the primary resistance low for good low-frequency response. The secondary winding resistance may be any value. The maximum number of secondary turns may be determined by the smallest practicable wire size rather than by winding resistance.

As the frequency increases, the primary inductive reactance X_N also increases until it has almost no effect on the frequency response. This is true for median frequency in Fig. 5.8. It is also true for higher frequencies. The OCL has an influence only on the low-frequency end of the frequency response curve. The ratio of R_2 to R_1 still limits the voltage ratio. If the amplifier works at only one frequency, OCL is determined by the deficiency in voltage gain that can be tolerated in the amplifier design. This can be found in Fig. 5.8.

In an amplifier with a band of operating frequencies, a well-designed transformer has uniform voltage ratio for the desired frequency range. This range extends from the frequency at which X_N ceases to exert any appreciable influence to a frequency designated as the high-frequency cutoff of the transformer.

5.5.2   High-Frequency Response

The factors that influence the high-frequency response of a transformer are leakage inductance, winding capacitance, source impedance, and load impedance. A new equivalent diagram [Fig. 5.7(d)] is necessary for the high-frequency end. Winding resistances are omitted or combined as in Fig. 5.7(b). Winding capacitances are shown across the windings. If primary and secondary leakage inductances and capacitances are combined, X_N is omitted as if it were infinitely large, and a^2 is dropped as before, the circuit becomes that shown in Fig. 5.7(e). X_L is the leakage reactance of both windings, X_C the capacitive reactance of both windings, and R_2 the load resistance, all referred to the primary side on a 1 : 1 turns ratio basis.

At any frequency, the equivalent voltage ratio in the circuit of Fig. 5.7(e) can be found by the ratio of impedances, as it was for the low-frequency response. The scalar value is

$$\frac{E_2}{E_1} = \frac{1}{\sqrt{(R_1/X_C + X_L/R_2)^2 + (X_L/X_C - R_1/R_2 - 1)^2}} \tag{5.6}$$

In equation 5.6 the term X_L/X_C may be written $4\pi^2 f^2 LC = f^2/f_r^2$, where $1/(2\pi\sqrt{LC}) = f_r$, the resonant frequency of the leakage inductance and winding capacitance, considered as lumped and without resistance. Also, $X_L/R_2 = X_C f^2/R_2 f_r^2$. Assign to the ratio X_C/R_1 a value B at frequency f_r. Then at any frequency f, $X_C/R_1 = Bf_r/f$. In the three cases considered at the low frequencies:

$R_1 = R_2$:

$$\frac{E_2}{E_1} = \frac{1}{\sqrt{(f/Bf_r + Bf/f_r)^2 + (f^2/f_r^2 - 2.0)^2}} \qquad (5.6a)$$

$R_2 = 2R_1$:

$$\frac{E_2}{E_1} = \frac{1}{\sqrt{(f/Bf_r + Bf/2f_r)^2 + (f^2/f_r^2 - 1.5)^2}} \qquad (5.6b)$$

$R_2 = \infty$:

$$\frac{E_2}{E_1} = \frac{1}{\sqrt{(f/Bf_r)^2 + (f^2/f_r^2 - 1)^2}} \qquad (5.6c)$$

Equations 5.6a to 5.6c are plotted in Figs. 5.9 to 5.11. If X_C/R_1 has certain

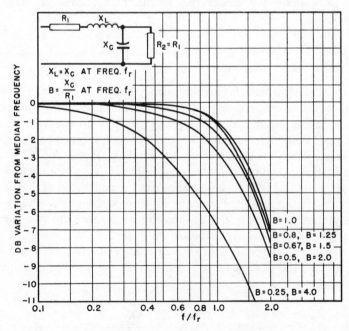

Figure 5.9. Transformer characteristics at high frequencies ($R_1 = R_2$).

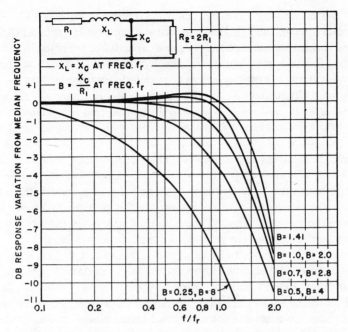

Figure 5.10. Transformer characteristics at high frequencies ($R_1 = 2R_2$).

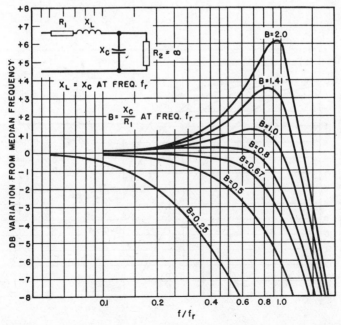

Figure 5.11. Transformer characteristics at high frequencies ($R_2 = \infty$).

values at frequency f_r, the frequency characteristic is relatively flat up to frequencies approaching f_r. Performance is good at $B = 1.0$ in all three figures.

When leakage inductance and winding capacitance are regarded as "lumped" quantities, current distribution in the windings is assumed to be uniform throughout the range of frequencies considered. This assumption is valid up to the resonant frequency f_r. At frequencies higher than f_r, there may be appreciable error in Figs. 5.9 to 5.11. But good frequency characteristics lie mainly below the frequency f_r, where the curves are correct within the assumed limits.

To use these curves in design work, choose the most desirable characteristic curve. From a knowledge of the source impedance, find the proper value of capacitive reactance X_C at frequency f_r. The value of f_r should be such that the highest frequency to be covered lies on the flat part of the curve. X_C and f_r determine the values of winding capacitance and leakage inductance that must not be exceeded in order to give the required performance.

In Fig. 5.7(e) the capacitance is shown across the load. This is correct if the main body of the capacitance is greater on the secondary than on the primary side. Normally, this is true if the secondary winding has the greatest number of turns. Figures 5.9 to 5.11 are thus plotted specifically for step-up transformers. Modifications are necessary for step-down transformers. The equivalent circuit is shown in Fig. 5.12. Analysis shows the scalar voltage ratio to be

$$\frac{E_2}{E_1} = \frac{1}{\sqrt{\left(\frac{R_1}{X_C} + \frac{X_L}{R_2}\right)^2 + \left(\frac{R_1}{R_2} \cdot \frac{X_L}{X_C} - \frac{R_1}{R_2} - 1\right)^2}} \tag{5.7}$$

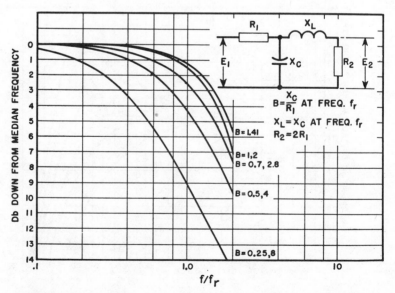

Figure 5.12. High-frequency response of step-down transformers.

Notice the similarity to equation 5.6. In fact, if $R_1 = R_2$, equation 5.7 reduces to equation 5.6; for this case the response is the same for step-down and step-up transformers, and is given by Fig. 5.9.

For $R_2 = 2R_1$, equation 5.7 becomes, after substitution in terms of frequency

$$\frac{E_2}{E_1} = \frac{1}{\sqrt{(f/Bf_r + Bf/2f_r)^2 + (f^2/2f_r^2 - 1.5)^2}} \tag{5.8}$$

which is plotted in Fig. 5.12. Nonuniform response comes at somewhat lower frequency than in Fig. 5.10.

The case of $R_2 = \infty$ for step-down transformers is not important. By inspection it can be seen to be the response of R_1 and X_C in series, because X_L carries no current. This case rarely occurs in practice.

5.5.3 Harmonic Distortion

Wide-band response may be good according to Figs. 5.9 to 5.12, but at the same time the output may be badly distorted because of changes in load impedance or phase angle. This possibility is considered here for the case in which the load impedance is twice the source impedance.

The phase angle of the equivalent circuits of Fig. 5.7(a) and (e) is found by taking the angle whose tangent is the ratio of imaginary to real components of the total circuit impedance in each case. This angle is plotted in Figs. 5.13 and 5.14 for the low- and high-frequency ranges, respectively, with the same abscissas as in Figs. 5.8 and 5.10. It is the angle between the voltage E_1 and the current entering the equivalent circuits of Fig. 5.7(c) and (e) and therefore represents the angle between ac input voltage and output amplifier current. A positive phase angle indicates a lagging amplifier current.

The phase angle exhibited by a transformer over the range considered in Figs. 5.13 and 5.14 does not exceed 30°. For the most favorable curve in Fig. 5.14 ($B = 1.0$) it does not exceed 15°. This phase angle by itself does not introduce significant distortion in the output of the transformer.

The influence of load impedance on distortion will be considered next. In Fig. 5.7(c) the load impedance, to the right of the dashed line, is

$$Z = \frac{jR_2X_N}{R_2 + jX_N}$$

Hence

$$\frac{Z}{R_2} = \frac{\sqrt{1 + (X_N/R_2)^2}}{R_2/X_N + X_N/R_2} \tag{5.9}$$

Equation 5.9 is plotted in Fig. 5.15. It shows the change in load Z from its median frequency value R_2 as the frequency is lowered. Abscissas are X_N/R_2 instead of X_N/R_1 as in Fig. 5.8.

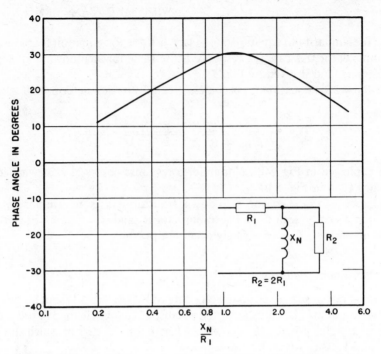

Figure 5.13. Variation of amplifier phase angle at low frequencies.

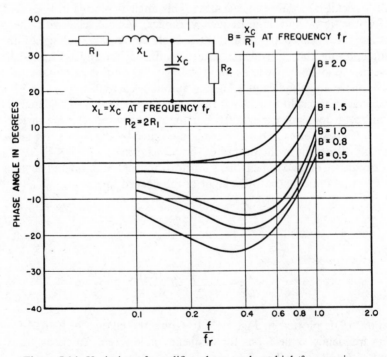

Figure 5.14. Variation of amplifier phase angle at high frequencies.

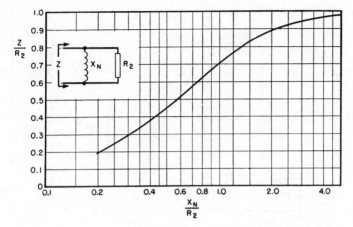

Figure 5.15. Variation of load impedance with transformer characteristics at low frequencies.

For the higher frequencies the load impedance at the right of the dashed line in Fig. 5.7(e) is

$$Z = \frac{jX_L R_2 + X_L X_C - jX_C R_2}{R_2 - jX_C}$$

$$\frac{Z}{R_2} = \frac{\sqrt{(X_C/R_2)^2 + (X_L X_C/R_2^2 + X_L/X_C - 1)^2}}{R_2/X_C + X_C/R_2} \tag{5.10}$$

If we let $X_C/R_2 = D$ at frequency f_r, then at any frequency f, $X_C/R_2 = Df_r/f$. If this substitution is made in equation 5.10 and also if $X_L/X_C = f^2/f_r^2$,

$$\frac{Z}{R_2} = \frac{\sqrt{D^2 f_r^2/f^2 + (D^2 + f^2/f_r^2 - 1)^2}}{f/Df_r + Df_r/f} \tag{5.10a}$$

Equation 5.10a is plotted in Fig. 5.16 for several values of D. The impedance varies widely from its median-frequency value, especially at lower values of D.

From Figs. 5.15 and 5.16 it is possible to compare the change of impedance with the frequency response curves in Figs. 5.8 and 5.10. When this comparison is made it should be remembered that $B = 2D$ for the conditions assumed here. If the amplifier response is allowed to fall off 1.0 dB at the lowest frequency, the corresponding value of X_N/R_1 from Fig. 5.8 is 1.3. This means that X_N/R_2 is 0.65. The corresponding load impedance in Fig. 5.15 is only 0.55 of its median-frequency value. For 0.5-dB droop of the frequency characteristic, the load impedance falls to $0.7R_2$. For a good load impedance of $0.9R_2$ the frequency characteristic can fall off only 0.1 dB. It is thus evident that the load impedance may vary widely even with comparatively flat frequency characteristics.

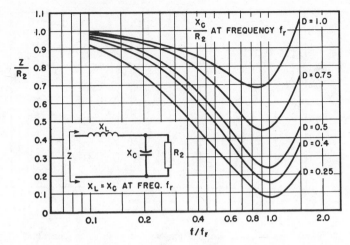

Figure 5.16. Variation of load impedance with transformer characteristics at high frequencies.

At high frequencies the divergences are still greater. Suppose, for example, that a transformer has been designed so that X_C/R_1 is 1.0 at f_r (i.e., $B = 1.0$ in Fig. 5.10). Suppose further that the highest frequency at which the transformer operates is $0.75f_r$. The amplifier then has a relatively flat characteristic, with a slight rise near its upper limit of frequency. In Fig. 5.16 the curve corresponding to $B = 1.0$ is marked $D = 0.5$, for which at $0.75f_r$ the load impedance has dropped to 32% of R_2, an extremely poor match for the amplifier gain element.

It might be thought that since $0.75f_r$ is the upper-frequency limit, the harmonics resulting from the low value of load impedance would not be amplified and no harm would be done. But at the frequency of $0.375f_r$, whose second harmonic would be amplified, the load impedance is only $0.69R_2$.

Between $0.375f_r$ and $0.75f_r$ (over half of the amplifier frequency range) the load impedance gradually drops from $0.69R_2$ to $0.32R_2$. Thus distortion is large over a wide frequency range. It would be much better to design the transformer so that $B = 2.0$; the change in impedance is much less and the rise in response is slight.

Although these curves were intended to be used with voltage-driven devices such as vacuum tubes, they were derived on the basis of a constant sinusoidal voltage at the source. Figure 5.16 demonstrates one important fact: For amplifier stages operating into loads of twice the output resistance, it is better to design transformers so that $B = 2.0$ or more. Then the output voltage and distortion are less affected by impedance variations at high frequencies. The actual frequency characteristics for triodes lie somewhere between the curves of Fig. 5.10 and the corresponding curves of Fig. 5.16.

Designing transformers for $B = 2.0$ means keeping the effective capacitance lower. The leakage inductance may be proportionately greater than for transformers having $B = 1.0$.

Variations in impedance at high frequency shown in Fig. 5.16 are for step-up transformers. Similar variations for step-down transformers may be found from equation 5.11.

$$\frac{Z}{R_2} = \frac{1 + (D^2 f^2 / f_r^2)}{\sqrt{(f/Df_r + Df^3/f_r^3 - Df/f_r)^2 + 1}} \qquad (5.11)$$

Equation 5.11 is plotted in Fig. 5.17. Impedance rises to peaks in the vicinity of f_r, in contrast to the valleys in Fig. 5.16. For the same variation of impedance, the frequency range is greater for step-down transformers, especially with values of $D = 0.5$ and 0.7.

Besides the harmonic distortion caused by variations in load impedance, at low frequencies additional distortion is caused by nonlinear magnetizing current. If a transformer is connected to a 60-Hz supply line, the no-load current contains large harmonics, but the voltage waveform remains sinusoidal because the line impedance is low. If distorted magnetizing current is drawn from a higher-impedance source, a distorted voltage waveform can appear across the transformer primary winding, caused mainly by the third harmonic of the magnetizing current. If the harmonic current amplitude I_H in the magnetizing current is found by connecting the transformer across a low-impedance source, the amplitude of harmonic voltage appearing in the output with the higher-

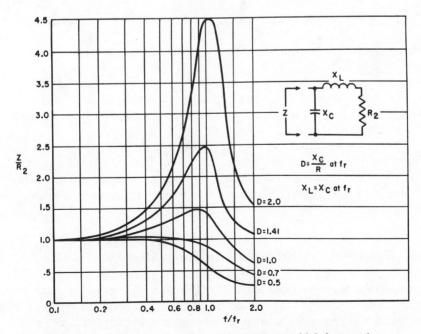

Figure 5.17. Step-down transformer impedance at high frequencies.

TABLE 5.3 Typical Silicon-Steel Magnetizing Current Harmonic Components With Zero-Impedance Source

B_m (G)	Percentage of Third Harmonic	Percentage of Fifth Harmonic
100	4	1
500	7	1.5
1,000	9	2.0
3,000	15	2.5
5,000	20	3.0
10,000	30	5.0

impedance source is

$$\frac{E_H}{E_f} = \frac{I_H R}{I_f X_N}\left(1 - \frac{R}{4X_N}\right) \tag{5.12}$$

where E_H = harmonic voltage amplitude
E_f = fundamental voltage amplitude
I_H = harmonic current amplitude
I_f = fundamental current amplitude
$R = R_1 R_2 / R_1 + R_2$; R_1, R_2, and X_N are shown in Fig. 5.7(c)

For a discussion of equation 5.12 and magnetizing currents in general, see Partridge (1942).

If the flux density is below the knee of the saturation curve, and if $X_N = 3R_2$ at the lowest operating frequency, the harmonic amplitude is less than 5%. An air gap in the core reduces this figure. Table 5.3 gives typical harmonic currents for silicon steel.

Output voltage distortion may be analyzed to find the harmonic content by the usual method of Fourier analysis. In general, if the recommended amplifier-stage load impedances are maintained, harmonic percentages will be determined primarily by the amplifying elements. If other load impedances are used, it will be necessary to perform a harmonic analysis.

5.6 PUSH-PULL AMPLIFIER TRANSFORMERS

The analysis of single-ended amplifiers in Section 5.5.3 applies to class A push-pull amplifiers, except that the second harmonic components in the amplifier output are due to unlike sources rather than to low-impedance distortion.

The internal impedance of a tube type class B amplifier varies so much with the instantaneous signal voltage on the input, power output, and output voltage that it is not practical to draw curves similar to Figs. 5.10 and 5.12 for class B operation. Bipolar transistors have a much lower internal impedance. The internal impedance variation does not have as much influence on the circuit operation. Qualitatively, the characteristic curves may be expected to follow the same general trend as for class A amplifiers. A basis for class B amplifier design is to make the transformer constants such that the load impedance does not fall below a given percentage of the load resistance R_2.

Usually, the decline with frequency response is greater for class B than for class A amplifiers, because the effect of internal output impedances is greater. In the extreme, frequency response falls off proportionately with load impedance.

A change in mode of operation occurs in a class B amplifier as the output passes from one half of the stage to the other in the region of amplifying device cutoff. This changeover may cause transient voltages in the amplifier which distort the output voltage waveform. If the two halves of the transformer primary winding are not tightly coupled, primary-to-primary leakage inductance causes nicks in the output voltage in somewhat the same way as leakage inductance in a rectifier transformer. In a class B amplifier, the change from one half of the amplifier to the other is less abrupt than in a rectifier, but perceptible nicks in the voltage waveform occur if the ratio of primary-to-primary leakage reactance to average output resistance is 4 or more (see Pen-Tung Sah, 1936).

Balanced operation in a push-pull amplifier, that is, equal current and voltage swings on both sides, is possible only if the amplifying devices are alike and if transformer winding turns and resistances per side are equal. Shell-type concentric windings do not fulfill this condition because the half of the primary nearer the core tongue has lower resistance than the other half. Balance is easier to achieve in the core type of arrangement shown in Fig. 5.18. In class A amplifiers close primary-to-primary coupling is not essential, and balance may be attained by arranging part coils as in Fig. 5.19.

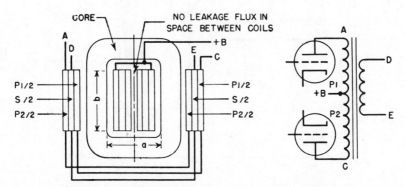

Figure 5.18. Core-type push-pull balanced windings.

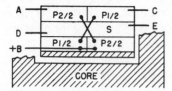

Figure 5.19. Shell-type push-pull balanced windings.

Because only half of the primary winding of a class B amplifier carries current during a half-cycle, the leakage flux and therefore the primary-to-secondary leakage inductance have approximately half the values, with both active all the time. With capacitive currents, both windings are active, at least partially. Transformers with $D > 1.0$ have low capacitive currents, low leakage inductance, high resonance frequency, and extended frequency range, in addition to the load-impedance advantage given in Section 5.5.3. At high frequencies a class B amplifier transformer presents a circuit to the amplifying devices like that in Fig. 5.20. Let L_1 be leakage inductance between the halves of the primary winding and L_2 that between each half of the primary and the secondary. L_1 is the inductance of one half of the primary winding, measured with the other half

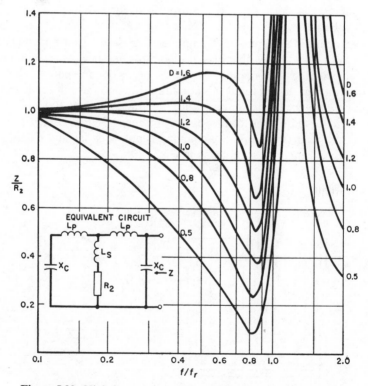

Figure 5.20. High-frequency load impedance of class B amplifiers.

primary short circuited and the secondary open. L_2 is the inductance of one half of the primary winding, measured with the other half primary open and the secondary short circuited. In Fig. 5.20, $L_1 = 2L_P$ and $L_2 = L_P + L_S$. Resonant frequency f_r is determined by X_C and $X_{L1} = 2\pi f L_1$. In this figure $D = X_C/R_2$ at f_r.

First one amplifying device delivers power into the equivalent circuit at one end. During the next half-cycle, this device is cut off and the other device delivers power into the circuit at the other end. Thus the transformer equivalent impedance Z seen looking into the circuit, first at the end shown and then at the other end, is fed by one of the active devices at all times. Impedance ratio Z/R_2 varies with frequency as in Fig. 5.20. For some values of parameter D, impedance falls more rapidly than for class A amplifiers (Fig. 5.14), but frequency f_r in Fig. 5.18 is determined by L and C having half the values of these elements in class A amplifiers. Hence class B impedance stays flat at higher frequencies, although response may droop at lower frequencies than for class A.

Figure 5.20 is drawn for a ratio of $L_1/L_2 = 1.5$, which is a practical design ratio. A lower L_1/L_2 ratio results in deeper valleys in the impedance curve; higher L_1/L_2 is more likely to cause nicks in the voltage wave. Good practice consists in designing class B amplifier transformers so that the highest operating frequency is less than $f_r/2$ and $L_1/L_2 = 1.5$. Then harmonic distortion at high frequencies should not exceed 5% (see Lord, 1950). Class B transformer impedance is influenced by circuit elements, so that maintenance of constant impedance over a wide frequency band becomes an overall amplifier problem.

Capacitive currents also cause unbalance at high frequencies, even with winding arrangements like Figs. 5.18 and 5.19. This is evident if the secondary winding in these figures is grounded at one end. The effective capacitances to the two primary windings are then unequal. This problem may be solved by keeping the capacitances small with liberal spacing. This practice increases leakage inductance and cannot be carried very far. Coil mean turn length should be kept as small as possible by the use of the most suitable core material. Core-type designs have smaller mean turns than does the shell type. Also, the two outer coil sections have low capacitance to each other and to the case if liberal spacing is used, without an increase in leakage inductance. Flux in the space between the outer sections links all the windings on one leg and hence is not leakage flux. Consequently, this space is not part of a term a in equation 5.14. In push-pull amplifiers the winding arrangement of Fig. 5.18 is advantageous because of the low capacitance between the points of greatest potential difference, A and C.

5.7   EFFECTS OF TRANSFORMER IMPEDANCE CHANGES ON DEVICE CURRENTS

In a lightly loaded amplifier the frequency characteristics remain flat at high frequencies, even with a droop in the load impedance. The current through the amplifying device and through the transformer primary must increase in inverse

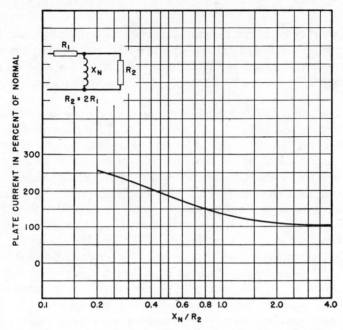

Figure 5.21. Rise in amplifying device current due to transformer impedance change at low frequencies.

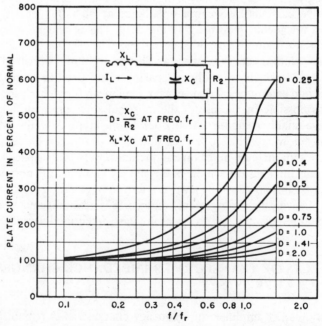

Figure 5.22. Rise in amplifying device current due to transformer impedance change at high frequencies.

proportion to the impedance change to maintain the same voltage. This type of change is most significant in vacuum-tube amplifiers. It may also be significant in bipolar transistor or FET amplifiers if the amplifying devices are being operated into a nearly matched impedance.

If the current rises high enough to maintain a constant output voltage, it may cause overheating of either the amplifying device, the transformer, or both. Values of current rise for constant output at both low and high frequencies are given in Figs. 5.21 and 5.22.

Amplifiers that operate at single frequencies over a wide band must have transformers designed so that this current rise is not sufficiently large to cause damage to either the transformer or the amplifying device. Audio transformers may have very high current rises at these extremes of frequencies. However, the time that high powers are handled at these extreme frequencies are usually short enough that the current rise is not a serious problem.

5.8   PENTODES, FETs, AND BIPOLAR TRANSISTORS

The vacuum tubes described in Section 5.7 had three elements. Tetrode tubes have an additional grid between the anode and control grid to reduce the grid to anode capacitance. This grid is operated at a positive potential. The anode voltage swing is limited to a minimum voltage determined by the voltage on this additional grid. A third grid between the screen grid and anode permits the same anode voltage swing as in a triode. The two-grid tube is called a tetrode. The three-grid tube is a pentode. A tube design for high-power applications is the beam power tube, a special case of a pentode.

Figure 5.23 shows the characteristics of a high-power beam tube. A typical load line of $2500\,\Omega$ is shown. The internal impedance of pentodes and beam tubes is very high.

Pentodes are essentially constant current devices. If we compare the characteristics of a bipolar transistor (Fig. 5.24) and a power FET (Fig. 5.25) to the characteristics of the beam power tube, the similarities are obvious. Bipolar transistors and FETs are also constant-current devices. The same design techniques that are used for pentode amplifiers can therefore be used with a change of impedance.

The value of load impedance is an indication of the output voltage for low frequencies. Response of a transformer coupled amplifier using constant-current devices to low frequencies can be found from Fig. 5.15.

At high frequencies, the leakage inductance of the transformer is between the amplifying device and the load. The primary voltage and secondary or load voltages are not identical. In Fig. 5.26 the change of output voltage for a constant input voltage at high frequencies is shown. In this figure the equivalent circuit of a pi filter is used to simulate the circuit. This type of circuit gives the best approximation to the transformer when a constant-current input is used.

The harmonic content of constant-current amplifiers can be high, especially

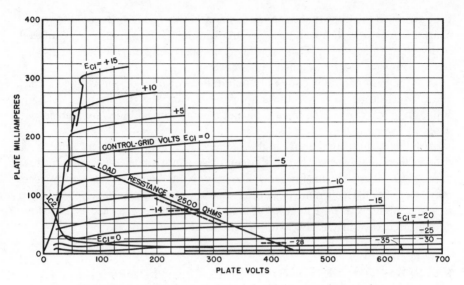

Figure 5.23. Plate characteristics of a high-power beam tube.

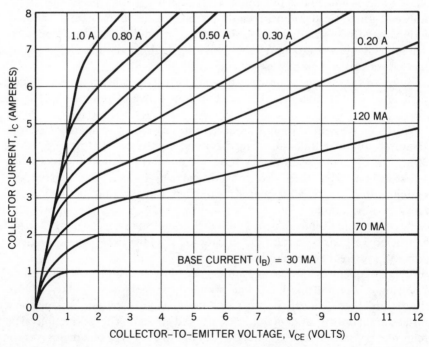

Figure 5.24. Collector characteristics of a high-power bipolar transistor (common-emitter connection).

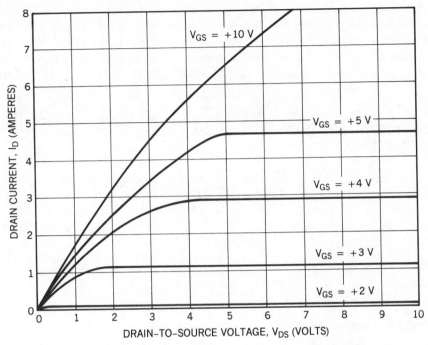

Figure 5.25. Drain characteristics of a power MOSFET.

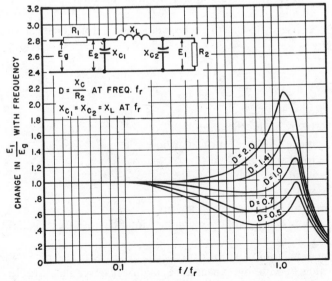

Figure 5.26. Constant-current amplifying device frequency response with pi-filter output circuit.

in single-ended amplifiers. A large phase angle and low load impedance also produce undesired distortion. It is best to use values of X_N/R_2 greater than 2 in Fig. 5.15, to reduce distortion at low frequencies. Holding the input and output voltage swings in single-ended amplifiers to very low levels will reduce distortion. Distortion is also reduced by using such devices in push-pull circuits.

5.9   CALCULATION OF INDUCTANCE AND CAPACITANCE

Transformer-coupled amplifier performance is dependent at low frequencies on transformer OCL, and at high frequencies on leakage inductance and winding capacitance. Calculation of these quantities is essential in design and useful in tests for proper operation. Inductance formulas are repeated here for convenience, along with capacitance calculations.

$$\text{OCL} = \frac{3.2N^2 A_c}{10^8(l_g + l_c/\mu)} \qquad \text{henrys} \qquad (5.13)$$

where N = turns in winding
$\quad A_c$ = core area, in.^{2}
$\quad l_g$ = total length of air gap, in.
$\quad l_c$ = core length, in.
$\quad \mu$ = permeability of core (if there is unbalanced direct current in the winding, this is the incremental permeability

For concentric shell- or core-type windings the total leakage inductance to any winding is

$$L_S = \frac{10.6N^2(\text{MT})(2nc + a)}{n^2 b \times 10^9} \qquad \text{henrys} \qquad (5.14)$$

where N = turns in that winding
$\quad \text{MT}$ = mean length of turn for whole coil
$\quad a$ = total winding height
$\quad b$ = winding width
$\quad c$ = insulation space
$\quad n$ = number of insulation spaces
$\quad$ = number of primary-secondary interleavings (see Fig. 3.11)

Winding capacitance is not expressible in terms of a single equation. The effective value of winding capacitance is almost never measurable, because it depends on the voltages at the various points of the winding. The capacitive current at any point is equal to the voltage across the capacitance divided by the capacitive reactance. Since many capacitances occur at different voltages in even the simplest transformer, no one general formula can suffice. The major components of capacitance are from

1. Turn to turn
2. Layer to layer
3. Winding to winding
4. Windings to core
5. Stray (including terminals, leads, and case)
6. External capacitors
7. Electrode capacitance

These components have different relative values in different types of windings. Turn-to-turn capacitance is seldom preponderant because the capacitances are in series when referred to the whole winding. Layer-to-layer capacitance may be the major contributor to the capacitance in high-voltage single-section windings, where thick winding insulation keeps the winding-to-winding and winding-to-core components small. Items 5 to 7 need to be watched carefully or they may spoil otherwise low-capacitance transformers and circuits.

If a capacitance C with E_1 volts across it is referred to some other voltage E_2, the effective value at reference voltage E_2 is

$$C_e = C(E_1^2/E_2^2) \tag{5.15}$$

By use of equation 5.15 all capacitances in the transformer may be referred to the primary or secondary winding. The sum of these capacitances is then the transformer capacitance which is used in the various equations and curves of preceding sections.

In an element of winding across which voltage is substantially uniform, capacitance to a surface beneath is

$$C = (0.225A\epsilon/t) \qquad \text{pF} \tag{5.16}$$

where A = area of winding element, in.^{2}
   ϵ = dielectric constant of insulation under winding, 3 to 4 for organic materials
   t = thickness in inches of insulation under winding (this includes wire insulation and space factor)

If the winding element has uniformly varying voltage across it, as in Fig. 5.27, the effective capacitance is the sum of all of the incremental effective capacitances. This summation is

$$C_e = C(E_1^2 + E_2^2 + E_1E_2)/3E^2 \tag{5.17}$$

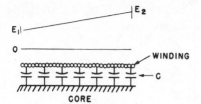

Figure 5.27. Transformer winding with uniform voltage distribution.

where C = capacitance of winding element as found by equation 5.16
E_1 = minimum voltage across C
E_2 = maximum voltage across C
E = reference voltage for C_e

If E_1 is zero and $E_2 = E$, equation 5.17 becomes

$$C_e = C/3 \qquad\qquad (5.18)$$

The capacitance to ground of a single-layer winding with its low voltage end grounded is one-third of the measured capacitance of the winding to ground. Measurement should be made with the winding ungrounded and both ends short circuited together, to form one electrode. Ground forms the other electrode.

In a multilayer winding, E_1 is zero at one end of each layer and $E_2 = 2E/N_L$ at the other, where E is the winding voltage and N_L is the number of layers. The effective layer to layer capacitance of the whole winding is

$$C_e = \frac{4C_L}{3N_L}\left(1 - \frac{1}{N_L}\right) \qquad\qquad (5.19)$$

where C_L is the measurable capacitance of one layer to another. The first and last layers have capacitance to other layers on one side only, and this is accounted for by the term in parentheses in equation 5.19.

Because the turns per layer and volts per layer are greater in a winding with many turns of small wire, such windings have higher effective capacitance than windings with few turns. In a transformer with a large turns ratio, whether step up or step down, this effective capacitance is often a barrier to a further increase in the turns ratio. With a given load impedance across the low-impedance winding, there is a maximum effective capacitance C_m which can be tolerated for a given frequency response. If layer and winding capacitances have been reduced to the lowest practicable value for C_1, the maximum turns ratio is $\sqrt{C_m/C_1}$. Appreciable amounts of capacitance across which large voltages exist must be eliminated by careful design.

Effective capacitance is greater at higher voltages. In step-down transformers the capacitance may be regarded as existing mainly across the primary winding. In step-up transformers the capacitance may be regarded as existing mainly across the secondary winding. The effect of this on frequency response has been discussed in Section 5.5.2.

5.10   CORE SELECTION

The selection of a core for a wide-band transformer is more complicated than the selection of a core for a power transformer or an inductor. The important parameter in the selection of a power transformer core is the VA that the transformer must handle. This parameter is very easily related to the D, E, F, and G dimensions of the core. In an inductor carrying dc current, the size can be determined to be a function of the ratio of the energy stored in the inductor to the power loss in the inductor. This too can easily be related to the D, E, F, and G dimensions of the core.

In a wide-band transformer, the important parameters are the power handled, the load impedance, and the open-circuit inductance of the winding. The basic core size can be determined from the power to be handled by the same equation that was used to select a core size for a power transformer. The basic core chosen by this equation may not have the other characteristics necessary for a wide-band transformer.

For a transformer to have a minimum size, the core and copper losses should be approximately equal. The ratios of core area, window area, and flux density must be adjusted so that the losses are equalized. At very low frequencies, the ratio of core area to window area may be increased. At higher frequencies, a reduction in maximum flux density may be required to reduce core loss.

The equation for leakage inductance helps in determining the core proportions. From equation 5.14 we can see that for minimum leakage inductance, the transformer should have the shortest possible mean turn with the longest possible winding space. These conditions are ideally met with a toroid of square core cross section. Such a core is difficult to wind economically. To get a minimum mean turn with a maximum cross section, a core with a square cross section should be used. As the winding length is increased for a given number of turns, the winding height would decrease.

When we combine all of the requirements of maximum core area, minimum mean turn, and maximum winding length with the basic requirement for handling the required power, we arrive at a basic core configuration of a square cross section of core having a large G dimension and a small F dimension. The ratio of these two dimensions must be determined for the particular case. Some help in arriving at the optimum dimensions can be derived by applying the method used by Petersen and Zener (1964) for power transformers.

5.11 AMPLIFIER TRANSFORMER DESIGN

In amplifiers that operate at a single frequency, transformers are similar in design to rectifier transformers. The size of core is determined by the required value of OCL (see Section 5.10). If the winding carries unbalanced direct current, an air gap must be provided to keep B_m within limits. Winding resistances are limited by permissible loss in the output, or in larger units by heating.

If the amplifier operates over a frequency range, the start of the design is with the OCL to ensure proper low-frequency performance. After ample core area and turns have been chosen, attention must be given to the winding configuration. Leakage inductance and winding capacitance are calculated and, from them, f_r and B. If the high-frequency response does not meet the requirements, measures must be taken to increase f_r or change B to a value nearer optimum. Sometimes these considerations increase size appreciably.

Below frequency f_r, the leakage inductance per turn is constant and equal to the total coil inductance divided by the number of turns. Capacitance per turn is constant and may be large because of the close turn-to-turn spacing. But the LC product per turn is smaller than the LC product per layer, because the layer effective capacitance is greater. Therefore, the frequency at which the turns become resonant is higher than that at which the layers become resonant. Similarly, if there is appreciable coil-to-coil capacitance, the layer resonant frequency is higher than the coil resonant frequency f_r. If the coil design is such that resonance of part of a coil occurs at a lower frequency than f_r, the transformer frequency response is limited by the partial resonance. This condition is especially undesirable in very wide range designs, but with reasonable care it can be avoided.

The resonant frequency f_r must be known to determine high-frequency properties. A resonance chart may be used to determine f_r, but often it is just as simple to calculate f_r from the equation $f_r = 1/2\pi\sqrt{LC}$.

Two examples of audio transformer design are given here to illustrate low- and high-frequency response calculation. Figure 5.28 shows an amplifier with a single-ended transistor stage driving a push-pull class AB amplifier stage. The load, R_L, is an 8-Ω loudspeaker, which will be considered a linear resistor for the calculations to follow.

Example 1: Input Transformer. The input transformer couples the ac signal from the class A single-transistor stage, Q_1, to the bases of the power transistors Q_2 and Q_3. V_{cc} for both stages is 100 V. The frequency range is 20 Hz to 20 kHz, with the response being ± 3 dB. Direct current flows in the primary of T_1. Fifty percent nickel-iron is used for the core to take advantage of the higher permeability, even though the effectiveness of the high permeability is greatly reduced by the necessary gap for the dc in the primary. For a 2N6514 transistor, a peak collector voltage swing of 72 V with a current peak of 6 A produces 200 W of output power. This is the equivalent of a 12-Ω load on the transistors.

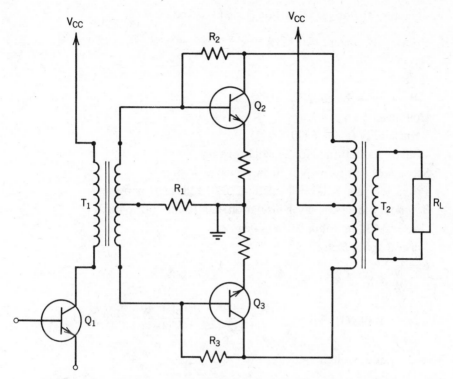

Figure 5.28. Input transformer driving class B push-pull transistors.

To drive the transistors to 7 A (this allows for losses) requires a peak base current of 300 mA, and a base-to-emitter voltage of 1.2 V. If this voltage is applied directly to the base of the push-pull transistors, severe distortion could result from the relatively high impedance of the class A amplifier stage. The base impedance of the push-pull stage at peak current is 4.0 Ω. If we add resistance in series with the bases, the effects of the nonlinearities of the base-emitter junction are reduced. If the series resistance is made equal to five times the base-emitter resistance, the total series resistance is 24 Ω. The power dissipated in the base resistors is 1.8 W. This is negligible compared to the power output. The voltage required from the base driver transformer is now 7.2 V. If a primary voltage swing of 36 V peak to peak is assumed, a transformer step-down ratio of 2.5 : 1 is required.

Primary impedance = 150 Ω
Required OCL (Fig. 5.8)$X_N/R_1 = 1$
$X_N = 150\,\Omega$
OCL = 1.19 H

Assume that the driver transistor Q_1 is biased to 50 V dc.

Primary dc current = 0.333 A
A_c = 0.351 in.^{2}
l_c = 6.25 in.
Core window height (F) = 0.625 in.
Coil mean turn = 4.5 in.
Permeability (μ_Δ) = 6500
Primary 1888 turns No. 29 single enamel wire
Secondary 754 turns No. 30 single enamel wire
Primary layers, 8; layer insulation, 0.0015-in. kraft paper
Secondary layers, 3; layer insulation, 0.0015-in. kraft paper
Gap = 0.032 in. total; 0.016 in./leg
Vertical space factor 0.9
b = 3.32 in.
c = 0.005 in.

$$\text{primary OCL} = \frac{3.2 \times (1888)^2 \times 0.351 \times 10^{-8}}{0.032 + 6.25/6500} = 1.21 \text{ H}$$

This is a good enough value for the OCL.

$$\text{primary leakage inductance} = \frac{10.6 \times (1888)^2 \times 4.5 \times (4 \times 0.005 + 0.221)}{4 \times 3.33 \times 10^9}$$

$$= 3.16 \text{ mH}$$

The primary layer-to-layer capacitance is

$$\frac{0.225 \times 4.5 \times 3.32 \times 3}{0.0015} = 6723 \text{ pF}$$

Then

$$\text{referred to total primary} = 6723/8 = 841 \text{ pF}$$

For the secondary layer-to-layer capacitance:

Referred to primary	359 pF
Winding to core	375 pF
Winding to winding	4034 pF
Total primary capacitance	5549 pF
External capacitance	1000 pF
Total capacitance	6549 pF

The total load resistance reflected to the primary is 150 Ω.

$$X_{pri} = 1.212 \times 2 \times \pi \times 20 = 152.3 \, \Omega$$

$$X_N/R_1 = 1.00$$

From Fig. 5.8, the response is down 1 dB at 20 Hz. The resonant frequency of 3.16 mH and 6549 pF is 35 kHz.

$$X_c = 695 \, \Omega$$

$$B = 695/150 = 4.63$$

From Fig. 5.12 the response is down 2 dB at 20 kHz.

Example 2: Output Transformer. The frequency requirements are the same as for the input transformer. There is no net dc current in the windings. For a 200-W output at 20 Hz, the design can be started by scaling the core from 60 Hz.

Core = 0.014-in.-thick silicon-iron laminations
Primary impedance = 12 Ω
Secondary impedance = 8 Ω
A_c = 1.38 in.^{2}
l_c = 9.57 in.
Core window height = 1.00 in.
Vertical space factor = 0.88
b = 5.12 in.
c = 0.010 in.
Primary: 390 turns AWG No. 18 center tapped, 97 turns/section
Secondary: 316 turns AWG No. 18, 158 turns/section
Primary: 2 layers/section, layer insulation 10 mils of kraft paper
Secondary: 3 layers/section, layer insulation 10 mils of kraft paper
Coil mean turn = 6.94 in.
Permeability = 500

For the output voltage,

$$E = \sqrt{W \times Z} = \sqrt{200 \times 8} = 40 \text{ V rms}$$

$$N_s = \frac{3.49 \times 40 \times 10^6}{20 \times 1.38 \times 16,000} = 316 \text{ turns}$$

permeability at 16,000 G = 500

$$L_{\text{OCL}} = \frac{3.2 \times (316)^2 \times 1.38 \times 10^{-8}}{0.001 + 9.57/500} = 0.219 \text{ H}$$

$$X_N/R_1 = 3.44$$

From Fig. 5.8 the response is still flat at 20 Hz.

$$\text{secondary leakage inductance} = \frac{10.6 \times (316)^2 \times 6.94 \times (4 \times 0.01 + 0.438)}{4 \times 5.12 \times 10^9}$$

$$= 1.71 \text{ mH}$$

For the capacitances,

$$\text{Secondary layer to layer} = \frac{0.225 \times 6.94 \times 2.56 \times 3}{0.01} = 1119 \text{ pF}$$

Total secondary capacitance	= 714 pF
Primary layer to layer referred to the secondary	= 1360 pF
Winding to winding	= 533 pF
Total distributed capacitance	= 2607 pF

$$f_r = 1/2\pi\sqrt{LC} = 10^9/2\pi\sqrt{2607 \times 179} = 214 \text{ kHz}$$

$$B = X_c/R_L = 10^9/2 \times \pi \times 214 \times 2607 \times 8 = 35.7$$

This is off the curve, so the response will be calculated from equation 5.6a. At 20 kHz the response is down 2.4 dB.
   Following are the losses:

Primary copper loss	= 12.9 W
Secondary copper loss	= 30.0 W
Core loss	= 2.5 W
Total losses	= 45.5 W

The temperature rise for a transformer this size is 2.0°C/W.

Total temperature rise	= 90.8°C
Total temperature 90.8 + 35	= 125.8°C

This is a satisfactory temperature for the design.

6 AMPLIFIER CIRCUITS

Amplifier applications may impose severe limitations on transformer design. These limitations may appear as a very wide usable bandwidth, limited phase shift over a wide frequency range, high Q, discontinuous current conduction, or other special requirements.

The required performance can sometimes be achieved by careful transformer design and construction. Other times it can only be achieved by considering the amplifying device and transformer together as part of the system and compromising the individual performance of the parts to obtain the desired total performance.

With the use of vacuum tubes, bipolar transistors, FETs, as well as integrated circuits, the potential applications of transformers in amplifying circuits are almost limitless. In many circuits where transformers could be used, circuit efficiency is sacrificed to eliminate the use of transformers. For example, most audio amplifiers at the present time are designed without transformers. An additional gain stage or two can be added to make up for inefficient coupling between stages caused by impedance mismatches far less expensively than using transformer coupling. Transformer-coupled amplifiers are still used in high-power applications, in very low frequency (VLF) transmitters, sonar equipment, and other broad-band applications where the state of the art in device technology is being pushed.

While the frequency range for transformer-coupled amplifiers has shifted away from the audio-frequency range, where most of the transformer design curves in Chapter 5 were developed, the design techniques developed for vacuum-tube audio amplifiers are still applicable. This upward shift in frequency has been made possible by the development of improved ferrite materials for

171

higher flux density applications and insulating materials with higher dielectric strength, low dielectric constant, and low dielectric loss.

In this chapter several different types of amplifiers will be discussed. The design of audio transformers will be considered since transformer-coupled audio amplifiers are still used in commercial and military applications. Several other types of amplifiers using the design techniques of Chapter 5 will also be discussed. These design techniques can be extended to frequencies of 200 MHz or higher. There are often other considerations in designing wide-band transformers in this frequency range. These are discussed in Chapter 7. For the purposes of this chapter, the techniques discussed are applicable up to 1 MHz and with great caution can be extended to 30 MHz.

The use of negative feedback for improving amplifier performance is also discussed. It is beyond the scope of this book to go into a thorough discussion of this subject. The basic principles are presented so that the transformer designer will have an appreciation of the impact of the transformer design on the performance of the equipment in which it is used.

A full discussion of wave filters is also beyond the scope of this chapter. The use of wave filters in high-power amplifiers is becoming more common. The design of high-Q inductors will be considered. Therefore, an elementary discussion of simple wave filters is presented.

6.1 WIDE-BAND AMPLIFIERS

Many amplifiers are required to work over a very wide band of frequencies. It is sometimes required that an amplifier pass several frequencies at the same time with very little distortion of the waveform or interaction of the frequencies. In other applications, the amplifier may be required to operate over a range of frequencies without tuning but pass only one frequency at a time even though these frequencies may cover a wide range.

Transformer coupling is very seldom used today in solid-state audio amplifiers. Older tube amplifiers commonly used transformer coupling for the output stage. Newer bipolar transistor amplifiers are less expensive to build without the transformer coupling. There are still amplifiers that use transformer coupling. Some of these are discussed in this chapter. Some examples of transformer-coupled amplifiers are found in high-power VLF transmitters for teletype transmission, sonar transmitters, amplitude modulation systems, and some servo systems.

6.2 INVERSE FEEDBACK

If part of the output of an amplifier is fed back to the input in such a way as to oppose it, the ripple, distortion, and frequency response variations are reduced. The amplifier gain is also reduced, but with the availability of high-gain amplifying devices an extra forward gain stage compensates for the reduction in gain caused by the inverse feedback. The improvement in performance justifies the extra stage. In the amplifier of Fig. 6.1, a network is shown connected to the output voltage E_0; part of this output is fed back so that the input to the amplifier is

$$E_2 = E_1 - \beta E_0 \tag{6.1}$$

here β is the portion of E_0 that is fed back. Let α be the voltage amplification of the amplifier, and let E_R and E_H be the ripple and harmonic distortion in the output without feedback. Let α', E_R', and E_H' be the same properties with feedback. Then the following equations hold, if α, E_R, and E_H are assumed to be independent:

Without feedback:

$$E_0 = \alpha E_2 + E_R + E_H \tag{6.2}$$

With feedback:

$$E_0 = \alpha' E_1 + E_R' + E_H' \tag{6.3}$$

From these equations it can be shown that

$$\alpha' = \frac{\alpha}{1 + \alpha\beta} \approx \frac{1}{\beta} \tag{6.4}$$

$$E_R' = \frac{E_R}{1 + \alpha\beta} \approx \frac{E_R}{\alpha\beta} \tag{6.5}$$

$$E_H' = \frac{E_H}{1 + \alpha\beta} \approx \frac{E_H}{\alpha\beta} \tag{6.6}$$

With high-gain amplifiers and large amounts of feedback, the output ripple and harmonic distortion can be made very small. The frequency response can

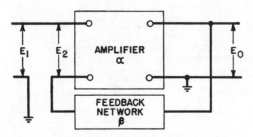

Figure 6.1. Voltage feedback.

likewise be made flat, even with mediocre transformers. Since the output and input are not linearly related, inverse feedback is not used in class C amplifiers.

Incidental effects in the amplifier, like distributed capacitance and leakage inductance, have to be carefully matched in the inverse feedback network so that the phase shift around the loop does not become too large. If it approaches 180°, feedback is regenerative. The amplifier may become an oscillator with the frequency determined by the circuit constants. There are several criteria for determining the stability of the feedback loop. These criteria differ only in the method used to determine the stability. They can all be related to the same set of conditions. Nyquist (1932) has shown that oscillation does not take place as long as the gain × feedback product $\alpha\beta$ is less than unity at the frequencies for which the phase shift is 180°. In a plot of $\alpha\beta$ made on the complex plane, the requirement for stability is that the curve of $\alpha\beta$ must not enclose the point 1,0, with the sign of β opposite that of α. Both gain α and feedback β are ratios of voltages. Both may be expressed in decibels and both are complex quantities.

Another method for determining feedback amplifier stability was developed by Bode (1950). This is the most commonly used method for determining amplifier stability. Readily measured parameters are used directly. These parameters are also relatively simple to calculate from the reactive elements (including transformers) in the circuit. Gains and phase shifts can be added directly. Sample Bode plots showing stable and marginally stable systems are shown in Fig. 6.2. The open-loop gain of the amplifier is plotted in decibels from

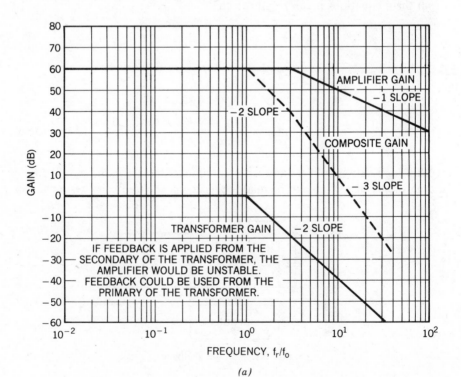

(a)

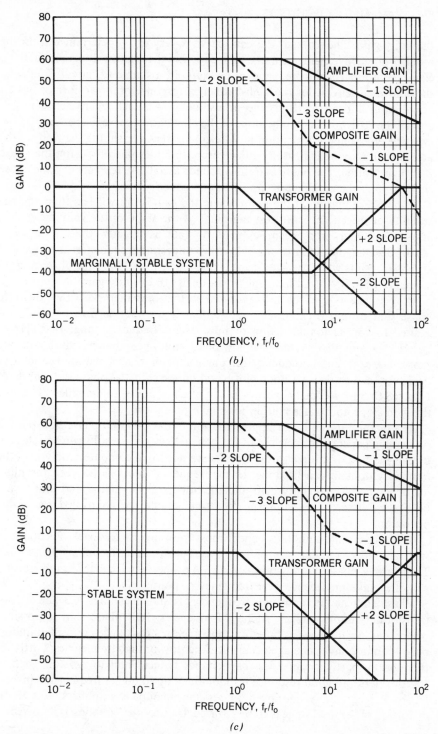

Figure 6.2. Bode plots. (*a*) Unstable system, (*b*) marginally stable system; and (*c*) stable system.

the unity gain. The slope of the gain curve is very easily determined from the number of reactive elements in the gain path and the relative reactances of these elements at a particular frequency. For example, a simple LC circuit similar to the equivalent circuit for the high-frequency response of a transformer has an attenuation slope of 40 dB/decade, or a slope of -2. Each reactive element will contribute 20 dB/decade to the slope of the attenuation curve. Depending on the type of element and the location of the element (series or shunt), the slope of the attenuation may be increased or decreased. In addition to the gain, the phase margin is plotted. The phase margin is defined as $180° - \Phi$, where Φ is the phase shift through the amplifier at a particular frequency.

If the phase margin at the point where the gain is unity is less than 45°, the stability of the circuit is at best marginal. If the slope of the gain curve as it approaches and crosses through unity gain is greater than a slope 1 (20 dB/decade), the amplifier may also have undesirable characteristics. A stable amplifier results when the gain slope is no greater than 1, and the phase margin is greater than 45° at all frequencies within 10 dB of unity gain.

Proper care in application is required so that amplifiers with 180° or more phase shift do not oscillate at some frequency outside the pass band. If it is desired to correct for distortion or hum over a frequency range of 20 Hz to 20 kHz, the amplifier should have low phase shift over a much wider frequency range. The exact cutoff frequency and phase shift for a particular application can be determined by doing a Bode plot based on the gain and phase shift curves of Figs. 5.8 to 5.14. RC or LC networks can often be added to amplifiers to adjust the gain slope and phase margin for stable operation.

The method uses a linear approximation of the gain and phase shift plotted on semilog paper in decibels versus frequency. Very good results can be achieved by this method. If more accurate results are needed, a computer analysis in the frequency domain can be done. In the frequency intervals between the high-frequency cutoff and unity gain and the low-frequency cutoff and unity gain, both the forward gain and the feedback should be controlled for stable operation.

Low-phase shift amplifiers benefit most from inverse feedback. Feedback in such amplifiers reduces size or improves performance, including phase shift. Transformer phase shift, therefore, is a vital property in feedback amplifiers and will often take precedence over frequency response.

Phase shift at low and high frequencies is shown in Figs. 6.3 and 6.4 for transformer-coupled stages. At high frequencies, 180° phase shift is possible, whereas at low frequencies but 90° is possible. In a resistance-coupled amplifier, only 90° phase shift per stage occurs at either low or high frequencies. Partly for this reason, partly because less capacitance is incidental to resistors than to transformers and good response is maintained up to higher frequencies, inverse feedback is generally employed in resistance-coupled amplifiers. But if the distortion of the final stage is to be reduced, transformer coupling is involved. It

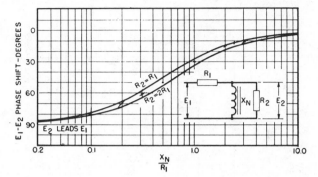

Figure 6.3. Transformer-coupled amplifier low-frequency phase shift.

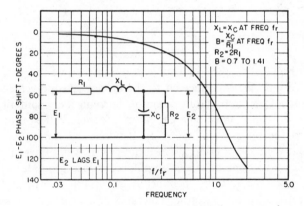

Figure 6.4. Transformer-coupled amplifier high-frequency phase shift.

is preferable to derive the feedback voltage from the primary side of the output transformer. This is equivalent to tapping between R_1 and X_N in Fig. 6.3, where the phase shift is much less. Connecting to the secondary of the transformer will reduce the distortion due to the transformer. If the feedback is to be taken from the primary side, the transformer must be designed for minimum distortion. The transformer must still present a fairly high impedance load to the output-amplifying device throughout the marginal frequency intervals to permit gradual decrease of both amplification and feedback.

Current feedback is obtained in the circuit of Fig. 6.5 by removing capacitor C. This introduces degeneration in the amplifying device negative element, which accomplishes the same thing as the bucking action of voltage feedback. It is less affected by phase shift and consequently is often used with transformer-coupled amplifiers.

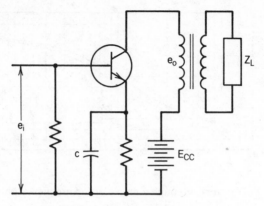

Figure 6.5. Biased amplifier stage.

6.3 CURRENT AMPLIFIERS

The circuit of Fig. 6.6 is typical of a current amplifier. Here the positive element of the amplifying device is connected to the dc power supply E_{cc} without any intervening impedance, so that for alternating currents it is essentially grounded. The input bias voltage e_b must be great enough to include the output voltage e_0 in addition to the normal voltage drop between the input elements at the minimum input voltage swing E_{min}. The input power is still the same as it would be were the negative element of the amplifying device grounded. This type of circuit is used when the output impedance Z_L is variable or low, so that it would be difficult to drive it from a normal voltage amplifier stage. The circuit has a low effective internal impedance as far as the output is concerned. It is approximately equal to the internal output impedance of the amplifying device Z_0 divided by the gain of the amplifying device. Current amplifiers have been used to drive highly variable loads such as the inputs of class B amplifiers. The circuit produces nearly constant output voltage which is somewhat less than the input voltage swing. The input power requirements may change with a change in load impedance, but this change is small because the output impedance of the

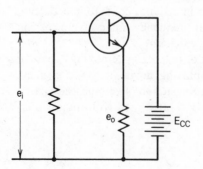

Figure 6.6. Current amplifier/emitter follower.

current amplifier is usually much lower than the load impedance. If the current amplifier feeds a very low impedance load, the voltage swing may be increased by transformer coupling. The response in current amplifier output transformers is usually flat over a very wide range of frequencies because of the low effective source impedance.

6.4 WAVE FILTERS

In the preceding sections dealing with transformer frequency response, means for extending frequency range have been considered. In some amplifiers and transmitters this is a vital problem. In other applications amplifiers are used over a limited frequency range. One example of this are transmitters used for sending teletype messages. These transmitters send one frequency for a mark and another for a space. Each frequency transmitted has a narrow bandwidth, but the change in frequencies must be made rapidly. It is desirable to allow the frequencies which are transmitted to pass through the amplifier at full amplitude but to suppress as much as possible any harmonic frequencies. The means usually employed to accomplish this is a wave filter. In any such filter, the band of frequencies which it is desired to transmit is known as the transmission band, and that which it is desired to suppress is known as the attenuation band. At some frequency, known as the cutoff frequency, the filter starts to attenuate. Transition between attenuation and transmission bands may be gradual or sharp, depending on the filter circuit and the design of the components. When a filter is used in conjunction with a transformer-coupled amplifier, the frequency response of both filter and amplifier must be coordinated. In a later section it will be shown how transformer response may be improved through the use of wave filter principles.

To avoid introducing losses and attenuation in the transmission bands, reactances as nearly pure as practical are used in the elements of the wave filter. For example, in the low-pass filter T section of Fig. 6.7, the inductance arms shown as $L/2$ and the capacitance C are made with losses as low as possible. The capacitors ordinarily used in filters have low losses, but it is a problem to make inductors that have low losses. Values of inductor Q ranging from 10 to 200 are common, depending on the value of inductance and frequency of transmission. Therefore, in wave filters the loss is mostly in the inductors. It can be shown (see Shea, 1929) that for pure reactance arms the values of reactance are such that in

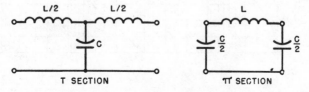

Figure 6.7. Low-pass filter sections.

the transmission band

$$0 > Z_1/4Z_2 > -1 \tag{6.7}$$

where Z_1 is the reactance of the series arm and Z_2 is the reactance of the shunt arm. In the T section of Fig. 6.7, Z_1 is $2\pi f[(L/2) + (L/2)] = 2\pi fL$ and Z_2 is the reactance of C. The attenuation for sections of filter like Fig. 6.7 is shown in Fig. 6.8, for a pure reactance network starting at the cutoff frequency. The attenuation is shown in decibels, and the abscissas are one-fourth of the ratio of series-to-shunt reactance in a full section.

In the transmission band, it is important to terminate the sections of the filter in the proper impedance. Like a transmission line, a wave filter will deliver its full energy only into an impedance that is equal to its characteristic impedance. Many wave filters are composed of multiple sections that simulate transmission lines. For the purposes of this chapter it is assumed that a filter exhibits the same impedance at either end when terminated at the opposite end with an impedance equal to its characteristic impedance. The impedance seen at any one point in the filter is called its image impedance; it will be the same in either direction provided that the source and terminating impedances are equal. In general, however, the impedance will not be the same for all points in the filter. For example, the impedance looking into the left or T section of Fig. 6.7 (if it is assumed to be properly terminated) will not be the same as that seen across capacitor C. For that reason, another half-series arm is added between C and the termination to keep equal input and output impedances. The terminating sections at both the sending and receiving ends of a filter network are half sections, whereas the intermediate sections are full sections. A full T section of the type shown in Fig. 6.7 includes an inductance L equal to $L/2 + L/2$. The

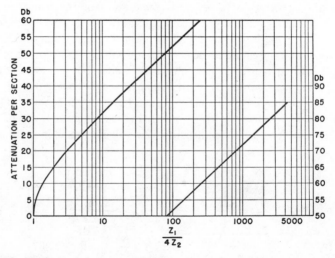

Figure 6.8. Single-section constant-k filter attenuation with pure reactance arms.

image impedance seen at the input terminals of the T section of Fig. 6.7 is known as the midseries impedance, and that seen across capacitor C is known as the midshunt impedance.

Similarly, in the pi-section filter shown at the right of Fig. 6.7, the midshunt image impedance is seen at the input or output terminals. The midseries impedance is seen at a point in the middle of coil L. This section terminates properly in its characteristic impedance at either end. Note that adjacent sections have $C/2$ for the shunt arm, so that a full section would again be composed of a capacitor C and an inductance L. The choice of T or pi sections is determined by convenience in termination, or by the kind of image impedance variation with frequency that is desired. If these precautions are not observed, wave reflections are likely to cause a loss of power transfer in the transmission band.

6.5   LIMITATIONS OF WAVE FILTERS

Wave filters are characterized mathematically by the solution of one of several types of equations. The solutions of these equations lead to values for the various elements of the filter. The solutions of the different equations may lead to different values for the elements as well as different connections of the elements to meet the same basic performance criteria. The basic pass-band response and cutoff frequencies may be the same, but the slope of the attenuation curve beyond cutoff, reflections, and phase shift may be totally different. When a filter is used with a transformer-coupled amplifier, the effect of the changes of the filter impedance on the transformer as well as the changes of the transformer impedance on the filter must be considered. A complete discussion of filter analysis and synthesis is left to others (Zverev, 1967). The discussion of filters presented here is based on the performance of the constant k and m derived filters. The limitations discussed in this section apply to all filter types. One may find that with the use of Butterworth or Tchebychev filters the limitations are often more severe than with the constant k or m derived filters.

Several factors modify the performance of wave filters, especially in the cutoff region. One is the reflection due to mismatch of the characteristic impedance. The load impedance is usually of constant value, whereas the image impedance changes to zero or infinity at cutoff for lossless filters. The resulting reflections cause rounding of the attenuation curve in the cutoff region instead of the sharp cutoff of Fig. 6.8.

Another cause of gradual slope cutoff is the Q of the filter inductors, or ratio of reactance to resistance. Figure 6.9 gives the attenuation at cutoff in terms of Q for a section of the so-called constant k filter (e.g., Fig. 6.7).

Still another cause of the gradual slope of the cutoff is the practice of inserting a resistor to simulate source impedance in attenuation tests. In typical cases the source and terminating impedances are equal. The correct prediction of filter response near cutoff requires a good deal of care. It cannot be taken directly

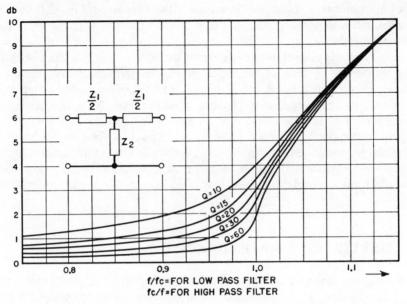

Figure 6.9. Insertion loss near cutoff of a constant-k filter section.

from the usual attenuation charts. With a digital computer and a good circuit analysis program in the frequency domain, a very good prediction of filter response near cutoff can be made.

The phase shift is nearly linear with frequency up to approximately 50% of the cutoff frequency for constant-k low-pass filters in the transmission band. This is an important consideration in connection with networks used for the transmission of steep wavefronts, as in video amplifiers. It is proved in books on network theory that when a nonsinusoidal voltage wave is applied to the input of a network, it appears at the output without distortion of its original shape if the phase shift of the network is proportional to frequency and if the amplitude response is flat for all frequencies. In no actual network are these conditions fulfilled completely. The closer a network approximates these phase-shift and frequency requirements, the smaller the distortion it causes in nonsinusoidal waveforms. Linearity of phase shift is usually more essential to a good waveform than is flatness of response. For this reason, where a nonsinusoidal wave passes through a filter, distortion is minimized if the major frequency components of the wave all lie in the linear region of the phase-shift curve. Considerable judgment must be exercised in the choice of cutoff frequencies. Higher-order harmonics are usually of smaller amplitude. Often too few harmonics are included in the pass band to give a good reproduction of the input waveform.

The effects just noticed are also present in band-pass filters. The filter designer must choose a bandwidth of transmission such that high attenuation is afforded at unwanted frequencies and low attenuation at desired frequencies. This is often not a simple choice. For a given frequency separation from the midfrequency, attenuation decreases as the filter bandwidth is made wider. Impedance variation is much less with a wider bandwidth. Choosing too narrow a bandwidth attenuates frequencies in the transmission band because of reflections.

6.6 ARTIFICIAL LINES

Transmission lines are used to couple devices where the impedances must be matched. Sometimes a certain amount of time delay must be interposed between one circuit and another. It may be that the physical distance between the units being connected is not correct for the line not to have reflections. Some means must be found to increase the length to the next-higher multiple of 90°. For either of these purposes, artificial lines are used. Artificial lines are a grouping of discrete components to approximate the distributed constants of a normal transmission line. They may operate at a single frequency or over a range of frequencies. They may be tapped for adjustment to suit any frequency in a given range, so that impedance and line length are correct. One common application is the network for matching the impedance of a transmitting antenna to the output of the transmitter. The configuration may be either T or pi, high or low pass. Figure 6.10 shows these four combinations for any electrical length θ of line section in degrees. It is assumed in this figure that the line operates at a single frequency and is terminated in a pure resistance equal in value to the line characteristic impedance Z_0. Figure 6.11 is the vector diagram for a leading phase shift pi-section line of 90° electrical length. Proportions of L and C are somewhat different in these line sections than in wave filters.

To obtain approximately constant time delay over a range of frequencies, several constant-k low-pass filter sections may be used, each having a cutoff frequency high enough so that the phase shift is proportional to frequency. The time delay per section is then $\theta/2\pi f$ at any frequency in the range, and

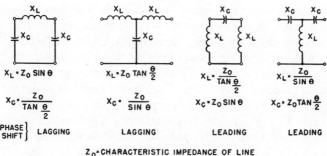

Figure 6.10. Artificial line relations.

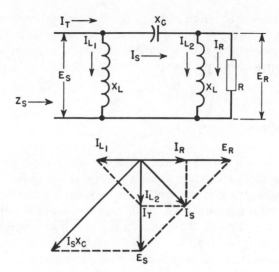

Figure 6.11. Vector diagram for 90° line length.

$\theta = 2\pi f \sqrt{LC}$, where θ is the phase shift in radians, L is the inductance per section, and C is the capacitance per section. In Fig. 6.11, $E_R = E_S$. If the section were terminated in an impedance higher than Z_0, $E_R > E_S$. The line section is then a kind of transformer, although the ratio E_R/R_S varies with frequency. Ninety-degree line sections are often used at high frequencies to obtain transformation of voltage.

6.7 INDUCTOR DESIGN FOR WAVE FILTERS

In Sections 6.4 and 6.5 it was shown that inductors for wave filters must have Q great enough to provide low attenuation in the pass band and proper performance at the cutoff frequency. In designing these inductors, attention must be given as much to Q as to inductance.

Low-loss core material is essential for high Q. Nickel-iron alloys are widely used, the lamination thickness depending on frequency. At frequencies up to 400 Hz, 0.014-in.-thick laminations are used. At higher frequencies, laminations 0.005 in. thick or less are used. Type C cores of nickel-iron alloys with lamination thicknesses of 0.002 and 0.001 in. thick are available. This is an approximate practical guide. Figure 6.12 shows how loss varies with thickness, frequency, and flux density. At frequencies higher than 1000 Hz, flux density must be small for low core loss. In the majority of applications, low-flux-density conditions prevail. Under such conditions, core loss is largely eddy current loss and may be treated as a linear resistance.

Core gaps are used in filter inductors to obtain better Q. For any core, inductance per turn, and frequency, there is a maximum value of Q. The ac

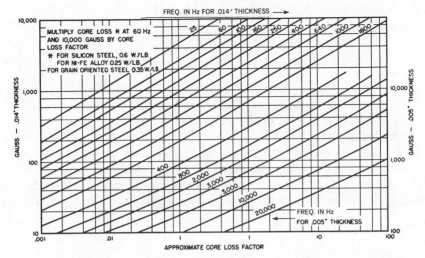

Figure 6.12. Core loss in laminations 0.014 and 0.005 in. thick.

resistance is composed of at least two elements: the winding resistance and the equivalent core loss. In previous chapters the core loss has been considered as an equivalent resistance across a winding. But it can also be regarded as an equivalent resistance in series with the winding. Figure 6.13 shows this equivalence, which may be stated

$$R_{\text{ser}} + jX = \frac{jR_{\text{sh}}X}{R_{\text{sh}} + jX}$$

For values of $Q > 5$,

$$R_{\text{sh}} \approx X^2/R_{\text{ser}} \tag{6.8}$$

where R_{sh} = equivalent shunt resistance
  R_{ser} = equivalent series resistance
  X = winding reactance = $2\pi f L$

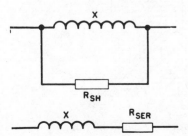

Figure 6.13. Shunt and series equivalent core loss.

The equivalence depends on frequency. The equation for large Q may then be changed to

$$Q = R_{sh}/X = R_{sh}/2\pi f L \qquad (6.9)$$

or Q is proportional to shunt resistance, the winding resistance being neglected. Thus Q can be increased by lowering L, and L may be lowered by increasing the core gap. There are limits on the increase of Q that can be obtained this way.

First the winding resistance is not negligible. With small gaps, maximum Q occurs when winding resistance and equivalent series core loss resistance are equal. For a given air gap there is a certain frequency f_m at which this maximum Q is obtained. At higher and lower frequencies, the manner in which Q falls below the maximum is found as follows: Let R_c be the coil winding resistance. Then

$$Q = X/(R_c + R_{ser})$$

If for R_{ser} we substitute the value obtained from equation 6.8, we have approximately

$$Q = \frac{X}{R_c + X^2/R_{sh}} = \frac{1}{R_c/X + X/R_{sh}} \qquad (6.10)$$

Equation 6.10 therefore gives the relation of Q to frequency. When it is plotted on log-log coordinates with frequency as the independent variable, it is symmetrical about the frequency f_m for which Q is a maximum. [See McElroy and Field (1942). This article also discusses choke design from the standpoint of similitude.] If the core gap is changed, frequency f_m changes. Figure 6.14 shows how the Q of a small inductor varies with frequency for several values of air gap in the core. All these curves have the same shape, a fact that suggests the use of a template for interpolating such curves.

Another phenomenon that limits Q is the flux fringing of the core gap. The influence of the core gap on inductance was discussed in Chapter 3. As the air gap increases, the flux across it fringes more and more, as shown in Fig. 6.15, and

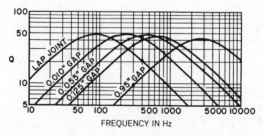

Figure 6.14. Frequency variation of Q for an iron-core inductor with air gaps.

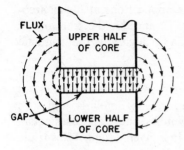

Figure 6.15. Magnetic flux fringing at core gap.

L ceases to be inversely proportional to the physical gap. Some of the fringing flux strikes the core perpendicular to the laminations and sets up eddy currents that cause additional loss. Accurate predictions of gap loss depends on the amount of fringing flux. For laminated cores this loss can be calculated from

$$W_g = K_g l_g dfB_m^2 \qquad (6.11)$$

where K_g = numerical constant from Table 6.1
l_g = core gap, in.
d = core width, in.
f = frequency, Hz
B_m = maximum core induction, G

Lee and Stephens (1973) stated that the accuracy of this equation is within $\pm 25\%$ at least up to a ratio of $l_g/l_c = 0.04$ for gaps to 0.17 in. Tests have shown that the gap loss in cut cores varies almost linearly with frequency up to

TABLE 6.1   Gap Loss Constant

Configuration	K_g
Shell	10×10^{-9}
Simple	5×10^{-9}
Core	2.5×10^{-9}

K_g multipliers for different magnetic materials

Material Type	Multiplier
Oriented silicon-iron	1.00
50% nickel-iron	1.11
80% nickel-iron	0.91
79% nickel-iron (Supermalloy)	0.77
Cobalt-iron (Supermendur)	1.72

2500 Hz, where it begins increasing rapidly. For applications at higher frequencies, particularly those covered in Chapters 12 and 13, the transformer designer should develop specific core gap loss data through tests that simulate actual operating conditions.

In Chapter 3 it was shown that under certain conditions maximum transformer rating for a given size is obtained when core and winding losses are equal. The same would be true for inductors with zero gap. Similarly, it may be shown that maximum Q is obtained with core, winding, and gap losses equal. This is only true when gap losses are significant. In a given design, if this triple equality does not result in the required Q, size must be increased. Losses may be compared by finding either the equivalent series resistances or the equivalent shunt resistances.

6.8 OTHER CORE MATERIALS

As frequency increases above a few kilohertz, gap loss becomes excessively large. At such frequencies, nonlaminated core materials developed for high frequencies are used for large Q. Cores for high frequencies are made from iron or nickel-iron alloys which have been finely powdered, or from magnetic ceramics known as ferrites. The powdered materials are bonded with low-dielectric-loss resins. Proportions of insulating bond material and powder are varied to obtain cores with permeabilities ranging from 10 to 250. The ceramics are fired into a solid mass at temperatures around 1200°C. Permeability in the powdered cores is far less than the inherent permeability of the magnetic material used because of the many small gaps throughout the core structure. Finely divided materials have low eddy current loss and virtually zero gap loss. Equation 6.9 indicates how Q varies with frequency. Low-permeability cores should be used to reduce inductance and maintain large Q at high frequencies. At frequencies higher than audio, coil eddy current losses make specially stranded wire necessary. This is discussed further in Chapter 7.

Ferrites are combinations of bivalent metal oxides which when chemically joined to trivalent iron oxide have magnetic properties. These materials have inherent volume resistivities ranging from 10^6 to 10^{10} Ω-cm. This keeps eddy current losses low. The material permeabilities range from a few hundred to 10,000. High-Q coils are designed by gapping the cores. Because of the high resistivity of the material, gap loss is usually negligible.

One of the problems of filter design is the maintainance of cutoff and attenuation frequencies under conditions of varying temperature. This may be so important as to dictate the choice of the core material. Powdered cores are available with very low temperature coefficients. Usually, these cores have less than the maximum Q for a given kind of magnetic powder. With low-temperature-coefficient cores, low-temperature-coefficient capacitors must be used in order to obtain the requisite overall frequency stability.

Powdered and ferrite cores are made in many forms and sizes. Table 6.2

TABLE 6.2   Shapes of Ferrite and Powdered-Iron Cores

Core Shape	Use
Shell	Low-voltage rf transformers and inductors
Cup	Adjustable and fixed low and medium rf transformers and inductors
Slug	Adjustable rf inductors
Toroid	Audio and low rf transformers and inductors
C and EI	High-voltage transformers and inductors

Figure 6.16. Some typical ferrite and powdered iron cores.

indicates some of the areas of usefulness of these core types. Figure 6.16 illustrates some of the core shapes available. Each core shape available is designed for a specific purpose. A study of available core shapes and materials is worthy of the designer's time.

6.9   CLASS D AMPLIFIERS

Vacuum tubes are commonly used in class A, B, or C amplifiers. These amplifiers are characterized by having essentially sinusoidal output voltages. Except for the class A amplifier, these amplifiers are usually push-pull. The efficiencies of these amplifiers is less than 70% due to the high plate losses in the tubes. Bipolar transistors operating in these classes of amplifiers are limited in power output since their collector dissipation capability is usually much less than the plate dissipation capability of high-power tubes.

The class D amplifier permits much higher power outputs from transistors than do the other classes of amplifiers. The input and output voltages of a class D amplifier are square waves. The input and output currents are approximately sinusoidal. If properly designed, class D amplifiers can achieve efficiencies in excess of 90%. Class D operation is achieved by connecting a resonant circuit in series with the load. The resonant frequency is adjusted to the fundamental frequency of the square wave. The value of C is determined by the current in the load. The value of L is then chosen to be resonant with the capacitance.

Transformer leakage inductance must be considered when determining the value of L. It is also important that the Q of the resonant circuit be high so that the impedance at resonance is low. If the Q is low, much power will be dissipated in the inductor. Unless some provision is made to vary L or C or both, the operating frequency of the amplifier is fixed. In this respect a class D amplifier is similar to a class C amplifier. The class C amplifier has a sinusoidal voltage and nonsinusoidal current.

To make the amplifier broad band, a low-pass filter is often used in place of the series resonant circuit. With a fixed cutoff frequency, the bandwidth is limited to a little more than 2 : 1 since the filter must have significant attenuation at the third harmonic of the square wave. If it is necessary to have a very pure sine wave out, a combination of a low-pass filter and a series resonant circuit are used. If a very wide band amplifier is desired, both the filter and resonant circuit are made with switchable or tunable elements or a combination of switchable and tunable elements.

A simplified schematic of a bipolar transistor class D amplifier is shown in Fig. 6.17. Most vacuum-tube amplifiers were designed as push-pull amplifiers. They were rarely used in bridge circuits because of the difficulty in balancing the effects of filament transformer capacitance between the tubes operating with the cathodes at or near ground and the tubes operating with the cathodes above ground. Bipolar transistors and FETs are usually used in bridge or half-bridge configurations to keep the collector voltages lower.

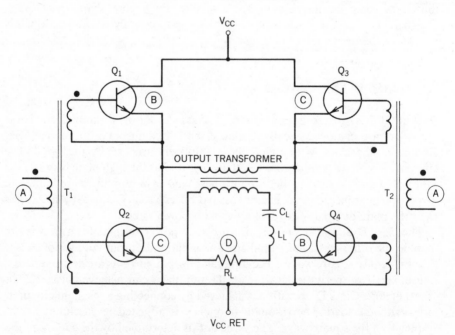

Figure 6.17. Class D bipolar transistor full-bridge amplifier.

In the full bridge shown in Fig. 6.17, transistors B are turned on for one half-cycle of the output. Transistors C are turned on for the other half-cycle. In a properly designed full-bridge amplifier, the voltage across a transistor never exceeds V_{CC}.

Base drive for this amplifier can be derived from either (1) one transformer with four secondaries operating push-pull, or (2) two identical transformers with one primary and two secondaries each, or (3) four transformers with one primary and one secondary each.

The input to the drive transformer is usually derived from single-ended switching amplifiers. Even if the amplifier itself is working at a fixed frequency, the input transformers must be capable of a wide bandwidth to pass the square wave of voltage. The design is further complicated since the transformer driving the lower transistor is essentially at ground potential. The transformer driving the upper leg of the bridge is swinging between ground and V_{CC}. It might be thought that a drive transformer could be designed with low primary-to-secondary capacitance to drive the base of the upper transistor, and then used to drive the base of the lower transistor. If the design is such that the ringing and slight differential in pulse delay can be tolerated, this could be done. In most class D amplifiers, the design is too critical to permit this ringing.

The voltage dependency of the effective capacitance was described in Chapter 5. Since the voltages are different on the base drive transformers in the upper and lower legs of a bridge amplifier, having the same actual capacitance in the two transformers is not desired. Having the same actual capacitance results in unequal effective capacitance. What is desirable is to have the capacitances different in a ratio inverse to the voltages. This can sometimes be achieved by winding two secondaries one over the other with a common primary. If this is done, it is necessary to adjust the leakage inductance to prevent the rise times and fall times of the two secondaries from being significantly different. A significant difference in rise time could cause one transistor in a leg to turn on before the other transistor in the leg. The transistor which has not turned on can see the full value of V_{CC} for a short time. A difference in fall time can cause one transistor to see a large ringing voltage caused by the sudden interruption of the current in the primary of the output transformer.

It is usually better to design the upper and lower transformers as two separate units. This way all of the critical parameters in one transformer are independent of the critical parameters in the other transformer. The same transformer design can be used for the two upper legs of the bridge. Similarly, the same transformer can be used for the two lower legs of the bridge. The base drive transformers for a half-bridge circuit are not as critical since only one transistor is turned on at one time.

The output transformer for this type of amplifier may be either step up or step down, depending on the desired load impedance. Since the output voltage is a square wave, this transformer must also pass a broad band of frequencies. In normal operation into a resistive load, the current in the transistor should cross zero at the time the transistor starts to turn off. The transistor should then turn off as rapidly as possible to prevent excessive shoot through when the other transistor turns on. Typical input and output voltage waveforms are shown in Fig. 6.18. If the output transformer has significant phase shift at the fundamental frequency, this condition of zero current at turnoff is not met. Low open-circuit inductance, in addition to creating phase shift, also increases the exciting current in the primary. High exciting current and too high a distributed capacitance can cause the voltage to decay slowly. One transistor may then still be on when the opposite one turns on. This causes a high peak current through the first transistor in the reverse direction. It takes very little current under these conditions to cause transistor failures. Even if the transistor turns off fast enough to prevent shoot-through damage, the current in the primary can cause a voltage spike which can destroy the transistor. *RC* networks can be incorporated to prevent these spikes from being excessive, but it is best to minimize the possibility of voltage spikes by good transformer design.

The other inductive component in the class D amplifier is the inductor in the output circuit. This inductor may be either in the tuned circuit or in the filter. The basic design of high-*Q* inductors was discussed in Section 6.7. An inductor for a low-pass filter may require a higher *Q* than an inductor for a series-resonant circuit to prevent excessive insertion loss in the pass band. If ferromagnetic cores are used in these inductors, the effect of the harmonic

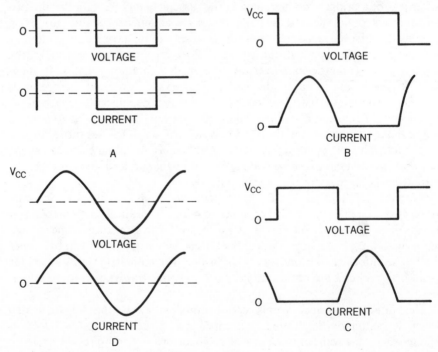

Figure 6.18. Input and output current and voltage waveforms: class D amplifier. (A, B, C, D correspond to Fig. 6.17.)

voltages on flux density and core loss must be considered. In a series circuit at resonance, the voltage across both the inductor and capacitor will be $Q \times E$, where Q is the loaded Q of the resonant circuit and E is the peak voltage across the output impedance. Because of this, the coil and capacitor both may have to be designed for high voltage even when the transformer secondary voltage is relatively low. The detailed design of coils for this type is discussed in Chapter 7.

6.10   DRIVER TRANSFORMERS

Requirements for class B, C, and D driver transformers are unusually difficult to satisfy. The transformer load is nonlinear. In tube-type amplifiers, the grid current is far from sinusoidal. In bipolar transistors, the load current changes rapidly with little change in voltage. Driver transformers for bipolar transistors are easier to design than those for vacuum tubes but still require careful design to handle the nonlinear load. The average power from a driver stage is relatively low, but the stage must deliver this power into a widely varying load impedance. The driver stage must be capable of supplying the requisite instantaneous current; otherwise, distortion will appear in the output stage of the amplifier. These currents contain harmonics of higher order, and to ensure faithful reproduction of the desired voltage waveform, the frequency range of the driver transformer at both ends must be extended well beyond the required frequency range of the amplifier. On the high end, reduced leakage inductance is required to allow the higher-frequency current components to be coupled to the input of the amplifying device. On the low-frequency end, the open-circuit inductance must be increased to prevent the transformer magnetizing current, itself nonlinear, from distorting the waveform. A resistance load across the transformer minimizes the magnitude of the load variation for a tube or FET driver. A series resistor will do the same for transformers driving class B or D bipolar transistor amplifier stages. Driver transformers are usually step down.

These conditions require transformers which are unusually large for their power level. For low (1 to 2%) overall harmonic distortion, driver transformer design becomes impractical, and it is advantageous to dispense with driver transformers entirely. This is accomplished by using a current amplifier stage (Fig. 6.6). For low-power amplifiers, RC coupling is commonly used. As the power becomes higher, the losses in the resistors becomes excessive if the resistor value is high enough for good low-frequency coupling with reasonable-sized capacitors. If the resistance value is lowered, the coupling capacitor value becomes too large to be practical.

To overcome the problems with RC coupling, LC coupling is used. With LC coupling, the dc impedance remains low while the ac impedance is high. When the coupling is done through an LC circuit for a push-pull amplifier, it takes the form of a symmetrical pi filter. The two input inductors carry the full driving element current. Coupling capacitors connect these inductors to the output-stage input element inductors, which carry the input element current. Sizes of inductors and coupling capacitors are chosen to give approximately constant impedance from the lowest frequency to be amplified up to the higher harmonics of the highest frequency. Inductor capacitance is minimized to preclude pronounced resonance effects throughout the frequency range.

The current amplifier circuit has another advantage. Leakage inductance in the driver transformer causes high-frequency phase shift between the driver output voltage and the next element input voltage. This does not exist in the coupling capacitor scheme. Since inverse feedback is often applied to amplifiers to reduce distortion, the absence of phase shift is a great advantage. The low frequency at which phase shift appears must be kept below the frequency that is to be amplified. This can be done without excessively large components.

6.11 TRANSFORMER-COUPLED OSCILLATORS

These circuits are similar to that in Fig. 6.19. Transformer OCL and C_1 form a resonant circuit, to which are coupled sufficient turns to drive the input circuit in the lower left-hand winding. The output circuit is coupled by a separate winding. For a good sinusoidal output in such an oscillator, class B operation is used. The turns ratio between the output and input circuits is determined by the voltage required for class B operation of the amplifying device as if it were driven by a separate amplifier stage. A single amplifying device may be used because the resonant circuit maintains the sinusoidal wave shape over the half-cycle during which the amplifying device is not operating. Bias on the input element is obtained from the circuit RC_2.

In such an oscillator, the load on the amplifying device equals the transformer

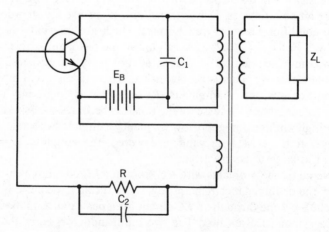

Figure 6.19. Transformer-coupled oscillator.

loss plus the input load plus the output load. In low-power oscillators, transformer loss may be the most significant part of the total load on the amplifying device. This loss consists of core, gap, and copper losses. Copper loss is large because of the large circulating current in the resonant tank. The wire size in the output element winding is larger than would normally be used in an amplifier transformer. The gap is necessary to keep the inductance to a value determined by the resonant circuit Q or volt-amperes. This in turn is dictated by the required harmonic content of the sine-wave output. The use and selection of core materials are approximately the same as indicated in Sections 6.7 and 6.8.

Class C oscillators are less desirable for very low harmonic requirements because of the difficulty of designing resonant circuits with sufficiently high Q. Where large harmonic values can be tolerated, the transformer can be designed for low Q, but the waveform becomes nonsinusoidal. Transformer input winding turns are large, approximately the same as the turns for the amplifying device output element. The input voltage would be high if the input current drawn did not limit the voltage swing. During the half-cycle when the amplifying device is operating, the voltage has a roughly rectangular shape. During the rest of the cycle it peaks sharply to a high amplitude in the opposite direction. Core losses are difficult to predict because core loss data for such waveforms is not usually available directly. Designs of this type of transformer usually involve cut-and-try for an optimum transformer design. The frequency of oscillation varies with changes in load; hence low-Q class C oscillators are to be avoided if good frequency stability is required.

Another type of transformer-coupled oscillator is shown in Fig. 6.20. In this circuit the amplifying device is operating class B. The close coupling through the tapped transformer helps to maintain excellent frequency stability.

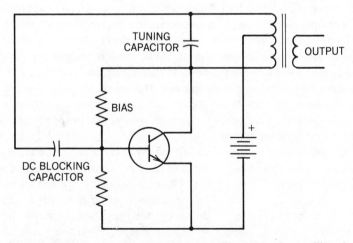

Figure 6.20. Autotransformer-coupled oscillator (Hartley oscillator).

6.12 SHIELDING

6.12.1 Magnetic Shielding

Many electronic circuits are sensitive to very small magnetic fields. It is important that only the desired signals are present in these sensitive circuits. Small amounts of extraneous voltage introduced at the input of a high-gain amplifier may spoil the quality of the signal or even make the amplified signal completely useless. One source of extraneous voltage or hum is the stray magnetic fields emanating from high-power inductive components in or near the sensitive circuit. The stray fields may enter the magnetic cores of low-level transformers and induce small voltages in the windings, which may be amplified to objectionably high values by the amplifier. Stray fields from transformers may also couple directly into wiring if the wires are located relative to the transformer so that the magnetic field can induce a voltage in the wire. Several means may be used to reduce this hum pickup:

1. The power transformer is located as far from the sensitive areas as possible.
2. The transformers and wiring are oriented for minimum coupling.
3. The transformers are magnetically shielded.
4. Core-type transformer construction is used.

A combination of two or more of the possible hum reduction methods are often required and used. The first expedient is limited by the space available for the amplifier. Space is particularly limited with high-gain operational amplifier integrated circuits. While input transformers are seldom used with this type of circuit, coupling into the wiring may be a severe problem due to the relatively high impedances of the circuits. The magnetic field varies as the inverse cube of the distance, so it is obviously helpful to locate the input circuitry as far from the power transformer as possible. The second method is to orient transformers so that the magnetic fields are perpendicular to each other. The power transformer is oriented to the wiring so that the magnetic field is parallel to the wiring. Magnetic shielding is a brute-force method of keeping out stray fields. It can be effective if properly done but does increase the space required for the transformer. Core-type construction is effective and does not increase the size of the transformer appreciably. All of these methods require the proper application of design and manufacturing techniques.

Magnetic shielding may be accomplished by a thick wall of ferromagnetic material or a series of thin, nesting boxes of high-permeability material encasing the windings and core. This type of shielding is normally applied to low-level transformers. It may be applied to power transformers if precautions are taken that the fields do not saturate the shield and for proper heat transfer from the transformer core and coil. The action of a thick shield in keeping stray flux out of its interior is shown in Fig. 6.21. Thick shields of magnetic material are not the most effective.

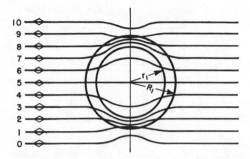

Figure 6.21. Refraction of magnetic field by iron shield.

Multiple shields increase the action just mentioned because eddy currents induced in the shields set up fluxes opposing the stray field. The most effective shielding at low levels uses alternate layers of high-permeability magnetic material alternated with layers of copper or other nonmagnetic electrically conducting material (see Gokhale, 1929; Gustafson, 1938). If the magnetic portion of the shield is made thick enough, this type of shield can be used for high flux levels. For high flux levels, thin magnetic materials can be used by placing a nonmagnetic layer next to the highest field. The flux density can be reduced to a level that will not saturate the magnetic portion of the shield.

In core-type transformers the flux normally is in opposite directions in the two core legs, as shown in Fig. 6.22. A uniform external field, however, travels in the same direction in both legs, and induced voltages caused by it cancel each other in the two coils. This type of winding reduces the external field, which can then be further reduced by shielding. If the flux density will not saturate the shield, the shield is most effective when applied to the source of the interference.

The relative effectiveness of these expedients is shown in Fig. 6.23. Pickup is given in decibels, with zero decibels equal to 1.7 V across 500 Ω and distance from a typical small power transformer as the abscissas. All curves are for 500-Ω windings working into their proper impedances and with no orientation for minimizing pickup. Using impedances much less than 500 Ω reduces the pickup. Orientation of the coil also reduces pickup. For all types of units there is a position of minimum pickup. With the unshielded shell type the angle between the transformer coil and the field is almost 90° and is extremely critical. With

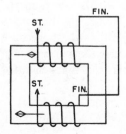

Figure 6.22. Flux directions in a core-type transformer.

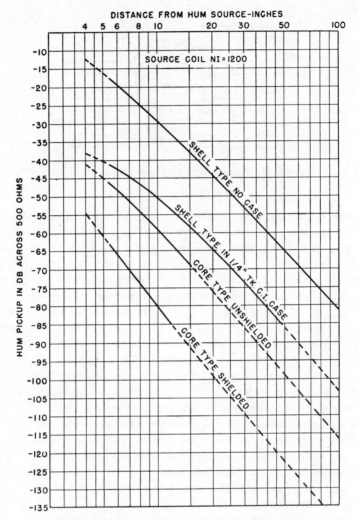

Figure 6.23. Hum pickup in low-level transformers.

shielding, this angle is less critical, but the minimum amount of pickup in this position is not noticeably reduced. The core type is less critical, especially with a shield. The minimum amount of pickup is from 10 to 20 dB less than the shielded shell type in its position for minimum pickup. Removing the shields from the core type may change its position of minimum pickup. This is because the shields reduce hum by a process of generating eddy currents, which in turn generate magnetic fields which are opposed to the magnetic fields that generated the eddy currents in the first place.

It is advantageous to have power transformers of the core type. Leakage fluxes from like coils on the two legs approximately cancel at a distant point.

The U and I shapes of the lamination are better than the L shape because of its symmetry. A type C core has the advantage that gaps are inside the coils. Thus fringing is reduced and stray flux from the core gap is minimized.

6.12.2 Electrostatic Shielding

Static shielding does not prevent normal voltage on a primary winding from being transferred inductively into a secondary winding. It is effective only against voltage transfer by interwinding capacitance. High-frequency currents from amplifier circuits are thus prevented from coupling into the 60- or 400-Hz power circuits through the power transformers. Without shielding, such currents may interfere with the operation of low-level amplifiers. The extent of static shielding depends on the amount of discrimination required. Usually a single, thin, grounded strip of metal between the windings is sufficient. The ends are insulated to prevent a short circuit. Magnetic flux in the interwinding space causes eddy currents to flow in such shields, and even shields with insulated ends look like a partial short circuit on test. This effect reduces the effective OCL of the transformer. If volts per layer are small compared to total winding voltage, a layer of wire is an effective shield. The start is grounded and the finish left free, or vice versa. A wire shield has none of the short-circuit effect of a wide strip shield.

Usually, if a transformer that requires static shielding has a low-voltage winding; a shield placed close to this winding needs little additional insulation and occupies but a small fraction of the total coil space. If shields are placed between high-voltage windings, the shields must be insulated from each winding with thick insulation. This materially increases the coil mean turn length, transformer size, and difficulty in obtaining good high-frequency response. Shields have questionable value in such transformers and are usually omitted.

6.13 HYBRID TRANSFORMERS

Hybrid transformers are used to isolate an unwanted signal from certain parts of a circuit and allow the signal to be used in other parts of the circuit. In the hybrid transformer shown in Fig. 6.24(a), the lower windings or primary sections are balanced with respect to each other, and the resistors R_2 and R_3 are equal. Voltage E_0 applied between the primary center tap and ground causes equal currents to flow in opposite directions through the two halves of the primary winding, and therefore produces zero voltage in the secondary winding. By this means, signal E_0 arrives at resistors R_2 and R_3 undiminished, but there is no voltage in R_4 connected across the secondary coil. Figure 6.24(b) shows what happens in this circuit if the voltage is applied across R_3 instead of across R_1. In this case the voltage E_3 appears across resistors R_1, R_2, and R_4, that is, in all parts of the circuit.

An inverted hybrid transformer is shown in Fig. 6.24(c). Here voltage E_s is applied across the upper coil, which is now the primary. The secondary sections

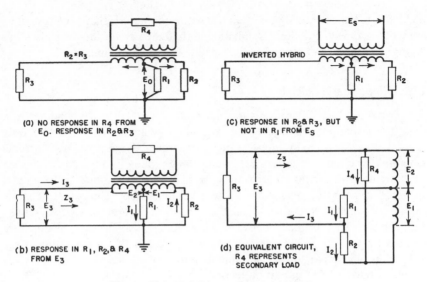

Figure 6.24. Hybrid transformer operation.

are assumed to be balanced. Therefore, there is zero voltage between the center point of the secondary winding and ground, and although a signal appears at R_2 or R_3 there is no signal across R_1. Thus a hybrid transformer works in both directions.

It has been assumed that R_2 and R_3 are equal and that the two primary half-windings are of an equal number of turns. This is not necessarily true. If the resistance R_2 is twice that of R_3, the number of turns connected to R_2 should be twice those connected to R_3. However, it is important that through the range of frequency in which the hybrid coil is desired to function, the balance between the two halves be maintained closely. The most exact balance is achieved for $R_2 = R_3$ by winding the two halves simultaneously with two different wires. This method gives good isolation of the undesired signal. Other methods introduce some ratio error, which reduces the isolation. For the same reason it is necessary to balance the circuit with regard to capacitance and leakage inductance. If a capacitance exists across R_3, such as line capacitance, for example, an additional equivalent amount should be added across R_2 to achieve the balance desired. Similarly, any inductive apparatus, adding either series or parallel inductance in one circuit, should be compensated for by inductance of like character in the other circuit. Adding series inductance, for example, in series with R_3 will not compensate for shunt inductance across R_2, or vice versa, as the two have opposite effects with regard to frequency and therefore balance is attained only at one frequency.

Assume a perfect transformer having no exciting current and no leakage inductance between the two halves, and a transformer with equal turns in the two halves of the primary winding. Assume currents in the directions shown in

Fig. 6.24(d). Then

$$I_1 = I_2 + I_3 \tag{6.12}$$

$$E_1 = I_2 R_2 + I_1 R_1 \tag{6.13}$$

$$E_3 = I_1 R_1 + E_2 \tag{6.14}$$

On the assumption of equal turns in the two half-windings, $E_1 = E_2$. If the magnetizing current is assumed to be zero, the ampere-turns and hence the volt-amperes of the two primary halves are equal. The secondary load can be considered as reflected into the primary winding as resistor R_4.

$$I_4 = (E_1 + E_2)/R_4 \tag{6.15}$$

$$(I_3 - I_4)E_2 = (I_2 + I_4)E_1 \tag{6.16}$$

If equations 6.12 to 6.16 are combined, an expression for Z_3 can be found:

$$Z_3 = \frac{E_3}{I_3} = \frac{4R_1 R_2 + 4R_1 R_4 + R_2 R_4}{4R_1 + 4R_2 + R_4} \tag{6.17}$$

If the secondary circuit is open, $R_4 = \infty$, and equation 6.17 becomes

$$Z_3 = 4R_1 + R_2 \tag{6.18}$$

Example 1: VLF transmitter line-matching transformer

Frequency range = 9 to 40 kHz; the power is reduced below 14 kHz
Maximum phase shift = 15°
Power outout = 120 kW
Secondary impedance = 300 Ω balanced
Primary impedance = 1500 Ω
Dc primary voltage = 11 kV
Voltage ratio primary to secondary = 2.24 : 1
Core: manganese-zinc ferrite
Core dimensions: D = 5.00 in., E = 5.00 in., F = 6.00 in., G = 12.00 in.
Turns primary/secondary = (74 + 74)/66
Primary wire size: 600 strands of AWG No. 38, Litz
Secondary wire size: 1200 strands of AWG No. 38, Litz
A_c = 24.6 in.^{2}
l_c = 56 in.
l_g = 0.01 in.

$$B_{AC} = \frac{3.49 \times 6000 \times 10^6}{14{,}000 \times 24.6 \times 66} = 921\,\text{G}$$

For 1% primary current unbalance,

$$B_{DC} = \frac{0.5 \times 74 \times 0.089}{0.85 \times 0.01} = 387\,\text{G}$$

$$\mu_\Delta = 1000 \text{ at this flux density}$$

For each half of the primary,

$$L_{OCL} = \frac{3.2 \times (74)^2 \times 24.6 \times 10^{-8}}{0.066} = 65\,\text{mH}$$

$X_L = 2 \times \pi \times 9000 \times 0.065 = 3676\,\Omega$ at 9 kHz. $X_L/R = 2.45$. From Fig. 5.8 the response is down 0.2 dB from midband. The phase shift (Fig. 6.3) = 12°.

$$\text{MT}_S = 2 \times (7.38 + 5.75) + \pi \times (0.25 + 0.26) = 27.9\,\text{in.}$$

$$\text{MT}_{P1} = 26.9 + \pi \times (1.10 + 0.19) = 30.2\,\text{in.}$$

$$\text{MT}_{P2} = 26.9 + \pi \times (1.79 + 0.19) = 32.5\,\text{in.}$$

The leakage inductance referred to primary is

$$L_s = \frac{10.6 \times (74)^2 \times 30.3 \times (2 \times 0.5 + 2.47)}{2 \times 7.5 \times 10^9}$$

$$= 204\,\mu\text{H}$$

The capacitances are

$$C_{S-P} = \frac{0.225 \times 29.1 \times 7.5 \times 2.2}{0.5} = 216\,\text{pF}$$

$$C_{P-P} = \frac{0.225 \times 31.4 \times 6 \times 2.2}{0.5} = 187\,\text{pF}$$

total capacitance = 403/3 = 131 × 2 = 262 pF

Further,

$$f_0 = \tfrac{1}{2} \times \pi \times \sqrt{204 \times 262 \times 10^{-18}} = 688\,\text{kHz}$$

$$X_C = \tfrac{1}{2} \times \pi \times 688 \times 262 \times 10^{-9} = 883\,\Omega$$

$$B = 883/1500 = 0.59$$

$$f/f_0 = 40/487 = 0.058$$

From Fig. 5.9 the response is still flat at 40 kHz. The phase shift is 5° at 40 kHz (Fig. 6.4).

The design more than meets the requirements. The frequency response exceeds the requirements. This is necessary to meet the phase-shift requirements.

Example 2: Hartley Oscillator Coil. A tapped inductor is needed for a Hartley oscillator coil at 1 kHz. The unloaded Q requirement is for a Q greater than 10.

Inductance = 2.63 H

R_{ser} = 795 Ω max.

R_{sh} = 320 kΩ min.

Current = 126 μA

Either a nickel-iron core or a ferrite pot core could be used for this coil. If a ferrite core is used

(core size, 26 mm diameter),

$$A_L = 630 \, \text{mH}/1000 \text{ turns}$$

$$N^2 = \sqrt{2.63 \times 10^6/0.63} = 2000 \text{ turns}$$

Place the tap at 667 turns. $Q = 140$ at 1 kHz. The wire size AWG is No. 38. The core used is a pregapped pot core. The value of A_L is within 1% of the nominal value. If a laminated core were used, it would be difficult to hold the inductance value to the kind of accuracy desired. A Q of 10 would be on the borderline of what is possible in a laminated core.

6.14 AMPLIFIER TESTS

The types of tests that may be run on an amplifier vary greatly depending on the application of the amplifier. An audio amplifier may be tested for distortion, frequency response, noise, and stability with a step input signal. A VLF amplifier would be tested for frequency response, phase shift, and into loads where the output current is leading or lagging the output voltage by a specified amount. There are many other tests that can be run. Some of these tests would be run on each amplifier. Others would only be run as a type test or as a troubleshooting test if the amplifier does not operate properly.

If the transformers are adequately tested before being installed in the amplifier, the reasons for deviation from normal operation will be easier to identify. Typical tests run on all transformers are turns ratio, DCR, exciting current, and electric strength. Wide-band transformers are usually tested for OCL instead of exciting current. If the transformer design is verified on early units by testing high- and low-frequency response and leakage inductance, it is necessary to make these tests on a sampling basis for the majority of the production run. Only if the samples start showing deviations from the early units do the tests have to be run 100%. Some transformers can be wound in a way that the characteristics may change drastically from unit to unit. These transformers are tested 100%. Very small transformers with spaced turns in the winding, very thin winding insulation which is a large percentage of the winding space, or windings that can be placed off-center with respect to each other are in this category.

Transformers used in push-pull amplifiers must be measured for balance. This test must be done under load. The test must measure both phase and amplitude balance. These measurements are not too difficult to make at audio or relatively low frequencies. At megahertz frequencies care must be taken to balance not only the load resistance, but also the reactance components of the loads. Load and lead reactances must also be kept low, or the load the transformer sees may be higher than the planned load and make the measurements invalid.

Inductors must be measured for inductance, DCR, and Q as a minimum. It may seem like a duplication of measurements to measure both DCR and Q, but it is possible for the DCR to be within limits and the Q too low. This is particularly true of coils wound with many stranded Litz wire. Since each strand of wire may only represent a very small percentage of the DCR, a reasonably large number of strands may not be connected before the DCR changes enough to be detected by a resistance measurement. A Q test will show this very quickly.

Inductor Q may be measured directly on a Q meter or by calculating the equivalent Q from the loss arm reading on a standard inductance bridge. If the transformer has an air core, a Q meter measurement at or near the operating frequency is satisfactory for any size coil. On an iron-core coil, the measurement should be made at or near the operating flux density. If this is not done, the actual Q may differ greatly from the measured value. To get the operating flux density at the operating frequency may be difficult. In this case the bandwidth of the resonant circuit may be measured. In percent of center frequency, $BW = 1/(2 \times Q)$. If this is measured at a few levels and the typical relative change of permeability and core loss are known with respect to frequency and flux density, a fairly accurate prediction of Q can be made.

If proper testing is done on the transformers and inductors before they are put in the circuit, many potential problems with circuit operation are automatically eliminated. One must decide in each case whether it is more economical to do extensive testing on the component or take a chance and do any required troubleshooting at the amplifier level.

7 HIGHER-FREQUENCY TRANSFORMERS

The required low-frequency response of a wide-band transformer is the primary factor that determines the size of the transformer. If high transformer OCL is required to maintain good low-frequency response, many turns or a large core are necessary, either of which tends to limit the high-frequency response. The terms high- and low-frequency are relative. The ratio of high- to low-frequency response primarily determines the difficulty of design. For example, in certain VLF transmitters, the frequency range is from about 25 to 60 kHz. It is then only necessary that the OCL be high enough to effect good response and low phase shift at 25 kHz. This is a great help in designing for proper response at 60 kHz. Ferrite cores, which have lower permeability and losses than 80% nickel-iron cores, can be used at these frequencies, but thin-gauge nickel-iron often results in a smaller transformer with lower overall losses.

7.1 LAMINATED-CORE TRANSFORMERS

At frequencies up to 1 MHz, the principles discussed in the preceding sections apply. In terms of mean frequency, the band is usually more narrow than the audio band, but at 25 kHz the curves for low-frequency operation portray amplifier performance just as they do at 30 Hz if they are both the low end of the respective frequency bands. Care must be used to prevent leakage inductance and winding capacitance from interfering with proper high-frequency operation.

In VLF transmitters, transformers are normally used to couple between stages and for coupling the output stage to the antenna or transmission line. They sometimes must handle large amounts of power with low distortion and phase shift. In addition to handling large powers, these transformers must be designed for low intermodulation distortion so that spurious frequencies will not be transmitted. Where space permits, coils are wound with single layers, spaced well apart to reduce capacitance, and have but few turns. If the necessary turns cannot be wound in a single layer, a bank winding like that shown in Fig. 7.1 may be used. This winding has more capacitance than a single layer but much less than two layers wound as a layer winding. Since intrawinding layer-to-layer capacitance is zero in these transformers, the resonant frequency is usually determined by winding-to-winding capacitance.

In high-impedance circuits, the winding-to-winding capacitance may be reduced by winding "pies" or vertical sections side by side. In low-power transformers, these pies will be self-supporting. They are wound with one or more throws per turn and may be several turns wide. They have the general appearance of Fig. 7.2 (see Kantor, 1947). High, narrow core windows or several pies are desirable to reduce leakage inductance. Depending on the frequency range and core material used, the transformer losses are mostly core loss. Thin-gauge grain-oriented silicon-steel (1 or 2 mil), nickel-iron (1 or 2 mil), or high-permeability ferrites are used advantageously in these transformers. In VLF transmitters, class B or class D amplifiers are commonly used, with or without modulation. The modulation may be amplitude modulation; however, VLF transmitters usually use frequency shift keying, in which the frequency is changed between two values to represent the mark and space of a teletype code. In a receiver, input and interstage transformers are also employed. These are used primarily to match impedances or to provide frequency selection. Similar transformers are used to match the antenna impedance to the input of the receiver.

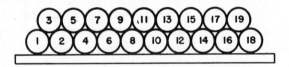

Figure 7.1. Two-layer bank winding.

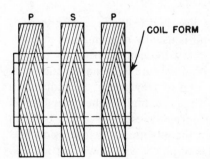

Figure 7.2. Pie-section winding.

Core data at these frequencies, of the type needed for transformer design, are often difficult to obtain except for a limited choice of materials. The power loss for several core materials is given in Fig. 7.3. It is usually necessary to pick a flux density and core loss and then calculate the temperature in the actual configuration in trying to estimate the transformer size. The thermal conductivity of ferrites is very low. Ferrite cores are often used in VLF transmitters in very large sizes, so that heat removal can become a serious design problem. Laminated materials have been used but have been largely replaced by ferrites. In some very large transformers, a core may be made up of several layers of thinner ferrite material with space between to improve core cooling. The apparent permeability of core material decreases at high frequencies. The amount depends on the lamination thickness. Oriented silicon-steel and nickel alloys have high permeability at low frequencies, but unless very thin laminations (2 mils or less) are used, this advantage disappears between approximately 10 and 50 kHz. This reduction in apparent permeability is shown in Fig. 7.4. Ferrites are less subject to this high-frequency reduction in permeability. Some of the recent developments in manganese-zinc ferrites have made their use advantageous in some applications below 10 kHz.

Transformers are used at frequencies into the megahertz region. Capacitance is usually the factor that limits the upper frequency at which an amplifier can operate. In a tuned amplifier circuit, the tuning includes the incidental and

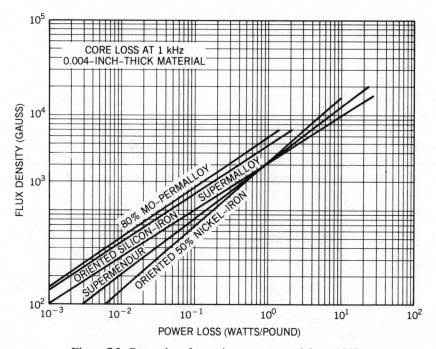

Figure 7.3. Power loss for various core materials at 1 kHz.

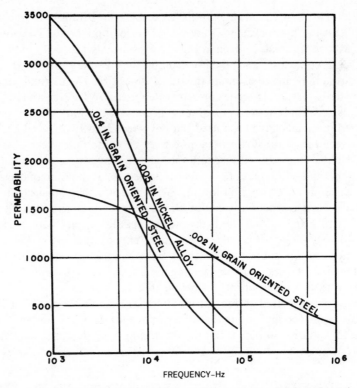

Figure 7.4. Approximate permeabilities of core steels at high frequencies.

amplifying device capacitance as well as the tuning capacitance in the resonant circuit. A wide-band transformer has no tuning to compensate for such capacitances. Even with zero winding capacitance there would be a frequency at which any amplifying device could not efficiently operate into an untuned transformer. The most favorable condition for the operation of transformers at higher frequencies is low circuit impedance. With solid-state devices and proper transformer design, it is possible to use conventional transformers to 200 MHz or above.

7.2   OTHER CORE MATERIALS

At the higher frequencies, ferrites cores have the advantage of high resistivity and practically no eddy current component of core loss. Several grades are manufactured, the different grades being useful for different applications. The manganese-zinc ferrites are the most commonly used for lower frequency or power applications. Figure 7.5 is a set of normal permeability curves for some types of ferrites. Figure 7.6 shows the initial permeability for the same materials.

PERMEABILITY

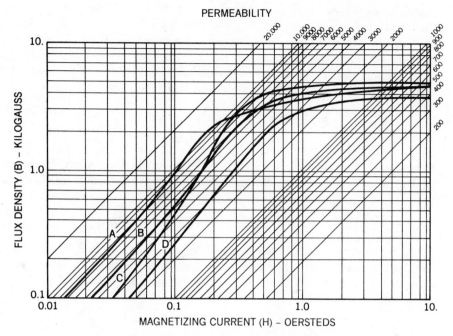

Figure 7.5. Ferrite normal permeabilities.

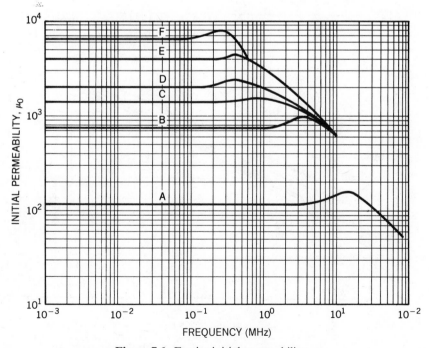

Figure 7.6. Ferrite initial permeability.

Figure 7.7 shows the core loss at various frequencies and flux densities for a Mn-Zn ferrite designed for high-power use. The nickel-zinc ferrites have lower permeabilities than the Mn-Zn and also much higher resistivities. These ferrites are used at frequencies up to several hundred megahertz.

At low flux densities, the losses in ferrites are usually stated as related to the product $\mu_0 Q$. This relationship is approximately as follows:

$$\frac{\text{core loss}}{\text{in.}^3} = \frac{0.41 \times 10^{-6} f B^2}{\mu_0 Q} \tag{7.1}$$

Instead of $\mu_0 Q$, the quantity $R_{\text{ser}}/\mu f L$ is sometimes plotted, where R_{ser} is the equivalent series resistance corresponding to core loss. Equation 7.1 then becomes

$$\text{core loss/in.}^3 = 0.065 \times 10^{-6} f B^2 (R_{\text{ser}}/\mu f L) \tag{7.2}$$

Neither of these expressions is valid when the ferrite is used at flux densities above a few gauss. While the mixture used and the processing of the materials both influence the core loss at higher flux densities, the core loss for Mn-Zn ferrites can be expressed

$$\text{core loss/in.}^3 = B_m^{2.4}(8.04 \times 10^{-10} f + 2.79 \times 10^{-15} f^2) \tag{7.3}$$

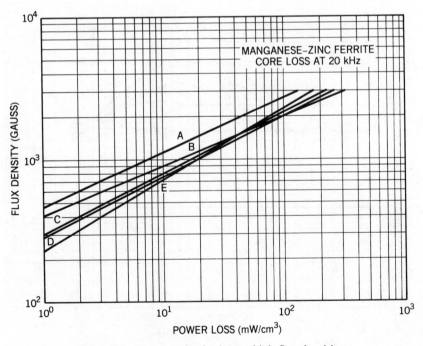

Figure 7.7. Core loss for ferrites at high flux densities.

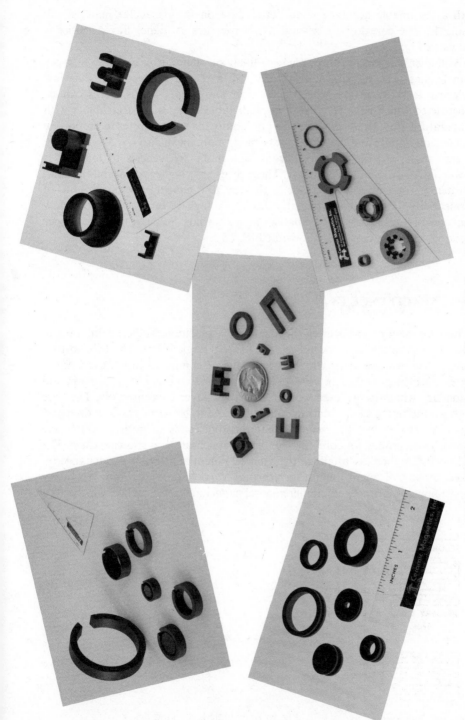

Figure 7.8. Ferrite core shapes. (Courtesy of Ceramic Magnetics)

211

with a reasonable accuracy if no better data on the particular material are available. The core loss on several Mn-Zn ferrites at higher flux densities is shown in Fig. 7.7.

At higher frequencies, cores made of finely powdered iron have lower losses than ferrites. The powdered iron cores use a plastic binder which has low dielectric loss. By proper selection of iron and binder ratios, cores can be constructed which have very low and predictable changes of permeability with temperature. Powdered iron cores do not require the temperatures in processing that ferrites require and therefore can be made more inexpensively. Another type of core material which is useful from high audio frequencies to low RF is powdered Molybdenum Permalloy. These cores have higher permeabilities than powdered iron and can be made with very low temperature coefficients of permeability. They are particularly useful in high-Q tuned circuits at low frequencies. These cores are available only as toroids. Some of the core shapes available in ferrite cores are shown in Fig. 7.8.

7.3 CAPACITANCE EVALUATION

In high-frequency transformers, the location of capacitances differs from those in audio transformers in that windings are usually single layers. The turn-to-turn capacitance of single-layer windings is negligible compared to the capacitance between windings and to the core. In the transformer in Fig. 7.9, the primary and secondary windings are each wound in a single layer concentrically. They are both wound in the same rotational direction and in the same traverse direction (right to left). It will be assumed that the right ends of both windings are connected to ground (or core) through large capacitances, as shown dashed. The right ends are then essentially at the same ac potential. Primary capacitance C_1 is composed of many small incremental capacitances C_p and secondary capacitance C_2 of many small incremental capacitances C_s, each of which has a

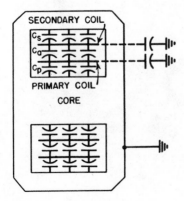

Figure 7.9. Single-layer windings.

different voltage across it. Similarly, many small incremental capacitances C_a exist between primary and secondary, which have different potentials across them. If the transformer is step-up,

$$C_1 = \frac{1}{3}\sum C_p \quad \text{and} \quad C_2 = \frac{1}{3}\left[\sum C_s + \frac{(N_s - N_p)^2}{N_s^2}\sum C_a\right] \quad (7.4)$$

If the transformer is step-down,

$$C_1 = \frac{1}{3}\left[\sum C_p + \frac{(N_p - N_s)^2}{N_p^2}\sum C_a\right] \quad \text{and} \quad C_2 = \frac{1}{3}\sum C_s \quad (7.5)$$

The ratio is 1:1,

$$N_s = N_p \qquad C_1 = \tfrac{1}{3}\sum C_p \qquad C_2 = \tfrac{1}{3}\sum C_s \qquad (7.6)$$

For transformers with opposite rotations of primary and secondary windings, or with opposite traverse directions (but not both), minus signs in the factors $(N_p - N_s)^2/N_p^2$ and $(N_s - N_p)^2/N_s^2$ in equations 7.4 and 7.5 become positive. There is no other change. For transformers with both angular rotations and traverse directions opposite, there is no change at all in these equations. If there is a shield between primary and secondary, omit the terms containing C_a in these equations, and $\sum C_s$ and $\sum C_p$ include the capacitance of secondary and primary to shield, respectively. $\sum C_s$ is the measurable capacitance of the short-circuited secondary to core, and $\sum C_p$ that of the short-circuited primary to core.

In push-pull amplifier transformers, the secondary winding is interleaved between two primary halves. The rotational directions of winding and traverse are important, as they affect not only effective capacitance but also the coupling between the outputs of the push-pull amplifying devices. It is usually best to have all windings with the same rotational direction and traverse, and to connect the primary halves externally.

Winding resistance increases with frequency because of eddy currents in the larger wire sizes. Copper loss increases proportionately. Equations for single-layer coils are given in the handbooks. The eddy current resistance of layer-wound coils is more difficult to calculate due to the proximity of one wire to another distorting the eddy currents in the wire and making the ac resistance of the middle turns of the coil different from the resistance in the end turns. Some very fundamental work in calculating the ac resistance of multilayer coils was done by Butterworth (1926). Using these equations or curves for hand calculations is very time consuming, but a simple computer program can be written to do the necessary calculations quickly. Eddy current resistance of layer-wound coils is plotted in Fig. 7.10 as a function of conductor size, frequency, and number of coil layers. It is a good approximation to the increase of resistance in a transformer winding.

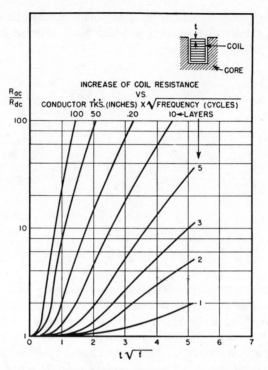

Figure 7.10. Increase of coil resistance at high frequencies.

Example. Line-Matching Transformer, 50 to 500 Ω

Frequency range = 50 to 150 kHz

Power output = 100 W

Primary voltage = $\sqrt{ZW} = \sqrt{50 \times 100} = 70.7$ V

Secondary voltage = $\sqrt{500 \times 100} = 224$ V

Core: 2-mil oriented silicon steel

$A_c = 0.45$ in.^{2}, $l_c = 6$ in., $l_g = 0.002$ in., core weight = 0.75 lb

Window 0.63 in. × 1.50 in.

Primary 31 turns No. 22 wire; mean turn 3.8 in.

Secondary 100 turns No. 30 wire; mean turn 4.6 in.

Windings arranged as in Fig. 7.9

Insulation between primary and core and between primary and secondary:
    0.12 in. of kraft paper

Secondary effective capacitance = 40 pF

$B_m = 350$ G

Secondary OCL = 20 mH

Secondary leakage inductance = 260 μH

Core loss = 8 W/lb $\times$ 0.75 = 6 W (this is practically the only loss)

X_N/R_1 at 50 kHz = $(2\pi \times 50,000 \times 0.020)/500 = 12.56$

$f_r = 1560$ kHz, $B = 5.00$

According to Figs. 5.8 and 5.9, this transformer has nearly flat response over the entire frequency range.

7.4 LEAKAGE INDUCTANCE AT HIGH FREQUENCIES

Provided that a transformer is operated at frequencies below resonance, the leakage inductance measured at low frequencies governs the response at high frequencies. Leakage inductance in concentric windings is lowest if the windings are symmetrically spaced in the traverse direction, as in Fig. 7.11(*a*). For a given number of turns, the leakage flux is least in Fig. 7.9, somewhat greater in Fig. 7.11(*a*), and much greater in Fig. 7.11(*b*). The increase in leakage flux is a function of core dimensions, winding-to-winding spacing, and marginal inequality. Figure 7.12 shows a typical increase of leakage inductance when one secondary margin is increased with respect to the other, as in Fig. 7.11(*b*). If the number of turns and wire size are not enough to fill the total space in the winding layer, the turns should be evenly spaced across the total distance between the assigned margins to minimize leakage inductance.

Leakage inductance is very low in toroids with windings that cover the whole magnetic path. Toroids are wound on special machines that thread wire into and out of the core. Carefully wound toroidal transformers can function at very high frequencies (see Maurice and Minns, 1947). If part of the core is not covered by the windings, as indicated by dimension *G* in Fig. 7.13, leakage flux escapes from the ends of the coils. This leakage flux determines the leakage inductance. The increase of leakage inductance reduces the frequency range.

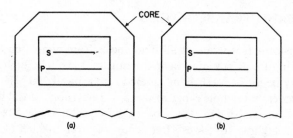

Figure 7.11. (*a*) Symmetrical and (*b*) asymmetrical spacing of concentric windings.

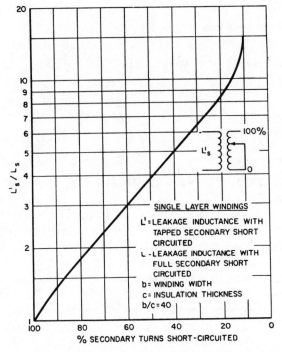

Figure 7.12. Leakage inductance of asymmetrical windings.

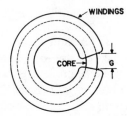

Figure 7.13. Toroidal core and coil.

7.5 WIDE-BAND TRANSFORMERS

Untuned transformers operate at all frequencies from almost dc to VHF. The lowest operating frequency may be a fraction of 1 Hz. The highest operating frequency may be in the VHF band, somewhere around 200 MHz. No known transformer covers this whole range at present. Television coaxial line terminating transformers have been made to cover the frequency range 50 Hz to 6 MHz, or a ratio of highest to lowest frequency of over 100,000:1. This is an exceptionally wide band. More common wide-band transformers are those in the audio band of 20 to 20,000 Hz, or 10 to 30,000 Hz, that is, about a 3000:1

frequency ratio. Wide-band transformers at other frequencies have frequency ratios of 100:1 or less.

In low-impedance circuits it is leakage inductance that primarily determines transformer behavior, whereas at high impedance the winding capacitance is the controlling factor. In most audio transformers the coupling coefficient is 0.9995 or higher. With bifilar windings, this figure may increase to 0.999995. Such a high coefficient of coupling requires the use of good core materials. It is difficult to achieve a 0.9995 coupling coefficient at high frequencies or at high voltage. It can only be done, if at all, by very careful arrangement of the windings and significant interleaving. For a given source impedance and transformer core material, the product of turns ratio and bandwidth is a rough indication of size. Quite generally, for low power the widest-band transformers use core materials of 80% nickel-iron.

In high-impedance circuits the matter of size is not merely one of space for mounting; it also has a direct bearing on the upper frequency limit, since transformer capacitance is roughly proportional to size. If capacitance is low, the bandwidth ratio (highest/lowest frequency) is approximately equal to the ratio of OCL/leakage inductance. This may be verified by comparing Figs. 5.8 and 5.9. It is most nearly true for low-impedance transformers. With a given primary impedance, core size, and material, there is a limit to the step-up turns ratio possible for any specified frequency response.

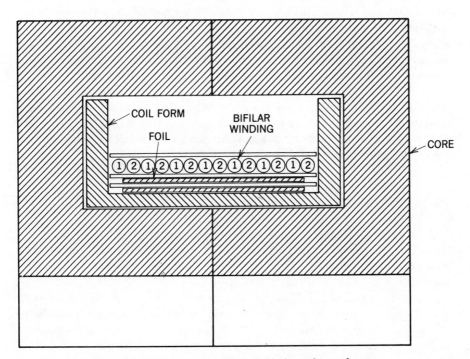

Figure 7.14. Low-power, wide-band balanced transformer.

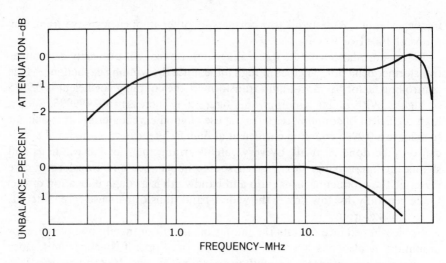

Figure 7.15. Frequency response and balance of the transformer in Fig. 7.14.

One way of building a transformer for very wide bandwidth at low powers is shown in Fig. 7.14. This type of construction is particularly good if good amplitude and phase balance is required over a wide frequency range. The OCL is determined by the required low-frequency response. The primary is wound with a very thin copper strip which is carefully centered on a thin insulating material. The start and finish leads are at or near the ends of the foil. The secondary is wound as a bifilar winding completely across the finish of the primary. The leads are brought out the same opening in the core. If the output is balanced, great care must be taken to keep the lead lengths equal.

Transformers wound this way have been designed with the frequency response to within 1 dB from 1 to 100 MHz. The balanced outputs have maintained combined phase and amplitude balance to better than 1% from 1 to 40 MHz. An example of the response and balance of a transformer wound as in Fig. 7.14 is shown in Fig. 7.15.

7.6 AIR-CORE TRANSFORMERS

Transformers considered previously have had iron or ferrite cores. A class of transformers widely used in higher-frequency circuits is either without cores or with small slugs of powdered iron or ferrite. In a transformer with an iron core, the exciting current required for inducing the secondary voltage is a small percentage of the load component of the current. In an air-core transformer all the current is exciting current and induces a secondary voltage proportional to the mutual inductance. Consider the circuit of Fig. 7.16 in which Z_1 is complex and includes the self-inductance of the primary coil. Secondary impedance Z_2 is also complex and includes the self-inductance of the secondary coil. With a

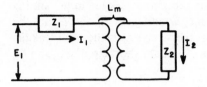

Figure 7.16. General case of inductive coupling.

sinusoidal voltage applied, Kirchhoff's laws give the following:

$$E_1 = Z_1 I_1 + j2\pi f L_m I_2 \tag{7.7}$$

$$0 = Z_2 I_2 + j2\pi f L_m I_1 \tag{7.8}$$

where L_m is the mutual inductance between the primary and the secondary coils.

From equation 7.8 we see that the voltage in the secondary coil is numerically equal to $2\pi f L_m I_1$, the product of primary current and mutual reactance at the frequency of the applied voltage E_1. The equivalent impedance of the circuit of Fig. 7.16 when referred to the primary side is given by

$$Z' = Z_1 + (X_M^2/Z_2) \tag{7.9}$$

where $X_M = j2\pi f L_M$.

In the equations above, the impedances Z_1, Z_2, and Z' are complex quantities whose real and imaginary terms depend on the values of resistance, inductance, and capacitance in the circuit. One common practical case arises when the primary resistance is zero, or virtually zero, and the secondary coil is tuned to resonance so that Z_2 is a pure resistance R_2. Under these conditions equation 7.9 reduces to

$$R' = X_M^2/R_2 \tag{7.10}$$

where R' is the equivalent resistance of the primary.

Equation 7.10 gives the value of mutual inductance required for coupling a resistance R_2 so that it will appear like resistance R' with a maximum power transfer between two coils. It states that the mutual reactance X_M is the geometric mean between the two values of reactance.

The ratio of mutual inductance to the geometric mean of the primary and secondary self-inductances is the coefficient of coupling:

$$k = L_M/\sqrt{L_1 L_2} \tag{7.11}$$

The value of k is never greater than unity, even when coils are interleaved to the maximum possible extent. Values of k down to 0.01 or lower are common at high frequencies.

Coupling coefficient is related to untuned transformer open- and short-circuit reactance by means of the transformer equivalent circuit shown in Fig. 5.7(a).

Assume that the transformer has a $1:1$ turns ratio, and leakage inductance is equally divided between primary and secondary windings. If L_1 and L_2 are the self-inductances of primary and secondary, respectively, L_s is the total leakage inductance (measured in the primary with the secondary short circuited), and L_m the mutual inductance,

$$L_1 = \frac{X_P + X_N}{2\pi f} = \frac{L_s}{2} + L_m$$

$$L_2 = \frac{X_S + X_N}{2\pi f} = \frac{L_s}{2} + L_m$$

From equation 7.11,

$$k = \frac{L_m}{\sqrt{L_1 L_2}} = \frac{L_m}{\sqrt{(L_m + L_s/2)^2}}$$

$$= \frac{1}{1 + (L_s/2L_m)} \qquad (7.11a)$$

If $L_m \gg L_s$,

$$k \approx 1 - L_s/2L_m \qquad (7.11b)$$

Equations 7.11a and 7.11b are useful in estimating approximate transformer bandwidth.

A tuned air-core transformer is often used for frequency selection in high-frequency narrow-band amplifiers and oscillators. An example is shown in Fig. 7.17. Here a sinusoidal voltage E_1 may be impressed on the primary circuit by an amplifying device. Resistances R_1 and R_2 are usually the inherent resistance of the coils, but occasionally, resistance is added to change the circuit response. The value of voltage E_2 obtained from this circuit depends on the impressed frequency. In Fig. 7.18 it is shown for three different values of coupling. If the value of coupling is such that

$$X_M = \sqrt{R_1 R_2}$$

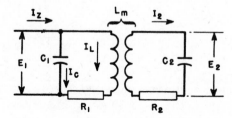

Figure 7.17. Tuned air-core transformer.

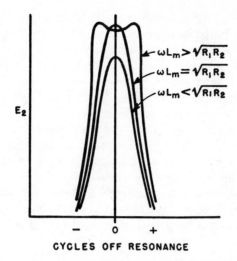

CYCLES OFF RESONANCE

Figure 7.18. Response curve for circuit of Fig. 7.17.

we obtain a condition similar to that of equation 7.10, in which the maximum power or current is produced in the secondary circuit. Maximum current through capacitor C_2 gives maximum voltage E_2. This value of coupling is known as the critical value. A smaller coefficient of coupling gives a smaller maximum value of E_2. A greater coefficient of coupling results in a "double hump," as shown in Fig. 7.18. The height of the resonant peaks and frequency distance between peaks depends on circuit Q and coefficient of coupling k. The double-hump curve of Fig. 7.18 is desirable when modulated signals are being amplified. Very little attenuation is offered to the modulating frequencies which add to or subtract from the carrier frequency normally corresponding to resonance. Frequencies in adjacent channels are rejected. Close tuning control and high Q are essential to good frequency response and selectivity.

If the primary circuit is made to resonate at a different frequency from the secondary, the response to the modulating frequencies is made worse, and considerable harmonic distortion is likely. The voltage output at the mean frequency is less then it would be if the circuits were tuned to the same frequency. Air-core transformers are often made adjustable for tuning and coupling.

7.7 MULTIPLE TUNED CIRCUITS

Double-hump resonance was obtained with higher-than-critical coupling in the circuit of Fig. 7.17. Frequency response with more humps is obtainable if there are more than two coupled loops. Such circuits are more difficult to tune and adjust than the circuit of Fig. 7.17 because of the reaction of each coupled loop on the others. Easier adjustment can be made with successive "stagger-tuned" band-pass amplifiers. Each amplifier stage is tuned to a slightly different frequency. Because of the isolation of the stages by the associated amplifying elements, tuning one stage does not influence the tuning of another.

Frequency response similar to that of multiple tuned coupled circuits may be obtained by filter sections. It does not matter whether the coupling is inductive or capacitive; the same shape of response is obtained from the same number of circuits tuned in the same manner. Since similar results are obtained from coupled circuits and filters, the choice between them may be made on the basis of convenience and cost. The amplifying function and frequency selectivity are often combined in circuits using high-gain integrated circuit operational amplifiers as the amplifying elements. These active filters result in the optimum economy in the use of parts. Considerable literature has accumulated on the design and adjustment of multiple tuned circuits, band-pass filters, and active filters. Special techniques have been developed for tuning them (see Dishal, 1951).

7.8 MUTUAL INDUCTANCE

It is evident from equation 7.8 that the secondary voltage depends on the mutual inductance between the coils. Mutual inductance can be calculated by equations that depend on the geometric configuration of the coils. If the coils are arranged concentrically, as shown in either part (a) or part (b) of Fig. 7.19, the mutual inductance of the coils can be found from

$$L_m = \frac{0.05a^2 N_1 N_2}{g}\left[1 + \frac{A^2 a^2}{8g^4}\left(3 - 4\frac{l^2}{a^2}\right)\right] \qquad \mu\text{H} \qquad (7.12)$$

where N_1 represents the primary turns and N_2 the secondary turns. All dimensions are in inches. For most purposes, the bracketed portion of this equation is approximately unity, and it has been plotted in Fig. 7.20 for a single-turn secondary. To find the mutual inductance for any given number of secondary turns, multiply the mutual inductance found from the curve by the number of secondary turns. The range of ordinates and abscissas can be extended indefinitely.

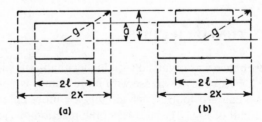

Figure 7.19. Concentric coaxial coupled coils.

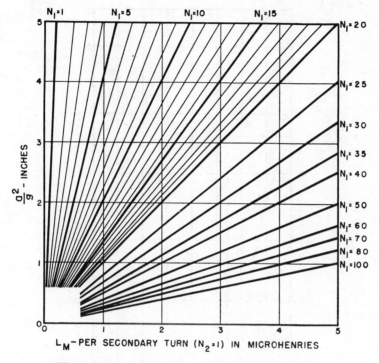

Figure 7.20. Mutual inductance of coils in Fig. 7.19.

If the coils are arranged coaxially as in Fig. 7.21, approximate values of mutual inductance are found as follows:

$$L_m = FN_1N_2\sqrt{Aa} \tag{7.13}$$

In this equation all dimensions are in inches and the mutual inductance is in microhenries. The factor F can be found conveniently in Fig. 7.22.

Self-inductance of single-layer coils is

$$L = 0.1a^2N^2K/l \tag{7.14}$$

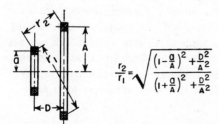

$$\frac{r_2}{r_1} = \sqrt{\frac{\left(1-\frac{a}{A}\right)^2 + \frac{D^2}{A^2}}{\left(1+\frac{a}{A}\right)^2 + \frac{D^2}{A^2}}}$$

Figure 7.21. Concentric noncoaxial coils of rectangular cross section.

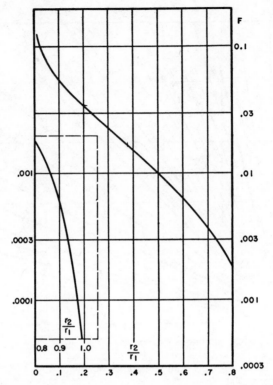

Figure 7.22. Factor F in equation 7.13 as a function of r_2/r_1 (see Fig. 7.21).

where a = mean coil radius, in.
   N = number of turns
   l = length of coil, in.
   L = inductance, μH

K may be found from Fig. 7.23. Equations 7.12, 7.13, and 7.14 are based on equations 192, 187, and 153 in National Bureau of Standards Circular 74. Simple computer programs have been written to calculate the inductance of air-core coils. The values for F and K may be found from a piecewise linear approximation to the values in the curves.

   Coaxial air-core coils are used for coupling and tuning at frequencies from 20 kHz to 400 MHz. If the wire is more than about 0.005 in. in diameter and the frequency is below 500 kHz, the wire is subdivided into strands of individually insulated wires. This type of wire is called Litzendraht and is available in sizes from three strands of AWG No. 44 to thousands of strands in cables measuring more than 1 in. in diameter. At higher frequencies, the current penetration is incomplete in even very fine wire, so solid or tubular conductors are used.

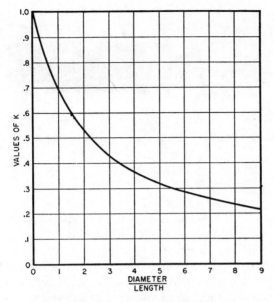

Figure 7.23. Factor K in equation 7.14.

7.9 TUNING SLUGS

Both self-inductances and mutual inductances of a coil may be increased by inserting a slug of powdered iron or ferrite inside the coil tube. Tuning a resonant circuit to a given frequency is often effected with a variable inductor with fixed capacitors instead of tuning with variable capacitors. A small coil of this type is shown in Fig. 7.24, with the tuning slug hidden by the coil form. At the left end is the screw and lock by which the inductance can be adjusted and maintained at a given value. The mutual inductance of a pair of coils can be changed similarly. This is preferable to attempting to vary the distance between the coils, since it requires no flexible connections. Powdered iron and ferrites are both available in several grades. The grade depends on the frequency range in which they are used. The powdered cores are available in both different alloys and powder sizes. Bonding compounds reduce the effective permeability of the core to values ranging from 10 to 125, depending on the grade of iron and the frequency. In a given coil, the insertion of a slug raises the inductance from 2 to 3 times the value it would have if no core were present. Circuit Q increases similarly. Higher Q results from a powdered iron or ferrite magnetic path, closed except for small air gaps. For an untuned transformer, where high Q is not essential, the air gap may be zero to reduce magnetizing current.

Figure 7.24. Coil inductance is varied by a magnetic slug.

7.10 RF INDUCTORS

When an inductor is used to pass direct current and present a high impedance to some frequencies, it may have a high voltage across it at the operating frequency. High inductor impedance at the operating frequency is necessary to reduce the copper loss caused by the high-frequency current, which reduces the useful power and overheats the inductor. If a single-layer inductor is connected to a high-frequency generator at a given voltage, and if the current is measured as in Fig. 7.25, the inductor impedance is the ratio of voltage to current measured.

By disconnecting the inductor from the circuit, the tuning reaction may be noted, and from this reaction, whether the reactance of the inductor is inductive or capacitive. The difference in watts input to the ac source, when the coil is removed and the tank resonating capacitor is retuned for minimum amplifying device current, is readily observable. This difference times the ac source efficiency is the loss in the coil at the particular voltage and frequency.

The impedance of a typical coil, found as described above, is plotted in Fig. 7.25 against frequency. At low frequencies (a), the curve follows a straight reactance line $X_L (= 2\pi fL)$. At a frequency somewhat below natural frequency b (determined by inductance and effective capacitance), the slope starts to increase and reaches a maximum point at a frequency c of $1.2b$ to $1.7b$. Above this frequency, the impedance decreases until a minimum value is reached at d, which is from 2.2 to 3.0 times b. At higher frequencies, the increase and decrease are repeated in a series of peaks and valleys at approximately equal frequency intervals. The second, fourth, and sixth peaks are of lower value than the first, third, and fifth, respectively. The seventh peak is followed by a flattened slope which suggests a submerged eighth peak. The points of minimum impedance rise in value, so that at higher frequencies the valleys appear to be partially filled in and the peaks to be leveled off. The watts loss are high at points of low impedance, and they rise sharply at the frequency d.

The change in reactance is shown in Fig. 7.25. The coil is inductive up to frequency b. From b to c it has no noticeable effect on the tuning and hence is a pure resistance or nearly so. Above c it is capacitive up to a frequency slightly below d, where it again becomes of indefinite reactance. Thereafter, it is capacitive, except for brief frequency intervals, where it is resistive or only

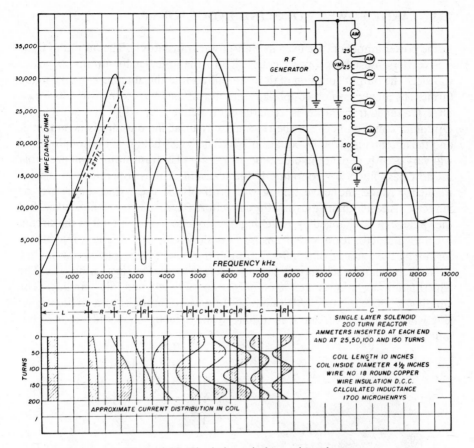

Figure 7.25. Single-layer inductor impedance.

slightly inductive. At all frequencies higher than the fifth peak, the coil is capacitive.

Since a coil has distributed constants it is subject to standing waves at the higher frequencies. The character of these waves may be found by tapping the coil at various points and inserting a high-frequency current-measuring device in series with the coil at these points. The current distribution is plotted in Fig. 7.25 against coil length. These diagrams show the kind of standing waves as the frequency increases.

Current distribution is uniform at all frequencies below *b*. Most inductors are used within the first impedance peak. The useful range for an inductor impedance of 20,000 Ω in Fig. 7.25 is 1700 to 2800 kHz. This inductor could be safely operated at 5500 kHz also, but the frequency range is narrower. The safe loss dissipation is less because it takes place over half of the coil surface. Pie-section inductors have similar impedance curves, but impedance peaks following the first are less pronounced.

7.11   LARGE POWER COUPLING INDUCTORS

Large power amplifiers are used in high-power transmitters. The output of these transmitters must be properly coupled into the antenna for optimum power transfer. The output coupling of these amplifiers into the antenna can take many forms. The most common coupling means are a tuned transformer or a low-pass filter. If a tuned transformer is used, the equivalent circuit is like that of Fig. 7.16. Optimum coupling between the resonant circuit at the output of the amplifier and the antenna is given by equation 7.10. The construction of the coupling coil will vary greatly depending on frequency range. VLF transmitters will probably use low-pass filters for coupling. If the transmitter must operate over a wide frequency range, inductive tuning will probably be used, since adjusting the inductors is easier than adjusting the capacitors. The capacitors will be fixed, but the values can be changed by switching the capacitors into and out of the circuit. A tuning inductor for a VLF transmitter is shown in Fig. 7.26. This inductor is for tuning the antenna of a 100-kW transmitter. Due to the voltages developed, it is insulated by enclosing it in sulfur hexaflouride, an insulating gas. The construction techniques are a combination of layer and pie winding. The winding is 2400 strands of AWG No. 38 Litz wire. Tuning is by a ferrite tuning slug which is motor driven. It is similar to the inductor shown in Fig. 7.24 but much larger.

Figure 7.26. Antenna tuning coil for a VLF transmitter.

Higher-frequency coupling inductors are usually similar to Fig. 7.20(b) with the coupling coil on the outside and spaced from the output tuned coil to reduce capacitive coupling. Taps are often provided on the coupling coil for frequency and antenna resistance adjustments.

When the secondary circuit is untuned and the secondary load is reactive, all the secondary volt-amperes (which may exceed the secondary watts many times) flow through the transformer windings. Tight coupling must be used between primary and secondary in order to prevent loss of power, due to current circulating in the primary without corresponding current flowing in the load. If the load power factor is less than 20%, currents and volt-amperes in the circuit may be considered independent of the winding and load resistances. In Fig. 7.16, let the load Z_2 be inductive, comprising L_3 and R_L. Also , let

$$R_1 = 0$$

$$L_2 = \text{secondary self-inductance}$$

$$L_m = \text{mutual inductance}$$

$$k = \text{coefficient of coupling} = L_m/\sqrt{L_1 L_2}$$

Then the secondary volt-amperes $= I_2^2 Z_2$, and primary volt-amperes $= E_1 I_1$. The ratio of maximum secondary volt-amperes transformed to the primary volt-

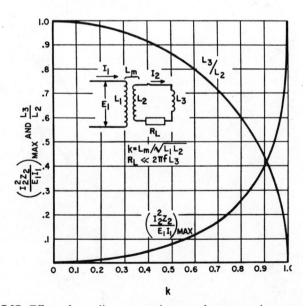

Figure 7.27. Effect of coupling on maximum volt-amperes in untuned load.

amperes is related to k as follows:

$$\left(\frac{I_2^2 Z_2}{E_1 I_1}\right)_{max} = \frac{k^2}{2(1 + \sqrt{1 - k^2}) - k^2} \tag{7.15}$$

This equation is plotted, together with values of ratio L_3/L_2 for maximum power transfer, in Fig. 7.27. If perfect coupling could be attained, all the primary volt-amperes could be transferred to the load. Iron-core transformers operate at the extreme right of Fig. 7.27. With air-core transformers it is often difficult to approach this condition, especially if voltages are high.

8 SATURABLE MAGNETIC CIRCUITS

The characteristic of magnetic cores to saturate when biasing ampere-turns are applied to the core has been applied to a variety of electronic circuits, switching elements, pulse generators, and amplifiers, for example. Amplifiers with saturable reactive elements are known as magnetic amplifiers. Such amplifiers have been built with power gains of over 1,000,000. Magnetic amplifiers are often more reliable and have longer life than some other types of amplifiers. In this chapter the operation and design of elementary magnetic amplifiers are described.

8.1 SATURABLE INDUCTORS

From the fundamental theory of transformers, it will be recalled that the voltage induced in a winding usually far exceeds the resistance drop in that winding. In other words, winding open-circuit reactance usually is much greater than winding dc resistance. Further, it will be recalled that a relatively small amount of direct current flowing into the winding of a transformer, in the core of which there is no air gap, causes the core to saturate. Thus the reactance of the transformer may be varied by a small amount of dc power. Now, if one winding of a transformer is connected between an ac supply and a load, the amount of power delivered to the load may be controlled by a small amount of dc power flowing in another winding. Because of the fact that open-circuit reactance ordinarily exceeds dc resistance, the possibility of power amplification is inherent in a transformer. When one winding of a transformer is used for dc control power and another for ac output power, the transformer is called a saturable inductor.

231

8.2   SIMPLE MAGNETIC AMPLIFIERS

A single inductor, with a battery-fed dc source controlling one winding and ac power fed through the other winding, would have ac voltage induced in the dc winding. If this dc winding were closed only on the battery, it would effectively short-circuit the ac voltage in the power winding. This difficulty might be overcome by using a high impedance in the dc control circuit. A more common solution is to use two inductors, one of the dc windings of which is reversed, while the ac windings add normally. Connections of this sort are shown in Fig. 8.1(a), with the ac windings in series; it is possible to connect them in parallel as in Fig. 8.1(b) to allow more load current to flow at lower ac voltage.

When there is zero direct current in the control windings of Fig. 8.1, both inductor impedances are large and prevent any load current except exciting current from flowing throughout the ac voltage cycle. When direct current is applied to the control windings, impedance remains large for the first part of a cycle, until saturation flux density is reached. Then inductor impedance is reduced and a large load current may flow. With rectangular B-H loop core material, such as that shown in Fig. 2.9, the change from high to low impedance is abrupt. If the loop were a true rectangle, the load current waveform would be as shown by i_L in Fig. 8.2(a). Only the exciting current flows in the load during the interval $0-\theta_1$. Then saturation is reached and load current suddenly rises to a large value. From θ_1 to π, i_L has sinusoidal shape. During the next half-cycle, this load current shape is repeated but in the reverse direction.

For a $1:1$ turns ratio in each inductor, current i_c in each control winding equals i_L minus the exciting current. In one inductor, because of the reverse connection, current i_c flows in the opposite direction. Total current in the control circuit is as shown by the lower trace of Fig. 8.2(a), the average value of which is the input direct current I_c. Thus load current contains fundamental and odd harmonics, whereas control current contains only even harmonics. If sufficient control current flows to saturate the cores over the full cycle, load current also flows over the full cycle and is sinusoidal in waveform. For turns ratios other than unity, load and control currents are inversely proportional to turns ratio.

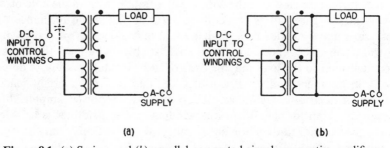

(a)   (b)

Figure 8.1. (a) Series- and (b) parallel-connected simple magnetic amplifiers.

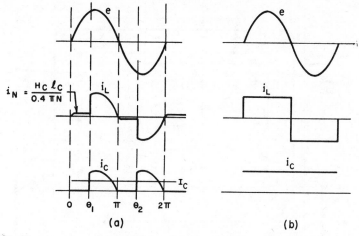

Figure 8.2. Simple magnetic amplifier voltage and currents with (a) $R_c \ll R_L$ and (b) $R_c \gg R_L$.

In the foregoing it was assumed that control current was free to assume the shape shown in Fig. 8.2(a). This is true on a 1:1 turns ratio basis only if the control circuit impedance is small. If total control circuit resistance is denoted by R_C and load resistance by R_L, for $R_C \ll R_L$, load and control currents are sine waves, or portions thereof. If the opposite is true, namely $R_C \gg R_L$, the control current wave shape is determined by R_C. For very large R_C, control current is continuous, and the current wave shapes approach those in Fig. 8.2(b). In this figure, the dc source impedance is large, even harmonics cannot flow, magnetization is "constrained," load current is flat-topped, and voltage across the reactor is distorted considerably. This distortion can be overcome by the use of a capacitor across the control coils as shown dashed in Fig. 8.1(a). When the reactors are parallel connected, as in Fig. 8.1(b), even harmonics may flow in the load coils, and capacitors are unnecessary for $R_C \gg R_L$.

Sometimes the two cores are combined into one, in the manner shown in Fig. 8.3. This is called a three-legged reactor, with one dc coil and two ac coils. Figure 8.3 shows the relative paths for the ac and dc fluxes. Equal turns in the ac coils set up equal ac magnetomotive forces which cancel in the center leg, and cause flux to flow as indicated by the solid line. No fundamental alternating voltage is induced in the dc coil, but dc flux flows in both outer legs as indicated by the dotted lines. A change of current in the dc coil causes a change in total flux linking the ac coils and hence a change of inductance. Ac coils may be connected in parallel instead of series, provided that equal turns in each coil and the flux polarity of Fig. 8.3 are maintained; for the same total number of turns the inductance is halved and the alternating current doubled. The middle core leg shunts the even harmonics of ac flux.

Rectangular B-H loops are obtained in grain-oriented core materials. It is

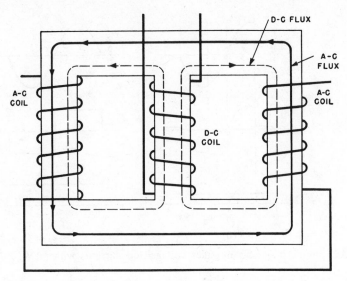

Figure 8.3. Windings and core flux paths in a saturable inductor.

only in these materials that the wave shapes of Fig. 8.2(*a*) are even approximated. In unoriented core steel, wave shape is much more rounded, and control current bears less resemblance to load current. Figures 8.4 and 8.5 indicate the contrast in saturation control afforded by unoriented silicon steel and oriented nickel steel. In grain-oriented steel cores there is an approximately linear relationship between dc ampere-turns per inch and ac ampere-turns per inch over a large range of flux density. Moreover, the ac NI/in. for a given dc NI/in.

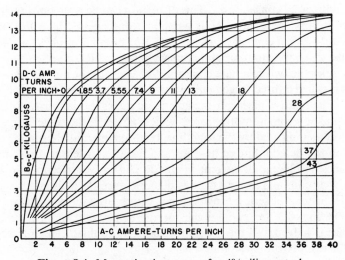

Figure 8.4. Magnetization curves for 4% silicon steel.

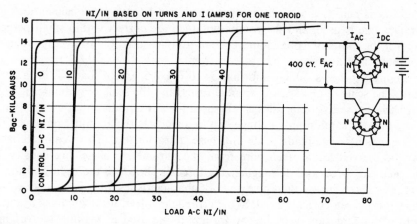

Figure 8.5. Typical magnetization curves for 0.002-in. grain-oriented nickel-iron toroidal cores.

change but little with ac flux density over this range. In Fig. 8.5, each of the lines for a given value of control magnetizing force is nearly vertical. For a given value of control NI/in., load current is almost independent of flux density and therefore of ac supply voltage. The sections following are based on the use of grain-oriented core steel.

8.3   GRAPHICAL PERFORMANCE OF SIMPLE MAGNETIC AMPLIFIERS

Since with a given core and supply frequency there corresponds a definite voltage for every flux density, and since for a given number of turns the ampere-turns per inch are proportional to current, the curves of Fig. 8.5 may be replotted in terms of voltage and current. We may then plot load lines on these curves in a manner similar to those in electronic amplifiers, so that the operation, efficiency, control power, and so on, may all be determined from a study of these load lines.

In Chapter 5 equation 5.1 indicates that the ac voltage in a vacuum-tube circuit is divided between the load and the tube. If a resistive load is used, a straight line can be drawn on the characteristics of a vacuum tube which will form the locus of plate current and plate voltage for any given load and supply voltage. This line is called the load line, and by use of it the gain and power output of the amplifier can be determined.

A similar method can be used with magnetic amplifiers. If a linear inductor were connected in series with the load, the voltage across the load and the voltage across the inductor would add at right angles. With rectangular loop core materials, currents are not reactive in the linear sense, so that the actual load line is neither a straight line nor an ellipse. For practical calculations the straight line is used, and the results obtained are correct within small percentages if the inductor voltage and current are measured on an average-reading voltmeter and ammeter.

Figure 8.6 shows similar information to Fig. 8.5, except that it is for a given core. The scale of abscissas is ampere-turns, and the scale of ordinates is volts per turn. These characteristics can be used for any amplifier that uses the same cores and the same supply voltage and frequency as the amplifier on which the measurements were made to obtain these characteristics. These characteristics can be derived for a given core from the parameters of ac flux density, ac magnetizing force, and dc magnetizing force as shown to the right and top of Fig. 8.6. These curves then give a set of characteristics for a given core rather than for a given core material. Some error is involved if these curves are used with a different supply frequency from the one used in making the original curves. Over a narrow range of frequency the curves of Fig. 8.6 using the scale at the top and to the right can be used to determine the operation of a magnetic amplifier for different loads. The curves of Fig. 8.5 may be used for magnetic amplifier calculations in this manner. For convenience of calculation, it is usually preferable to make a set of characteristic curves for several core sizes and for each supply frequency.

An example will show how these curves can be used in the design of magnetic amplifiers. Assume that it is necessary to design an amplifier with 30 W output using cores with E/N and NI of Fig. 8.6. The supply voltage is 100 V, the load is 200 Ω, and 0.01 A is available for use in the control winding. The characteristics show the E/N can be varied from about 1.4 to 0.2 and still stay on the linear part of the characteristic curves. The power output is equal to $\Delta E \times \Delta I$, which is also

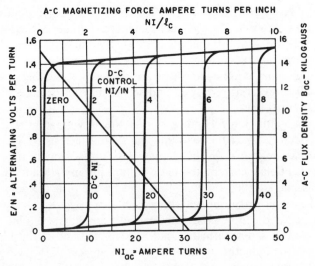

Figure 8.6. Generalized magnetic amplifier characteristics and load line.

equal to $\Delta(E/N) \times \Delta NI$, where E is the alternating voltage across the reactors, I is the alternating current through the reactors, and N is the number of turns in the load windings of the reactor. ΔNI needed for 30 W $= 30/(1.4 - 0.2)$ $= 25$ At. Load impedance is $\Delta E/\Delta I$. A load line on Fig. 8.6 is $(\Delta E/N_L) \div \Delta NI = (\Delta E/\Delta I)(1/N_L^2)$, where $N_L =$ turns in load winding. For 200 Ω, the load line passes through the points $E/N_L = 1.4$, $NI_{ac} = 2.5$, and through $E/N_L = 0.2$, $NI_{ac} = 27.5$. When this line is extended to the ordinate it intersects at 1.54. This is the point of zero alternating current or of infinite reactor inductance. At this point the total supply voltage would be across the reactor. Since the supply voltage is 100 V, $100/N = 1.54$, and $N_L = 65$ turns. By interpolation of the dc NI curves, we see that, for $E/N = 0.2$ and $NI_{ac} = 27.5$, 25 ampere-turns are necessary in the control winding. The turns in the control windings are $N_c I_c/I_c = 25/0.01 = 2500$ turns. Here I_c is the current in the control winding, and N_c are the turns in the control winding. Control winding resistance is determined by the wire size. For the purpose of this example, assume that the resistance of the control winding is 500 Ω. Then the power in the control winding is $500 \times I_c^2 = 0.05$ W. Power gain of the amplifier is power out/power in $= 30/0.05 = 600$. The impedance of either the input circuit or output circuit can be changed by changing the number of turns in the respective windings. Either impedance varies with the square of the number of turns used in the winding. For example, the load line which was used for 200 Ω in the preceding example could be used for 800 Ω. The load winding would then have $\sqrt{800/200} \times 65 = 130$ turns, and the supply voltage would be 200 V instead of 100 for $E/N = 1.54$ at zero current.

Power output is proportional to the area of the rectangle of which the load line forms a diagonal. More power output can be obtained by using a load line with less slope, but gain may increase or decrease, depending on the winding resistances and core material. In the preceding example, the load windings were assumed to be in series with the load, as in Fig. 8.1. This is the connection commonly used when the source is a 60-Hz ac line. With a high impedance source, it is preferable to connect the load windings in shunt with the load. Then the ordinates of Figs. 8.5 and 8.6 correspond to load voltage at all times.

If we choose three line voltages corresponding to flux densities within the linear portions of Fig. 8.5, and plot the dc control versus ac load ampere-turns per inch, the curves of Fig. 8.7 result. If, instead of $NI/$in., average load current is plotted, Fig. 8.7 gives the transfer curves for a simple magnetic amplifier. The curves are symmetrical about zero ampere-turns. The difference between the transfer curve and a straight line indicates the degree of nonlinearity in the amplifier for any load current. With grain-oriented core material the ac load current is nearly independent of supply voltage for ac inductions less than saturation. Provided that appropriate changes in scale are made, transfer curves may be plotted between load voltage and control current, or between load ampere-turns and control ampere-turns, or between combinations of these.

Load current is the result of flux excursions beyond the knee of the normal magnetization curve. In Fig. 8.5 the curve for zero control $NI/$in. is normal

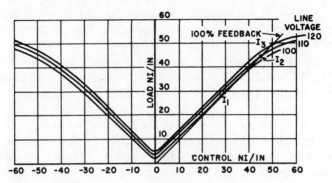

Figure 8.7. Simple magnetic amplifier transformer curves with line-voltage variations.

magnetization for the material. When direct current flows in the control windings, it sets up a constant magnetizing force in the core. Then superposed ac magnetizing force readily causes a flux excursion beyond the knee of the curve, permeability suddenly drops, and a large current flows through the load winding. The point in the voltage cycle at which this sudden increase in current occurs depends on the amount of direct current in the control winding. Magnetic amplifiers with steep current curves like those of Fig. 8.5 can be used as control relays.

Load current is usually measured with an average-reading ammeter, such as a rectifier-type instrument. This kind of ammeter is generally marked to read the rms value of sinusoidal current but actually measures the average value. Thus the ammeter reading is $0.707/0.636 = 1.11$ times the average current over a half-cycle. When the meter is used to measure nonsinusoidal current, it still reads 1.11 times the average.

Except for the slight amount of nonlinearity noted in Fig. 8.7, the *average* value of ampere-turns in the load winding of each reactor equals the dc ampere-turns in the control winding. But since the ac ammeter reads 1.11 times this value, the load ac NI/in. are 1.11 times the control dc NI/in., plus the differential due to core magnetizing current. Thus if a core had infinite permeability up to the knee of the magnetization curve and zero permeability beyond the knee, the transfer curve would be exactly linear. Oriented nickel-iron alloy cores approach this ideal and therefore are more nearly linear than other materials.

8.4 RESPONSE TIME

Because of the inductance of the reactor coils, when a change is made in the control winding direct current, load current does not change immediately to its final value. An interval of time, called *response time*, elapses between the change in control current and the establishment of a new steady value of load current. If the inductance were constant during the change, the response time constant would be the time required for a load current increase to rise to 63% of the final value after a sudden control current increase. Magnetic amplifier response time cannot be evaluated as an ordinary linear L/R time constant. Storm (1951) shows that the time of response of simple magnetic amplifiers is independent of core permeability. An average or equivalent control circuit inductance may be found from the relation

$$T_d = \frac{L_C}{R_C} = \frac{R_L}{4fR_C}\left(\frac{N_C}{N_L}\right)^2 \qquad (8.1)$$

where T_d = time for load current increment to reach 63% of final value
L_C = equivalent total control coil inductance, H
R_C = total control circuit resistance, Ω
R_L = load resistance
f = line frequency
N_C = turns in control winding
N_L = turns in load winding

An obvious method of decreasing magnetic amplifier response time is by increasing R_C, but this has the disadvantage of reducing overall power gain. Gain and response time are so related that the ratio of gain to time constant in a magnetic amplifier is usually given as a figure of merit.

8.5   FEEDBACK IN MAGNETIC AMPLIFIERS

If a rectifier is interposed between the reactor and load, and a separate winding on the reactor is connected to this rectifier as in Fig. 8.8, it is possible to obtain sufficient power from the rectifier to supply most of the control power. If the control power from the rectifier furnishes the ampere-turns represented by the straight line in Fig. 8.7, the amplifier is said to have 100% "feedback." It is then necessary for the control winding to supply only the amount represented by the

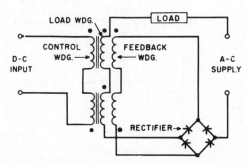

Figure 8.8. Magnetic amplifier with external feedback.

horizontal difference between the transfer curve and the straight line. This greatly increases the amplification of a pair of reactors.

Typical transfer curves for a simple magnetic amplifier are plotted in Fig. 8.7 for three ac supply voltages: 100, 110, and 120 V. A 100% feedback line intersects the transfer curves at I_1, I_2, and I_3, respectively. The control NI/in. are furnished by the feedback, except for the control current difference between the feedback line and the transfer curve. Positive control current is required when the transfer curve is at the right, and negative current when it is at the left, of the feedback line. Net control NI/in. for the three voltages are plotted in Fig. 8.9 with expanded abscissa scale. Now the transfer curve is asymmetrical. Most of the amplifier gain occurs with negative control current changes. On the steep parts of the transfer curves, gain is fairly linear and greatly exceeds the gain of simple amplifiers. Below the steep parts, output current reaches a minimum but remains small with relatively large excursions of negative control current. These current minima are called *cutoff points*. Reference to Fig. 8.5 shows that cut-off current is I_N, the normal exciting current at supply voltage E. With positive control, current output levels off to a nearly constant value, depending on the voltage. Feedback causes output current to be quite dependent on variations in ac supply voltage, because I_N has a greater effect than in simple amplifiers.

Computing control current for transfer curves with feedback as described in the preceding paragraph involves a small difference between two large quantities. Minor measurement errors in the original data cause large inaccuracies in the feedback transfer curves of Fig. 8.9. A more accurate derivation of 100% feedback transfer curves is given in Section 8.9.

To the left of the cutoff points, the transfer curve rises slowly toward the left along a straight line, as in Fig. 8.10(a). This line corresponds to 100% *negative* feedback; it is practically linear, but gain is much reduced. The transfer curves of Fig. 8.9 would, if continued to the left, merge into such a line.

Polarities in Fig. 8.8 are for positive feedback with positive direct current entering the control winding at the top. Negative feedback is obtained if the control current is reversed. If series feedback is derived as shown in Fig. 8.8, the

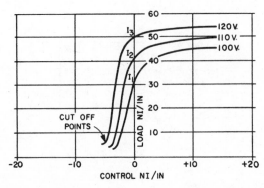

Figure 8.9. Transfer curves for magnetic amplifier with feedback.

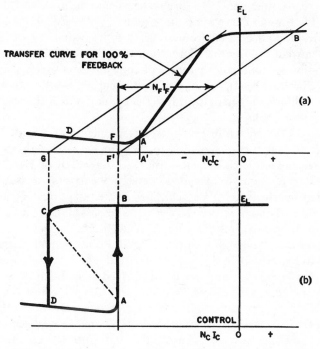

Figure 8.10. (*a*) Typical transfer curve and (*b*) bistable magnetic amplifier.

feedback current is E_L/R_L. It is possible to connect the feedback circuit across the load to obtain voltage feedback. To conserve power, the feedback resistance should be large relative to R_L.

8.6   BISTABLE AMPLIFIERS

Positive feedback in a magnetic amplifier can be increased to more than 100% by increasing turns in the feedback winding. Transfer curves may then become double-valued and give rise to abrupt load current changes with changing control current. Such amplifiers are called bistable. In Fig. 8.7, the effect of increasing feedback would be to *decrease* the slope of the feedback line. If the feedback were increased gradually, operation would remain stable until the feedback line had the same slope as the transfer curve. Then the load current would become some indefinite value along the transfer curve. If the feedback were increased further stable operation would be had at only one of two values of load current. Bistability is illustrated in Fig. 8.10(*a*). Here a transfer curve similar to those of Fig. 8.9 is shown except that it is with load voltage ordinates and expanded $N_C I_C$ abscissas. The amount of feedback in excess of 100% is drawn as line *AB* with slope less than that of the main part of the 100% feedback transfer curve. Another line, *CD*, is drawn parallel to the line *AB*. These lines are tangent to the transfer curve at points *A* and *C*. With feedback > 100%, let dc control current be decreased from some negative value toward zero. Load voltage or current follows the transfer curve until it reaches point *A*; then it jumps to point *B*, and further increase of control current results in very little load voltage increase beyond point *B*. If control current is subsequently reduced, load voltage follows the top of the transfer curve until it reaches point *C*; then it drops abruptly to point *D*.

Bistable action is shown in Fig. 8.10(*b*) as a function of control NI/in., with points *A*, *B*, *C*, and *D* corresponding to those in Fig. 8.10(*a*). Line *AB* in Fig. 8.10(*a*) represents feedback ampere-turns $N_f I_f$ in excess of 100%, which are proportional to E_L. Line *AB* extended intersects the axis of abscissas at *F′*, and *CD* extended intersects at *G*. Vertical lines erected at *A′* and *F′* intersect the transfer curve at *A* and *F*, respectively. *F′A′* represents ampere-turns $N_f I_f$ when control ampere-turns $N_c I_c$ are at point *F*. When decreasing negative $N_c I_c$ reach value *F*, the load voltage jumps from *A* to *B*. Points *F′* and *G* are projected downward to Fig. 8.10(*b*). In this figure the output jumps to final value *B*, but the increase actually takes place along the dashed line. Decreasing additional feedback $N_f I_f$ reduces the differential amount *F′G* of control $N_c I_c$ and reduces the width of the bistable loop. Conversely, increasing $N_f I_f$ widens the loop and provides a greater margin for variations in $N_f I_f$ due to voltage, temperature, etc. Bistable amplifiers are used in protective and control circuits to turn relays or indicators on or off when control power varies between narrow limits and the inherent lock-in action is desirable.

8.7   CURRENT TRANSDUCTORS

In some countries the term *transductor* is used to denote any magnetic amplifier. Here it denotes a saturable reactor circuit for measuring direct current. A current transductor is hardly an amplifier; it is a metering device. A transductor circuit is shown in Fig. 8.11. It is similar to that of Fig. 8.8 but with feedback windings and ac load removed. Operation is entirely different. Cores are circular or square, and are wound in-and-out toroidally in a manner resembling

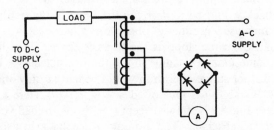

Figure 8.11. Current transductor circuit.

through-type current transformers. The heavy dc bus then may be inserted through the toroid to form a single turn on each core. In Fig. 8.11 the dc load windings are shown aiding, and the ac windings bucking; this accomplishes the same core flux polarities as for Fig. 8.3. Load direct current is determined by the load resistance, which is large compared to the reactance of the transductor. Control circuit impedance multiplied by the turns ratio is large; magnetization is constrained. It will be recalled from Section 8.2 that, under this condition, even current harmonics cannot flow. Therefore ac winding current is flat-topped. After this flat-topped current is rectified, it flows through the ammeter as smooth direct current. For a description of the current and flux conditions, see Tweedy (1948).

At any instant one inductor of the pair is saturated, and the other unsaturated. On each ac half-cycle the unsaturated inductor maintains the output current constant. Total output dc ampere-turns of course must equal twice the load direct current at all times. Transductors are like simple magnetic amplifiers as far as the relations of load and output currents are concerned. They have been built to measure currents of 10,000 A or more, with good linearity.

8.8   SELF-SATURATED MAGNETIC AMPLIFIERS

In Section 8.5 it was seen that the use of feedback windings greatly increases the gain of a magnetic amplifier. Several circuits have been devised to provide the feedback by means of the load circuit and thus eliminate the extra feedback winding. Such circuits are termed self-saturating. A "building-block" or elementary self-saturating component is the half-wave circuit of Fig. 8.12 from which several magnetic amplifiers may be formed. Impedance Z in the control circuit prevents short-circuiting the inductor. It may be the control winding of another inductor in a practical amplifier. Rectifier RX prevents current flow into the load in one direction, so that the core tends to remain in a continually saturated condition. This condition is modified by negative control winding NI/in., which opposes the load winding NI/in. and permits the core to become unsaturated during the portion of the cycle when there is no load current flowing. The greater the control NI/in., the less the average output current. Transfer characteristics are similar to those of Fig. 8.9. Ideally, the circuit has 100% feedback.

Assuming the core to be saturated at all times with zero control current, current flows into the load throughout the whole positive half-cycle and is zero for the whole negative half-cycle. With a given value of negative control current, reactor inductance is high at the start of the positive half-cycle and load current does not build up appreciably until an angle θ_1 is reached when the core saturates. Then it climbs rapidly and causes most of the supply voltage to appear across the load as shown by the curve marked e_L in Fig. 8.12(b) for the remainder of the positive half-cycle. As negative control current increases, so does angle θ_1. In the limit $\theta_1 = 180°$; that is, with large negative control current, virtually no load current flows. The similarity of load voltage wave shape to thyratron action

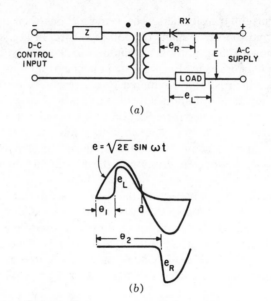

(a)

(b)

Figure 8.12. (a) Half-wave self-saturated magnetic amplifier circuit and (b) load and rectifier voltage wave shapes.

is at once evident. It has led to the use of the same terminology. Angle θ_1 is often called the firing angle of a magnetic amplifier. Load voltage is reduced as θ_1 increases, approximately as in Fig. 8.13. There are some important differences, too:

1. Inductance is never infinite, and magnetizing current is therefore not zero. This means that during the interval $0-\theta_1$ a small current flows into the load. The change in inductance at the firing instant is not instantaneous; the time required for the inductance to change limits the sharpness of load current rise.

2. Even with tight coupling between control and load windings, the saturated inductance is measurable. This saturated inductance causes the load current to rise with finite slope.

3. After load voltage reaches its peak and starts to drop along with the alternating supply voltage e, core flux continues at saturation density. An instant a is reached when the load voltage exceeds the supply voltage. Beyond a, the inductance increases and magnetizing current decreases, but at a rate slower than the supply voltage because of eddy currents in the core.

4. After supply voltage e in Fig. 8.12(b) reaches zero, the inductor continues to absorb the voltage until the core flux is reset to a value dependent on the control current, that is, until angle θ_2 is reached. Then part of the negative supply voltage rises suddenly across rectifier RX as shown by the waveform of e_R.

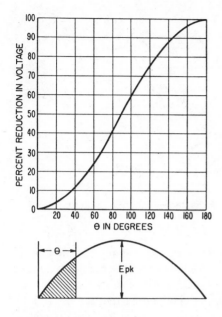

Figure 8.13. Relation of firing angle to voltage output.

During the interval $0-\theta_1$ the inductance is high and virtually all the supply voltage appears across it. The voltage time integral $\int e\,dt$ represented by the inductor flux increase during this interval is equal to $\int e\,dt$ during $\pi-\theta_2$. That is, the energy stored in the core before the firing instant is given up during the negative half-cycle of supply voltage.

Self-saturated magnetic amplifiers have transfer curves similar to that of Fig. 8.10(a). A small amount of additional positive feedback makes them bistable. Negative feedback makes the transfer curve more linear but reduces the gain. Ordinates and abscissas may be current, ampere-turns, or oersteds, as for simple magnetic amplifiers.

8.9 HYSTERESIS LOOPS AND TRANSFER CURVES

Several workers have observed (see Dornhoefer, 1949) that the transfer curves of Fig. 8.9 are similar in shape to the left-hand or return trace of the hysteresis loop. There is a connection between the two. In Fig. 2.8 it was shown that in a core with both ac and dc magnetization the minor hysteresis loop follows the back trace of the major loop in the negative or decreasing B direction, and proceeds along a line with less slope in the positive direction until it joins the normal permeability curve at B_m. Also, it was pointed out in connection with Fig. 3.21 that, if ΔB has the maximum value B_m, the result is the banana-shaped figure OB_mD'. Here again the loop representing flux excursion $O-B_m$ follows the left-hand side of the hysteresis loop in the downward or negative direction. In a rectangular hysteresis loop material with B-H loop shown in Fig. 8.14(a), the

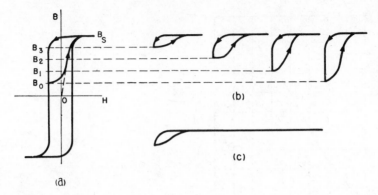

Figure 8.14. Minor loops in rectangular hysteresis-loop core material.

path traced over a flux excursion B_0B_s is more irregular in shape but still follows the left-hand trace of the loop. If magnetic amplifier cores are biased to a series of reset flux positions B_0 to B_3 the corresponding flux excursions and minor loops are those shown in Fig. 8.14(*b*). Usually, the load current far exceeds the control current necessary to reset the cores, so that these loops actually have a much longer region over which the loop width is practically zero, as shown in Fig. 8.14(*c*). This is true of all the loops regardless of flux excursion.

The foregoing is true of a slowly varying flux excursion, so that the locus of the lower end points of the minor loops is the left-hand trace of the dc hysteresis loop. Because of eddy currents most magnetic materials, including rectangular loop materials, have a wider loop when the hysteresis loop is taken under ac conditions. The difference between loops is as shown in Fig. 8.15. The locus of the end points of the minor loops under ac flux excursions is neither the ac nor the dc loop but an intermediate line such as that drawn dot-dash in Fig. 8.15.

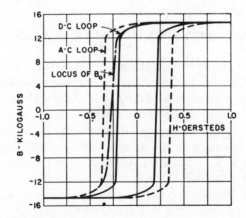

Figure 8.15. Dc and ac *B-H* loops for grain-oriented nickel-iron.

The slope of this line is less than that of either the ac or the dc loop, and the gain of the magnetic amplifier is accordingly reduced.

An analysis for the self-saturated magnetic amplifier of Fig. 8.12(a) is given below. Load current is assumed to have the same shape as e_L in Fig. 8.12(b), and the following assumptions are made:

1. Sinusoidal supply voltage and negligible ac source impedance
2. Negligible inductor and rectifier forward voltage iR drops
3. Negligible rectifier back leakage current
4. Negligible magnetizing current compared to load current
5. Negligible saturated inductance
6. High control circuit impedance
7. $E = 4.44fN\phi_s \times 10^{-8}$ ⠀⠀⠀⠀⠀⠀⠀⠀⠀⠀⠀⠀⠀⠀⠀⠀⠀⠀⠀⠀(8.2)

This will be recognized as equation 1.4 with peak flux at saturation value ϕ_s. Other terms are listed as follows:

θ_1 = firing angle as in Fig. 8.12(b)
$t_1 = \theta_1/\omega$
$\omega = 2\pi \times$ supply frequency f
E = rms supply voltage
ϕ_s = saturation flux = $B_s A_c$ (for B_s see Fig. 8.14)
A_c = core section, cm^2
ϕ_0 = reset core flux = $B_0 A_c$ (for B_0 see Fig. 8.14)
R_L = load resistance
I_{av} = average load current
i = instantaneous load current
N = turns in load winding

Under the assumptions, equation 1.1 becomes

$$\sqrt{2}E \sin \omega t = \frac{N}{10^8} \frac{d\phi}{dt} \qquad \text{for } 0 < \omega t < \omega t_1 \qquad (8.3)$$

Integrating equation 8.3 gives

$$10^{-8} \int_{\phi_0}^{\phi_s} N \, d\phi = \int_0^{t_1} \sqrt{2}E \sin \omega t \, dt \qquad (8.4)$$

and

$$\frac{\omega N(\phi_s - \phi_0)}{\sqrt{2}E \times 10^8} = 1 - \cos \omega t_1 \qquad (8.5)$$

During the interval $\theta_1 < \omega t < \pi$, load voltage is

$$\sqrt{2}E \sin \omega t = iR_L \tag{8.6}$$

where R_L is the load resistance. This may be integrated to give

$$\frac{\omega R_L}{\sqrt{2}E} \int_{t_1}^{\pi/\omega} i\, dt = 1 + \cos \omega t_1 \tag{8.7}$$

Combining equations 8.5 and 8.7 and substituting equation 8.2,

$$\int_{t_1}^{\pi/\omega} i\, dt = \frac{N}{R_L}(\phi_s + \phi_0) \times 10^{-8} \tag{8.8}$$

The left side of equation 8.8 is the average load current over the conducting interval $\pi/\omega - t_1$. Average load current over the whole cycle is

$$I_{av} = \frac{fN(\phi_s + \phi_0)}{R_L \times 10^8} \tag{8.9}$$

Equation 8.9 has two flux terms: ϕ_s, which is a fixed quantity for a given core material; and ϕ_0. The relation between ϕ_0 and control current is, as indicated in Fig. 8.14, the return trace of the major hysteresis loop. Thus equation 8.9 states that the average load current is the sum of a constant term and a term which has the same shape as the return trace of the hysteresis loop. Quantitatively, a self-saturated half-wave magnetic amplifier has a *current* transfer curve the same as the return trace of the core hysteresis loop, except that ordinates are multiplied by $fA_cN/10^8R_L$ and are displaced vertically by an amount $fB_sA_cN/10^8R_L$.

Comparison with equation 8.2 reveals that the ordinate multiplier and vertical displacement are $E/4.44R_LB_s$ and $E/4.44R_L$, respectively. As noted above, the return trace should be modified to mean the dot-dash line of Fig. 8.15.

8.10   SELF-SATURATED MAGNETIC AMPLIFIER CIRCUITS

In Fig. 8.16 three single-phase circuits are diagrammed which comprise two of the half-wave elements described in the preceding sections. These circuits are discussed briefly below.

1. *Doubler Circuit.* This is really two half-wave circuits working into a common load. Rectifier polarities are such as to cause ac voltage to appear across the load, as in Fig. 8.17(a). The wave shape departs somewhat from alternately reversed half-waves. In the doubler, the reactor which is carrying load current during a given half-cycle causes a reduction in the resetting voltage,

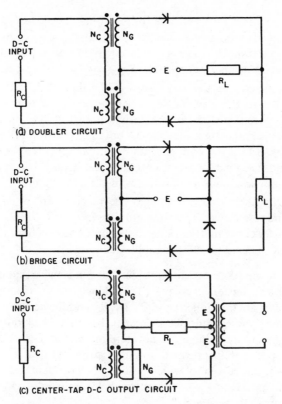

Figure 8.16. Self-saturated magnetic amplifier circuits.

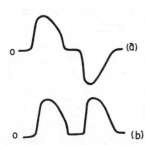

Figure 8.17. Single-phase magnetic amplifier output: (a) ac voltage across load; (b) dc voltage across load.

and therefore in the time rate of resetting flux change of the other reactor. This increases the output and gain for a given control current compared to the half-wave circuit but has no effect on current minima at the cut-off points (see Fig. 8.9).

When control circuit resistance R_C is large, the control current and associated magnetizing force are fixed, but, when R_C is small, even harmonic currents flow freely in the control circuit and influence the wave shape for a given control current further. Generally, low values of resistance R_C cause a slight increase in the control oersteds for a given output but virtually no change in slope. In other words, the whole transfer curve is displaced slightly to the right.

2. *Single-Phase Bridge Circuit.* Here two extra rectifiers isolate the two reactors at all times, and the wave form is like that of the half-wave rectifier, except that it occurs twice each cycle. Load current is dc; that is, both reactors produce load current of the same polarity, as in Fig. 8.17(*b*). Because of the isolation of the two reactors, the transfer curve closely follows a dot-dash line like that in Fig. 8.15 if the core is grain-oriented nickel steel, or a similar line between ac and dc loops for other core material. Control resistance R_c affects output in a manner similar to that mentioned for the doubler.

3. *Center-Tap DC Circuit.* Although the reactors are not isolated in this circuit, load and resetting currents are still the same as for the bridge circuit, and hence the transfer curve has the same shape, unless the rectifier reverse currents are appreciable. Then gain is appreciably reduced.

In all these single-phase circuits, the load current is twice that of the half-wave circuit. Therefore, transfer curves may be predicted from B-H loops as in Section 8.9, but ordinates are multiplied by $E/2.22R_L B_s$, and the vertical displacement is $E/2.22R_L$. From these multipliers it can be seen that output current is proportional to supply voltage E, and therefore power gain is proportional to E^2. In this respect, a self-saturated amplifier contrasts with a simple magnetic amplifier, the output current of which is nearly independent of E, for rectangular B-H loop core material. At least this is true below maximum current, or current flow over a complete half-cycle.

As an example of the manner in which a transfer curve is found from the B-H loop, suppose that, in a given self-saturated amplifier, Fig. 8.15 is the B-H loop, supply voltage $E = 230$ V, $R_L = 500\,\Omega$, $B_s = 14.7$ kG. The ordinate multiplier is $230/(2.22 \times 500 \times 14.7) = 0.0141$, and displacement is $230/(2.22 \times 500) = 0.207$.

Table 8.1 indicates the change in ordinates. The last two columns of the table are plotted in Fig. 8.18(*a*) as load current in milliamperes. Also indicated is the "normalized" value of unity for maximum output current. For any load impedance the same calculated transfer curve can be used, and all ordinates multiplied by $E/1.11R_L$. Abscissas may be normalized likewise, with cutoff $H = -1.0$.

Normalized output current at cutoff is $\Delta_N = I_N R_L/E$. Cutoff control current is most accurately found from H corresponding to $-B_r$. This is $H = -0.5$ in Fig. 8.15. These relations are, of course, idealized, but they are still very useful in practical work. For example, winding resistance R_G and rectifier forward resistance R_F reduce load current and output power, but these resistances may be added to the actual R_L arithmetically to obtain total resistance $R_T = R_G + R_F + R_L$. Then the transfer curve ordinates are

$$I_{\text{av}} = \frac{E(B\text{-}H \text{ loop ordinates})}{2.22R_T B_s} \tag{8.10}$$

TABLE 8.1   Derivation of Transfer Curve From *B-H* Loop (Fig. 8.15)

H (Oe)	B (kG)	Average 0.0141B (A)	Vertical Displacement (A)	Average Load Current, Fig. 8.18(a)	
				Amp.	Normalized
−0.5	−14.3	−0.202	+0.207	0.005	0.012
−0.4	−14.0	−0.197	0.207	0.010	0.024
−0.3	0	0	0.207	0.207	0.500
−0.15	13.0	0.183	0.207	0.390	0.943
0	14.0	0.197	0.207	0.404	0.975
0.5	14.7	0.207	0.207	0.414	1.000

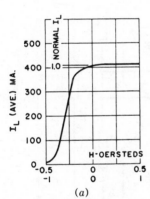 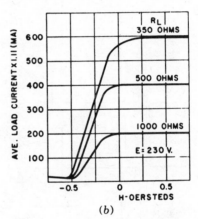

(a)   (b)

Figure 8.18. Self-saturated magnetic amplifier output: (a) calculated for 500 Ω from Fig. 8.15; (b) in actual amplifier.

displaced vertically by

$$E/2.22R_T \tag{8.11}$$

Output current and power are reduced somewhat by these inevitable resistances. This can be verified in Fig. 8.18(b) which is a plot of transfer curves for an actual doubler amplifier with $E = 230$ V, with 350-, 500-, and 1000-Ω load resistances, and with average load current × 1.11 as read directly on the output meter. The 500-Ω load resistance curve is approximately the same as Fig. 8.18(a); this means that $R_F + R_G \approx 0.11R_L$ in this particular amplifier. The accuracy of Fig. 8.18(a) is evidently poorest at cutoff. Upward slope at control currents more negative than cutoff is not shown at all. For the most practical region (i.e., to the right of cutoff) the calculated curve is eminently useful.

Additional windings are often used on the inductors for control purposes. One common winding, called a *bias* winding, carries negative control current. The function of this winding is to maintain low output in the absence of control current. Thus in Fig. 8.9, with $E = 120$, $-5NI/$in. of bias magnetizing force keeps the amplifier load $NI/$in. at 5. Then positive control current raises the load current to the desired value. Most of the gain is obtained with less than $+5NI/$in. control magnetizing force.

Additional control windings are used for adding or subtracting input signals. This provides a simple means of combining several control functions in one magnetic amplifier.

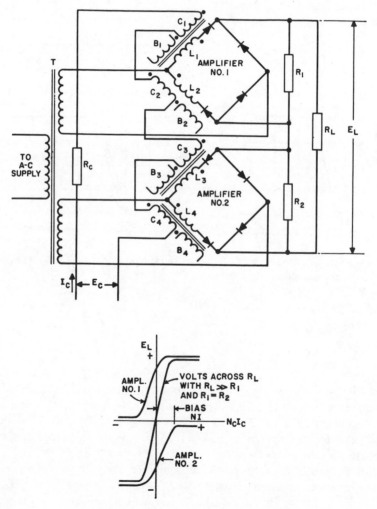

Figure 8.19. Dc push-pull magnetic amplifier.

Response time in a self-saturated amplifier is longer than in a simple amplifier, but the gain per second is much greater. The time constant is

$$T_d = \alpha_v/2f \qquad \text{seconds} \tag{8.12}$$

where T_d = time for 63% response to step input
$\quad \alpha_v$ = amplifier voltage gain for 1:1 turns ratio
$\qquad = (\Delta E_L/\Delta E_C) \times (N_C/N_L)$ for any turns ratio
$\quad f$ = supply frequency
$\quad \Delta E_L$ = change in load voltage
$\quad \Delta E_C$ = change in control voltage
$\quad N_C$ = turns in control winding
$\quad N_L$ = turns in load winding

Equation 8.12 is valid for T_d down to approximately four cycles minimum. Although smaller T_d may be obtained, it does not follow equation 8.12.

Push-pull amplifiers are used to provide ac or dc output, with the output polarity dependent on input polarity. A dc push-pull circuit which senses input polarity is shown in Fig. 8.19. Bias windings on each reactor carry current in such directions that amplifier outputs cancel for $E_C = 0$. For positive E_C, amplifier 1 produces positive E_L, and for negative E_C, amplifier 2 produces negative E_L. This circuit has low efficiency, owing to the power dissipated in the balance resistances R_1 and R_2 but has linear output.

Whenever two or more inductors are used in a magnetic amplifier the inductors must be alike in turns and in the cores used in the inductors. The flux-current characteristics of the cores must be closely matched since it is not feasible to compensate for core differences in balanced amplifiers by bias adjustments. Lynn et al. (1960) discuss the properties and methods of testing magnetic materials for magnetic amplifiers.

8.11   HALF-WAVE CONTROL OF MAGNETIC AMPLIFIERS

Through attention to ac voltages present in the control circuit, Ramey (1951) analyzes magnetic amplifiers in a manner that gives rise to new circuits with desirable properties. A half-wave building block of such circuits is shown in Fig. 8.20(a) for a 1:1 turns ratio inductor. The load circuit is the same as in the half-wave amplifier of Fig. 8.12. The control circuit comprises ac voltage E and rectifier RX_C in addition to variable rectified control coltage e_C of polarity indicated. Ac voltage polarities are for the positive or conducting half-cycle in the load circuit. During this half-cycle, RX_C blocks and the control voltage is zero. During the next half-cycle, ac line voltage $E - e_C$ appears across the reactor control coil. If e_C is zero, the core is not magnetized by control current flowing during the positive half-cycle, and the core is completely reset by E during the negative half-cycle. If the peak value of e_C is equal to $\sqrt{2}E$, it appears

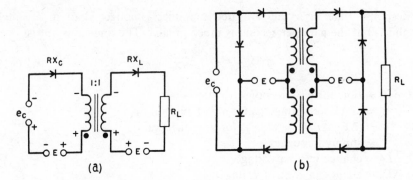

Figure 8.20. Half-wave controlled magnetic amplifiers.

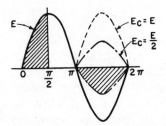

Figure 8.21. Resetting voltages with half-wave control.

across the reactor in opposite phase to the line voltage and completely cancels it during the resetting half-cycle. This is shown dashed in Fig. 8.21, with both voltage waves designated by capital letters. This cancellation results in zero resetting; therefore, full output current flows over 180° of the positive half-cycle. If $E_C = E/2$ it subtracts from E, resulting in the lower dot-dash line of Fig. 8.21. The area under $E - E_C$ (shown hatched) is just half of the area under E and therefore equals the hatched area under E during the interval 0 to $\pi/2$ of the positive half-cycle. That is, the inductor absorbs voltage E during the interval 0 to $\pi/2$ and allows current to flow from $\pi/2$ to π. But this half of full or maximum output. Thus the output current is

$$\text{zero for } E_C = 0$$

$$\tfrac{1}{2}\text{max. for } E_C = E/2$$

$$\text{max. for } E_C = E$$

Several advantages accrue from this type of control:

1. Output is proportional to control voltage.
2. Output depends only on control voltage and is independent of variations in line voltage or frequency.
3. Time of response is short (two cycles or less).
4. Filtered dc source of control power is not necessary.

Proportionality of output to input voltage is strictly true only for zero control circuit resistance or zero inductor exciting current. The lower the control circuit resistance and inductor exciting current, the more nearly is output proportional to input. Rectangular B-H loop core material is necessary for linearity. Control circuit resistance can be made small without causing slow response in this circuit. Exciting current and control circuit resistance give rise to load voltage output with zero control voltage. Raising control voltage e_C restores linearity of output. With half-wave control, voltage gain is more important than power gain; voltage gain is approximately equal to turns ratio. Mixing is not so readily accomplished in half-wave control circuits. Figure 8.20(b) shows how two half-wave sections are combined to form a full-wave bridge circuit with dc output. This circuit differs from the circuit of Fig. 8.16(b) in that the control windings are isolated from each other by the control circuit rectifiers. Voltage E in the control circuit is an ac bias voltage, and e_c is rectified ac signal voltage. Zero output voltage appears across R_L with $e_c = 0$. When e_c is increased, full-wave rectified voltage appears across R_L. The fundamental ac component of this voltage is zero.

8.12 MAGNETIC AMPLIFIER DESIGN

Of first concern in design is the inductor core material. Supermalloy or other high-percentage nickel alloys are best suited as core material for the low power input stage. Grain-oriented nickel steel is used in the stages where output power is appreciable, and grain-oriented silicon steel where power is large. Figure 8.22 shows the dc loops of two grain-oriented core materials, Hipersil and Orthonik. Although both materials have approximately rectangular B-H loops, the difference in rectangularity is marked. Grain-oriented nickel-steel strip such as Orthonik is usually wound into toroids, to ensure that the flux flows in the preferred direction. The toroidal cores are protected from mechanical damage and strain by encasing them, as in Fig. 8.23, after the core material is annealed to preserve the magnetic properties. Grain-oriented silicon-steel cores are much less sensitive to damage; type C cores may be used, with coils wound as described in Chapter 2.

In either type of core there is a small inevitable air gap. In a toroid, the flux must change from one lamination to the next as it flows around the core. If the insulating space between laminations is 0.0005 in. and the average core length is 5 in., the effective core gap is $0.0005/5 = 0.0001$ in. This gap is not negligible in high-permeability core material, but it is about one-tenth of the gap that manufacturers allow in type C cores. Effective core gap requires more control NI/in. and reduces gain because the gap causes a more sloping B-H loop (see Fig. 10.18). Special U-shaped punchings of grain-oriented steel are sometimes used with alternate stacking to reduce the effective core gap.

Another effect that reduces gain is rectifier "back" resistance, or current flow during the part of the negative half-cycle when inverse voltage exists across the

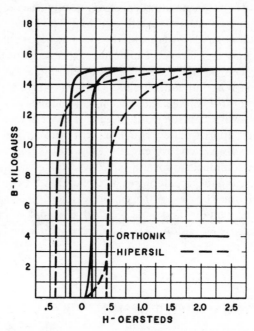

Figure 8.22. Typical dc magnetization curves and hysteresis loops for 2-mil Hipersil and 2-mil Orthonik toroidal cores.

diode. The peak value of inverse voltage divided by the corresponding reverse current is the rectifier back resistance. For a given peak source voltage $\sqrt{2}E$, the inverse peak rectifier voltage is $2\sqrt{2}E$ in the center-tap dc circuit, and it is $\sqrt{2}E$ in the bridge circuit, for zero winding and rectifier forward IR drops. In a doubler amplifier with zero forward drop, inverse peak voltage is zero, and increases with forward drop up to a maximum of $\sqrt{2}E$. The reverse current corresponding to these voltages resets the core more than control circuit current with no rectifier reverse current. This causes transfer characteristic slope to decrease; the unity ordinate of the normalized transfer curve is displaced to the right by the ratio of reverse current to cutoff control current I_C. Normal cutoff control current I_C and cutoff output current I_N are not affected, because I_C operates to reduce *load* current during the positive half-cycle. Good-quality rectifiers are as important as good core material. This applies equally well to leakage current and forward current IR drop. Losses may limit output in rectifiers as well as in inductors. Most of the I^2R loss in windings of self-saturated amplifiers is usually in the load windings. This loss occurs during the part of the cycle in which load current flows, or while the core is saturated and core loss is zero. I^2R loss is a maximum when $\theta_1 = 0$ in Fig. 8.12(b). When $\theta_1 = 180°$, I^2R loss is negligible and core loss is a maximum.

When the supply frequency is high, choice of rectifiers is limited to those with

Figure 8.23. Toroidal core of grain-oriented nickel-iron in a case and with the case top removed.

good high-frequency properties. At radio frequencies this may mean that suitable rectifiers are not available; simple magnetic amplifiers must then be used. To reduce core loss at high frequencies, ferrites are used.

Insulation of toroidal coils is difficult to apply. Insulation between concentric windings is taped in and out like the wire. If voltage is low, the wire enamel is sufficient insulation. For 115- or 230-V circuits, windings are laid on the core progressively, that is, with turns bunched so that adjacent turns have but a small ac voltage difference. Insulation difficulties increase with voltage, and high-voltage reactors are preferably layer wound, with type C or stacked cores.

Induced voltage in control windings requires careful attention, especially when control current is limited and many control turns are required. Although fundamental ac voltage cancels in the control circuit, the full magnitude of this voltage is induced in the control windings. In the example of simple magnetic amplifier given in Section 8.3, the voltage induced in the control windings is $2500/65 \times 100 = 3850$ V. With layer-wound coils and solventless resin coil impregnation the insulation is readily provided, but it would be difficult with toroidal coils.

Winding space in a toroid is limited by the minimum practicable hole size in the finished coil. This varies with the kind of winding machine and also with the size of toroid. If

$$d_1 = \text{hole diameter}$$

$$d_2 = \text{core case inside diameter}$$

$$d_3 = \text{core case outside diameter}$$

$$A_w = \text{total winding area}$$

then

$$A_w = (\pi/4)(d_2^2 - d_1^2) \qquad (8.13)$$

On the outside of the toroid, the winding builds to a smaller height than on the inside. Since A_w is fixed by the minimum hole size, the coil outside diameter is

$$d_4 = \sqrt{d_3^2 + 4A_w/\pi} \tag{8.14}$$

$$\text{coil axial length} = \text{core case height} + 2A_w/l_c \tag{8.15}$$

$$\text{mean turn of first winding} = \text{case periphery} + \pi A_{w1}/l_c \tag{8.16}$$

where A_{w1} is area occupied by first winding. Equation 8.16 is approximate because wire turns tend to become circular after several layers are wound on the core.

Example. Control Reactors for Single-Phase Rectifier. Assume the following conditions:

Power supply-400 Hz
Center-tap dc circuit per Fig. 8.16(c)
Control current available = 40 mA dc
Plate transformer $E = 125$ V per side
At full output $I_{dc} = 2$ A in R_L
Percent reduction in $E_{dc} = 33\%$ at minimum output

Assume grain-oriented nickel-steel core with $A_c = 0.1$ in.^{2}, $l_c = 5.5$ in., and $B_s = 14,700$ G; core-case dimensions $1\frac{1}{4}$ in. I.D., $2\frac{3}{16}$ in. O.D., $\frac{15}{32}$ in. high.

Each reactor must be capable of absorbing the voltage-time integral corresponding to 33% voltage reduction, or $0.33 \times 125 = 41$ V. From equation 3.4,

$$N_L = \frac{3.49 \times 41 \times 10^6}{400 \times 0.1 \times 14,700} = 244 \text{ turns}$$

With full output, load winding current $= 2\pi/(2 \times 2) = 1.57$ A rms. From Fig. 8.18(a), this can be controlled with $H = 0.5$ Oe

$$0.5 = 0.5N_C I_C/l_c$$

$$N_C = l_c/I_C = 5.5/0.04 = 138 \text{ turns}$$

This will be increased to 276 turns to allow for rectifier reverse current, variations in slope of the core *B-H* loop, and effective core gap. Using 650 cmil/A, and single enameled wire, yields $1.57 \times 650 = 1020$ cmil or No. 20 wire for N_L, and $0.040 \times 650 = 26$ cmil or No. 35 wire for N_C. With an average winding area space factor of 60%, the coil winding areas required are, from Table 2.4, $244/(860 \times 0.60) = 0.48$ in.^{2} for N_L and $276/(24,500 \times 0.60) =$

0.019 in.^{2} for N_C. If N_C turns are wound concentrically over N_L, the load winding inside diameter is, from equation 8.13,

$$d_1 = \sqrt{d_2^2 - 4A_w/\pi}$$

$$= \sqrt{(1.25)^2 - (4 \times 0.48/\pi)} = 0.975 \text{ in.}$$

N_C turns occupy but a single layer. Then, for N_C, $d_1 = 0.955 - 2(0.0064) = 0.94$ in. With 10-mil insulation over N_C, the hole diameter becomes $0.94 - 0.02 = 0.92$ in. Space required to insulate the ends of the windings and space for additional control windings reduce the hole diameter further.

Winding mean turn lengths are, for a core-case periphery of 1.88 in.,

$$\text{MT}_L = 1.88 + \frac{\pi \times 0.48}{5.5} = 2.16 \text{ in.}$$

$$\text{MT}_C = 2.16 + \pi([0.48/5.5] + 0.0064 + 0.020) = 2.44 \text{ in.}$$

$$\text{resistance of load winding} = \frac{244 \times 2.16 \times 10.3}{12{,}000} = 0.45 \, \Omega$$

$$\text{resistance of control winding} = \frac{276 \times 2.44 \times 338}{12{,}000} = 19 \, \Omega$$

load winding $IR = 0.71$ V $I^2R = 1.12$ W

control winding $IR = 0.76$ V $I^2R = 0.0305$ W

$$\text{power gain} = \frac{(125/1.11) \times [1 - (0.67)^2]}{0.0305} = 2050$$

$$\text{time constant} = \frac{41 \times 276}{1.11 \times 0.76 \times 244 \times 800} = 0.07 \text{ s}$$

with no external series resistance in the control circuit. With feedback applied to the control winding, this rectifier can be made self-regulating. If the feedback is further refined by comparison with a voltage reference, a stable voltage regulator results.

8.13 MAGNETIC-AMPLIFIER LIMITATIONS

Several limitations may affect the practical usefulness of magnetic amplifiers. Some of these limitations are beneficial in certain applications:

1. Residual output with zero input.
2. When more than one reactor is used in a circuit, reactor cores must often be matched.

3. Zero drift. At low input levels (of the order of 10^{-13} W for toroids of rectangular loop core material) magnetic amplifiers do not track because of hysteresis.

4. Amplifiers with feedback or high-gain self-saturated amplifiers are subject to instability when biased to cutoff and may change linear amplifiers into bistable amplifiers.

5. When the amplifier operates over a wide range of ambient temperature, variations in resistance of the reactors and rectifiers, and hysteresis loop width, cause changes in gain, output, and balance.

6. Response time of a magnetic amplifier is a limitation in comparison with an electronic amplifier.

7. Variations in supply frequency and voltage cause variations in gain and output, especially with self-saturated amplifiers.

8. Whereas the vacuum tube is a relatively high-impedance device, the magnetic amplifier is better adapted to low impedances, where the turns are fewer.

9. Saturation inductance is greater than the leakage inductance of the reactor, measured as in a transformer. The B-H curve slope at B_s, even with rectangular loop core materials, always gives μ greater than unity at the top. This effect reduces output and gain, and causes a sloping wavefront at the instant of firing.

8.14 MAGNETIC PULSE GENERATORS

In Chapter 14 it is seen that thyratron operation can be approximated by self-saturating magnetic amplifiers. This fact points to a saturable reactor to replace the hydrogen thyratron in the pulser of Fig. 11.3. Several factors militate against the direct substitution of saturable inductors for thyratrons:

1. Departure of core material from sharp rectangularity interferes with steep pulse voltage rise.
2. Saturated value of inductance interferes with large current flow needed during pulse.
3. Inductors are ac devices; hence ac charging must be used. This limits the choice of PRF.

Despite these difficulties, saturable inductors have been used successfully in pulsers. Low power pulses may be formed by use of the circuit of Fig. 8.24. Inductor L_1 is linear and resonates with C_1 at the supply frequency. Resonance therefore tends to maintain current i sinusoidal in wave form. Current i is large enough to saturate inductor L_2, which has rectangular B-H loop core material. Twice each cycle current i passes through zero, and near these current zeros L_2

inductance becomes large. This large inductance forces most of current i instantaneously into C_2 and R_L, and builds up a comparatively large peak voltage across R_L. Such pulses are peaked in wave shape and alternate in polarity twice each cycle. Pulse durations of less than 0.1 μs have been obtained with this circuit. Owing to the large interval of time during which i is large, and not producing pulses, the power output is limited to small values.

In Fig. 8.24 the voltage across L_2 at a given line frequency f is nearly proportional to the saturation flux density B_s of the core for rectangular loop material. If the core area is A_c and turns N in L_2, this voltage is, neglecting losses,

$$e_s \approx \frac{0.8 B_s f A_c N}{10^6 \theta_s} \tag{8.17}$$

where $\theta_s/2$ is the angle, starting from zero, at which saturation is reached, as in Fig. 14.6. If θ_s is very small, and $2\pi/\omega \gg R_L C_2 \gg \theta_s/\omega$, substantially all of e_s appears across R_L.

Higher power may be obtained from cascaded stages as in Fig. 8.25. Inductor L_1 is linear and resonates with C_1 at the supply frequency. Inductors L_2, L_3, and L_4 are saturable; bias windings are used, but not shown. Suppose that L_2 and L_4 are initially unsaturated, and L_3 is saturated. C_1 charges in series with L_1 and L_3. As the voltage across L_2 reaches maximum, L_2 saturates and discharges C_1. Discharge current causes L_3 to become unsaturated and L_4 saturated; then C_2 charges until L_3 saturates again. As the wave proceeds towards R_L both charge and discharge times become successively shorter. Pulse duration in each stage is determined by the saturated value of inductance and associated C. In the last

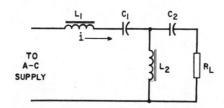

Figure 8.24. Magnetic pulse generator.

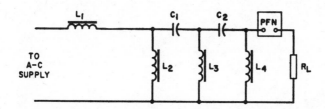

Figure 8.25. Cascaded stages in a magnetic pulse generator.

stage, the pulse is shaped by PFN to the desired duration and flatness at the top. As this sequence is repeated once each cycle, the line current is not sinusoidal, so a line capacitor is useful for supplying the current harmonics. One modification of this circuit is the use of saturating transformers instead of inductors in order to provide the stepped-up voltage necessary for magnetron operation. With a magnetron, R_L is replaced by a pulse transformer primary winding. In another modification saturable inductors are the series elements and capacitors the shunt (see Melville, 1951).

9 HIGH VOLTAGE

High voltage may be defined in different ways. Usually, electric potentials are considered from the standpoint of a hazard. How dangerous is the voltage to a human being?

Human factors and safety should be considered in the design of all electronic equipment and will have an effect on the layout of the circuits and the packaging of the equipment. However, high voltage, insofar as transformer insulation systems and electrical clearances are concerned, is considered any voltage at which partial discharge can be initiated. This is affected by the environment, physical spacing of electrodes, pressure, the insulating material(s), and the configuration of the electrodes and other parts. Partial discharge can initiate at less than 50 V (see Dakin and Berg, 1962).

9.1 HIGH-VOLTAGE MAGNETIC DEVICES

A variety of transformers and inductors may be utilized in high-voltage circuits. They may be the source of the high voltage, they may be required to isolate the high voltage from lower voltages or ground, they may be required to withstand high transient voltages, or they may be required to couple high voltages at different impedance levels. These magnetic devices may be required to withstand ac, dc, pulse, or transient voltages individually or in combination.

9.2 HIGH-VOLTAGE POWER TRANSFORMERS

High-voltage power transformers usually are the source of the high voltage in electronic circuits; they may be operated at the nominal power frequency, either single or polyphase, or they may be operated at a higher frequency which has been generated by an inverter circuit. These devices transform a lower voltage to a high ac voltage which is usually rectified, filtered, regulated, and utilized as a high dc voltage. The transformer must be designed to withstand the ac induced voltages, the high dc voltage, and any transients generated during circuit operation. Usually, a step-up three-phase high-voltage transformer is operated

Figure 9.1. Three-phase high-voltage transformer.

in a delta-wye configuration; this reduces the secondary winding voltages. A three-phase high-voltage power transformer is shown in Fig. 9.1. This transformer supplies an 86-kV dc power supply from a 400-Hz source. The major insulation between the high-voltage secondary windings and the primary and ground is SF_6 at 3 atmospheres. The secondary coils are sectored to reduce layer-to-layer stresses and also to reduce winding capacitance. The epoxy glass terminal supports have barriers and openings to increase surface creepage paths.

9.3 FILTER INDUCTORS

Filter inductors may be operated in one of two modes; in the first the winding insulation must provide isolation between the high dc voltage and ground. In the second the winding and the core are isolated from ground. Each mode of operation imposes different insulation requirements on the inductor.

In case 1 the coil must be insulated to withstand the full dc voltage to ground. In addition, the coil must be designed to withstand transients which may substantially exceed the dc voltage during faults and power supply turn-on and turn-off.

In case 2 one end of the coil is connected to the core so that the core and mounting bracket or case are always at the same voltage as the coil (Fig. 9.2).

Figure 9.2. High-voltage filter inductor.

Normally, the voltage across the coil is that developed by the IR drop of the inductor. However, under fault or switching conditions, the load end of the winding may be instantaneously grounded, this would place at least the full dc voltage on the input terminal of the inductor. This voltage does not distribute equally across the coil and may require special winding and insulating methods (see Chapter 10). Figure 9.2 is a 30-H filter inductor in an 86-kV power supply; it is the second type, the coil is connected to the core and mounting bracket. The coil is insulated with Kapton film and is cooled with circulating SF_6. Since virtually all of the losses are in the coil, copper heat sinks are wound into the coil so that losses can be conducted to external fins. The radiating fins are rounded so that the electric stress concentration at their edges is minimized. Similarly, the edges of the mounting bracket also have radii.

9.4   ISOLATION TRANSFORMERS

Frequently, a transformer may be required to provide relatively low voltages which power electronic circuits operating at a high dc voltage. The transformer

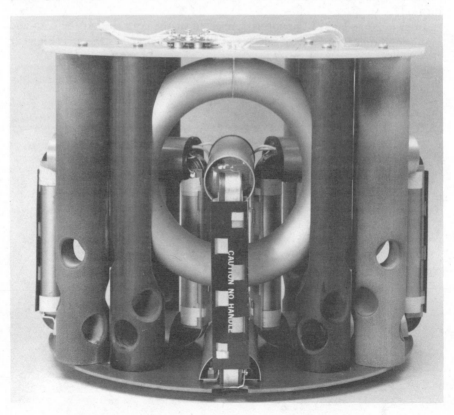

Figure 9.3. Low-capacitance three-phase isolation transformer.

must provide isolation between the high voltage and ground. Often, transformers such as this must also have very low capacitance between the secondary winding and ground and the primary winding. A low-capacitance three-phase isolation transformer is shown in Fig. 9.3. This transformer isolates circuits floating at 86 kV from ground. Each secondary winding is enclosed in an aluminum torus. Besides low capacitance this configuration provides virtually a uniform electric field. The epoxy glass supports are pierced in such a way that there is an SF_6 gas gap between the torus at 86 kV and ground.

9.5   PULSE TRANSFORMERS

Many electronic circuit applications require transformers which provide high pulse voltages, trigger applications, pulse radars, and video circuits, for example. The performance of these transformers is directly related to the circuit impedances, circuit and component capacitances and inductances, as well as the nature of the pulse waveform and the pulse voltage. Of prime importance in the selection of an insulation system is its performance at high frequency. The dielectric characteristics of materials often are not as attractive under pulse and high-frequency conditions as at dc and low-frequency ac operation. See Chapter 10 for a discussion of pulse transformers.

9.6   HIGH-VOLTAGE INSULATION

In the design and construction of high-voltage transformers, consideration must be given to a total insulation system, that which insulates the integral elements of the transformer (turns, layers, leads, windings, and the core) and that which provides an insulating environment for the complete transformer. Some of the materials may perform both functions.

These materials usually fall into two categories: (1) mechanical support, often with barrier action, and (2) filling or impregnation to displace air with higher-dielectric-strength materials. Liquids and gases fill only the second category, but solid insulations can function in both categories.

9.6.1   Liquids

Liquid dielectrics are very effective for filling the interstitial spaces within the transformer and impregnating the porous and permeable insulating materials, thereby increasing the dielectric strength and the corona threshold of a system. As can be seen in Table 9.1, the effective dielectric strength of liquids is about midway between that of gases and solids, but it can be approached by compressed gases (see Fig. 9.4). The practical dielectric strength of liquids varies with spacing, increasing with decreasing and decreasing with increasing spacing. Values of 350 V/mil for a 0.1-in. spacing are typical for most liquid insulants,

TABLE 9.1 Electrical Properties of Various Insulations

Type	Material	Dielectric Constant (100 Hz)	DC Volume Resistivity (Ω cm³)	Dielectric Strength [V(rms)/0.001]	Max. Stress [V(rms)/0.001]	Hot-spot max. (°C)
Gas	Air	1.0	1×10^{25a}	—See Sec. 9.6.3—	—See Sec. 9.6.3—	180
	Sulfur hexafluoride (SF$_6$)[b]	1.0	1×10^{25a}			105
Liquid	FC75[b]	1.9	8×10^{15}	350	150	105
	Coolanol 20[c]	2.5	9×10^{10}	350	150	105
	Askarel	5.5	1×10^{12}	350	150	105
Solid	Epoxy resin	4.2	1×10^{15}	>500	200	130
	Polyester resin	4.0	1×10^{14}	550	150	130
	RTV	3.0–3.5	2×10^{15}	>300	150	200
	Laminate, epoxy glass	5.2	1×10^{15}	500	150	130
	Dyallphthalate (DAP)	4.2	5×10^{16}	>300	150	130
	Urethane	5.2	2×10^{12}	>400	150	130
Film	Mylar[d] (0.001 in. thick)	3.3	1×10^{18}	>3000	200	130
	Kapton[d] (0.001 in. thick)	3.5	1×10^{18}	>5000	250	200
	Teflon (0.002 in. thick)	2.2	5×10^{13}	>900	200	180
	Polypropelene	2.0	3×10^{15}	>4000	250	105
Sheet	Kraft paper, unimpregnated	2.0	1×10^{16}	>250	75	105
	Kraft paper, resin impregnated	3.5	1×10^{15}	>500	200	130
	Nomex[d] unimpregnated	2.7	1×10^{16}	150	75	180
	Nomex[d], liquid impregnated	2.8	1×10^{16}	>850	200	180
	Nomex[d], resin impregnated	2.9	1×10^{16}	>350	150	155
	Pressboard	2.0	1×10^{15}	>200	100	130
Other	Mica	5.8	1×10^{15}	3000	250	200
	Porcelain	6.5	5×10^{12}	>350	150	200
	Alumina	8.5	5×10^{14}	400	150	200
	Wire enamel	3.2	1×10^{16}	>800	200	105–200

[a]The volume resistivities of insulating gases are essentially infinite until breakdown occurs, at which time the resistivity drops to almost zero.
[b]3M trademark for fluorocarbon liquids.
[c]Monsanto Co. trademark for silicate-ester-base fluid.
[d]Trademark of E.I. duPont de Nemours & Co., Inc.

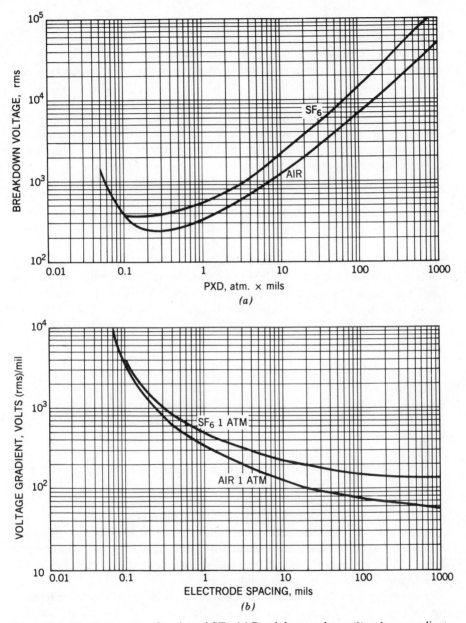

Figure 9.4. Paschen's curves for air and SF$_6$. (*a*) Breakdown voltage; (*b*) voltage gradient.

which includes chlorinated aromatic, mineral, silicone, and polyester oils, and fluorocarbon liquids.

Dc and ac crest dielectric strength are comparable, while impulse voltage strengths are two to three times higher than the 1-minute dielectric strength. The strength of liquids is reduced by the presence of conducting or semiconducting contaminants suspended in the fluid. Gas bubbles greatly reduce the dielectric strength of the liquid. Filling in a vacuum should be done to remove any entrapped air. If the liquid is saturated with a gas it will break down at lower stress when subject to lower pressure. The container should be designed to accommodate the changes in fluid volume with changes in temperature; this will eliminate the formation of a partial vacuum at lower temperatures.

Insulating fluids are also subject to a degradation of properties by oxidation or by a reaction between the liquid and the solid insulations which are a part of the system. The most common effect is an increase in the conductivity of the liquid which could lead to a thermal breakdown. Fluorocarbon liquids are the most inert. Mineral oils are slightly less inert but will oxidize in air. Silicone fluids are fairly inert and have high thermal stability. The chlorinated aromatic fluids have higher dielectric constants and are solvents for or may cause many resins to swell and dissociate ionic substances.

9.6.2 Solids

Solid insulations generally have much higher dielectric strengths than the fluid insulations. However, it is usually not possible to utilize this inherent strength because of the difficulty in maintaining or obtaining a void-free condition with the solid, particularly adjacent to metallic electrodes and insulated conductors. As is true with the gaseous and liquid insulators, the dielectric strength of solids decreases with increasing thickness.

In a high-voltage transformer the solid insulations of interest are the sheet insulations and tapes which are used between layers and windings, the resin or varnish used to impregnate the coils or embed the transformer, and any other materials that are used for structural or barrier purposes.

The selection of the sheet insulation, adhesive tapes, and other materials used in the coil should be based on expected hot-spot temperature, the voltage stress between layers and windings, the type and size of magnet wire, and whether the transformer insulation is liquid, solid, or gaseous.

The resin or varnish used as an impregnant or embedment should be selected on the basis of its compatability with the sheet insulations, tapes and magnet wire insulation, the hot-spot temperature, the voltages in the transformer, and also processing costs, which includes special tooling and processing equipment.

Processing as well as mechanical, electrical, and thermal properties must be considered when selecting materials for use as structural insulators. Usually, materials that can be molded are not as strong as laminated plastics; however, most molded materials are more homogeneous than laminated materials.

Laminated materials, which are often used for structural as well as insulating purposes, have much lower dielectric strength parallel to the laminate than perpendicular to the laminate. This is due to voids that occur at the bond between the layers of the reinforcing laminate, usually woven glass fiber sheet.

Some molded materials have inherent defects which are introduced during the molding operation. For example, acetal sheet often contains a "blend line," a layer of small bubbles through the plane of the sheet. These bubbles could well be sites of partial discharge if the bubbles are within a high electric field.

9.6.3 Gases

Gases are used primarily as the dielectric medium around the structural materials and the barrier insulation. As such their most important property is dielectric strength. Table 9.2 lists the relative dielectric strength of the more common insulating gases. As with the liquid insulating fluids, the dc and ac crest dielectric strengths are usually equal and as much as 25% higher for microsecond pulse voltages.

The dielectric strength in volts/mil, gradient, decreases with increasing spacing as shown in Fig. 9.4(b). The breakdown voltage varies as the product of the pressure (gas density) and the electrode spacing; this is shown in the familiar Paschen's curve in Fig. 9.4(a). All gases exhibit a minimum voltage below which voltage breakdown will not occur in a uniform field. At pressure-spacing values below the minimum, larger gaps break down at lower voltages than do smaller gaps. This characteristic of gases must be considered in all open electrical systems which are to be operated at very low pressures.

In Table 9.2 it can be seen that gases other than air or mixtures of gases can give substantial increases in dielectric strength. By increasing the pressure of the gas, even higher strengths can be obtained. For increases in the pressure up to about 5 atm, the strength increases about in proportion to the pressure, but at higher pressures this drops off.

Condensation temperature, thermal stability, and reactivity are other properties of gases which are important. The chlorinated gases are less thermally

TABLE 9.2 Relative Dielectric Strengths of Gases

Gas	ϵ	Gas	ϵ
Air	0.95	SF_6	2.3–2.5
N_2	1.0	CF_4	1.1
CO_2	0.90	C_2F_6	1.9
H_2	0.57	C_3F_8	2.3
A	0.28	C_4F_8	2.8
Ne	0.13	CF_2Cl_2	2.4
He	0.14	C_2F_5Cl	2.6
		$C_2F_4Cl_2$	3.3

stable, may cause corrosion, are absorbed by many resins, and may cause decomposition of resins above 100°C. The fluorocarbon gases are much more stable and may be operated at temperatures as high as 300°C with stable resins and inorganic materials. SF_6 is somewhat less stable and may dissociate at higher temperatures; some of the decomposition products are highly reactive and some are toxic. SF_6 is not recommended for operation at temperatures above 175°C.

9.6.3.1 Stress Concentrations.   Most curves of breakdown voltage versus the "pressure-spacing product" [the Paschen curve, Fig. 9.4(a)] are based on uniform electric fields. In practice, uniform fields in high-voltage equipment are virtually impossible to achieve. Consequently, it becomes necessary to apply utilization factors to the average stresses determined by dividing the voltage by the spacing. The layout of high-voltage equipment based on an average stress in volts/mil may well lead to the initiation of partial discharge if the locales of high voltage or low voltage (ground) create a concentration of stress in excess of the corona inception voltage (CIV).

Figure 9.5 shows utilization factors for some common electrode con-figurations. Factors for other configurations are available in the literature. Utilization factors generally are applicable to systems in which there is a single dielectric material; gas, liquid, or solid and are not directly applicable where the insulation system consists of several materials of different dielectric constant in series.

It is apparent that the critical stress is dependent on the ratio of the electrode spacing to the radius. To reduce the stress and increase the utilization factor it is

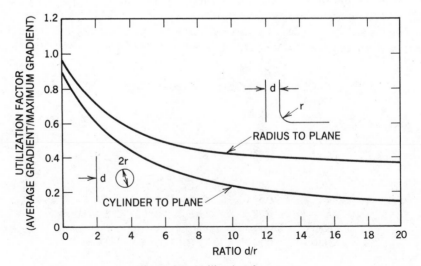

Figure 9.5. Utilization factors.

important to increase the effective radius. In some instances increasing the radius will decrease the peak stress more effectively than will increasing the spacing.

9.7 VOLTAGE STRESS ON DIFFERENT DIELECTRICS IN SERIES

Often, in high-voltage transformers and other high-voltage devices, a combination of insulating materials may be used in such a manner that in some locales more than one of these materials intervenes between points of different electric potential. The manner in which the insulating materials share this electric stress is determined primarily by the voltage; if it is ac, the stress distribution is capacitive, if it is dc, it is resistive (see Dakin, 1986).

9.7.1 Stress Distribution in ac Fields

When more than one insulating material is used in series between circuit elements with an ac potential difference, the electric stress in each is proportional to the thickness/dielectric constant ratio. Thus the insulation with the lowest ϵ has the highest stress, all other things being equal. Often this is an insulating gas or air which usually has the lowest dielectric strength of the series insulations.

The ac voltage stress on each insulation element in series may be calculated by

$$G_x = \frac{V}{\epsilon_x(T_1/\epsilon_1 + T_2/\epsilon_2 + \cdots + T_n/\epsilon_n)} \tag{9.1}$$

where V = total voltage across all insulation
$\quad G_x$ = electric stress, V/mil
$\quad T$ = insulation thickness, mils
$\quad \epsilon$ = dielectric constant
$\quad x = 1, 2, \ldots, n$

9.7.2 Stress Distribution in dc Fields

When more than one insulating material is used in series between circuit elements with a dc potential difference, the electric stress in each is proportional to the ratio of their resistivity-thickness products (assuming that the electric field goes through the same area of each). Thus the insulation with the highest ρ will have the highest stress, all other things being equal.

The dc voltage stress on each insulation element in series may be calculated by

$$G_x = \frac{\rho_x V}{(T_1\rho_1 + T_2\rho_2 + \cdots + T_n\rho_n)} \tag{9.2}$$

where V = total voltage across all insulation
$\quad G_x$ = electric stress, V/mil
$\quad T$ = insulation thickness, mils
$\quad \rho$ = volume resistivity
$\quad x$ = 1, 2, ..., n

9.8   DIELECTRIC STRENGTH

The usual figure given for the dielectric strength of an insulating material is the breakdown value in rms volts at 60 Hz in a 1-minute test. It is not possible to operate most transformer insulations anywhere near these values: some, such as kraft paper, because of their cellular structure in which corona can initiate, and others, such as films and resinous solids, in which voids can form and become sites of partial discharge. Table 9.1 provides dielectric strength design data for several insulating materials. It should be noted that the recommended voltage stresses are substantially lower than the intrinsic dielectric strengths found in the literature for these materials.

9.9   FLASHOVER AND CREEPAGE

Flashover is arcing in a gas or liquid between points of differing voltage. It is comparable to the dielectric strength shown for the liquids and gases in Table 9.1. Creepage or tracking is the breakdown across the surface of a solid insulation. Tracking occurs only on plastic and elastomeric insulation. Glass, porcelain, and other ceramic insulators do not track, but other insulating properties can be affected. Usually, a creepage arc is preceded by a treeing pattern on the surface of the insulation. The treeing is initiated by localized discharges on the surface of the insulation in regions of high electric stress forming a carbon path. The propagation of these trees is permitted by ionic contaminants on the surface and the presence of moisture. When the treeing pattern bridges two electrodes of different potential a creepage arc occurs and an arc track may be formed on the insulator. An insulating material's resistance to tracking when subjected to high electric stress should be considered when selecting a material as a barrier insulator. Some materials will track within seconds when subjected to ASTM tests, particularly the aromatic epoxies. However, there are coatings that can be applied to finished insulators which make them much more track resistant. Most of these coatings are a cycloaliphatic epoxy which has been filled with anhydrous alumina.

Creepage is a function of the dielectric constant of the insulation; the higher the dielectric constant, the lower the voltage to surface breakdown. Creepage has the least effect (as compared to flashover) when the dielectric constant is low and when the surface is parallel to the electric field. Figure 9.6 provides design limits for flashover and creepage in air.

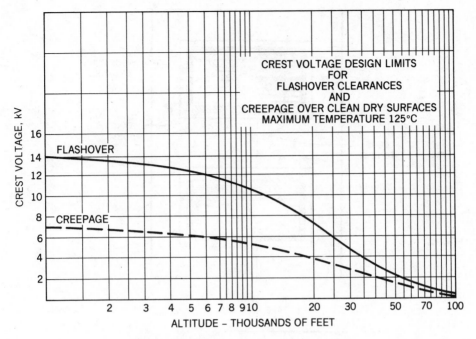

Figure 9.6. Flashover and creepage in air.

While the dielectric strength of the solid insulation is important in a high-voltage transformer, the most common insulating problems are maintaining suitable surface creepage distances, such as margins between the winding and the core along the layers of sheet insulation, or margins between leads and the frame or other grounded parts of the transformer. Insulating channels are often used to increase the direct creepage distance across the margins as in Fig. 9.7(a). This is especially helpful when the inner part of the coil adjacent to the core tongue is at a low voltage and the outer part at a high voltage. When the whole coil is at a high voltage, it may be insulated by taping; taping is expensive and should not be used if the creepage distance provides the necessary insulation strength.

Creepage distances over organic insulation in air are shown in Fig. 9.8 for breakdown voltages up to 100 kV. The primary purpose of these curves is to permit the determination of the proper margins for coils adjacent to the core.

Insulation between the start (or finish) turn of the first layer and the core consists of the margin plus the thickness of the coil form; however, if the coil lead is brought across the margin and up the side of the coil, the only creepage distance is the thickness of the coil form. In low-voltage coils this is probably adequate, but in high-voltage coils an insulating barrier should be placed between the coil form and the core beneath the place where the lead is brought out of the coil. This is shown in Fig. 9.9. The dimensions of the insulating barrier

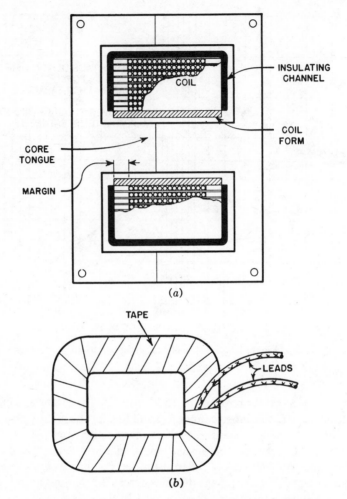

Figure 9.7. (*a*) Use of insulating channel; (*b*) taped coil.

should be such that a distance at least equal to the coil margin should intervene between the start lead and the core in all directions; it should be at least as thick as the coil form.

When the finish lead is at the top of the coil it will have a longer creepage distance to the core if the height of the coil is greater than the margin. It is necessary to avoid using materials on the sides of the coil which would result in a decrease of dielectric strength. Some materials with high dielectric strength may be more susceptible to creepage failure. The last layer of wire may be insulated from the core with a channel as in Fig. 9.7(*a*).

When practical coil margins, even with barriers, are insufficient to support the induced or applied voltage and where a change to a liquid or gaseous insulation cannot be made, it may be necessary to tape the coil or cast it in some

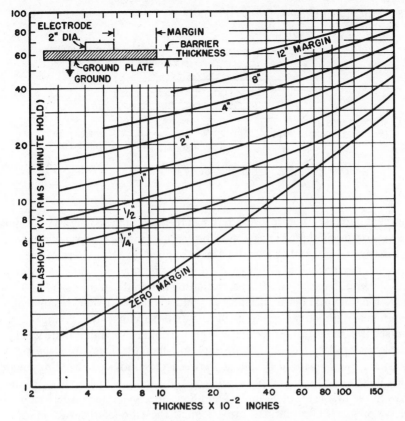

Figure 9.8. Creepage curves in air over smooth organic insulation.

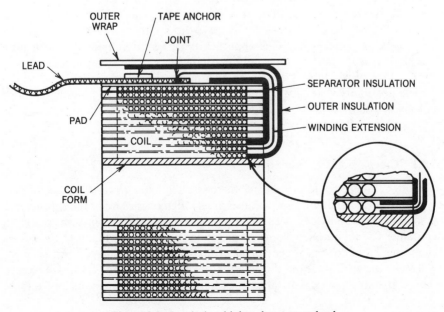

Figure 9.9. Insulating high-voltage start lead.

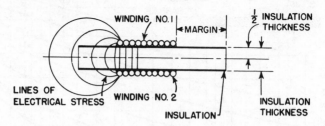

Figure 9.10. Adaptation of Fig. 9.8 for insulation between coils.

impregnating resin. Figure 9.7(*b*) shows a taped coil, if the coil were to be molded in an epoxy or similar resin care must be taken to ensure complete impregnation and that a coverage of at least 0.10 in. is applied on all surfaces of the coil.

Ordinarily, a winding is insulated from the winding over which it is wound by wraps of kraft paper or other sheet insulating materials. In the coil (Fig. 9.10) the insulation thickness between windings 1 and 2 is shown divided by an imaginary centerline. With equal margins in the two windings the voltage stress is symmetrical about this centerline. Margins should be such that there is sufficient creepage distance in conjunction with one-half the insulation thickness, to withstand one-half the test voltage between these adjacent windings. That is, when full test voltage is applied between the windings, only half of it appears between the first layer of winding 1 and the centerline of the insulation between the windings. If the margins are unequal, the sum of the two margins, in conjunction with the total insulation thickness, should be large enough to withstand the full test voltage, in accordance with Fig. 9.8.

When the layer-to-layer insulation stress is too great for safe operation, the coil can be wound in sections to increase the number of layers and thereby reduce the volts per layer. This approach must be carefully analyzed because transformer leakage inductance and capacitances can be changed significantly by this type of winding as compared to a nonsectioned winding.

The possibility that edge corona may form should also be considered. Edge corona is discussed in Section 9.10.1.

9.9.1 Creepage in Oil Insulation

Although, in electronic equipment, there is a tendency toward the use of dry-type transformers, frequently, voltages are so high or the size of the transformer is so limited that air clearances are impracticable and the unit must be placed in container filled with oil or some other insulating liquid. In Fig. 9.11 the curves show rms breakdown voltage versus creepage distance under oil. An example will show the advantage of oil insulation. From Figs. 9.8 and 9.11 it will be seen that a 10.0-in. creepage distance is required in air to withstand a 1-minute

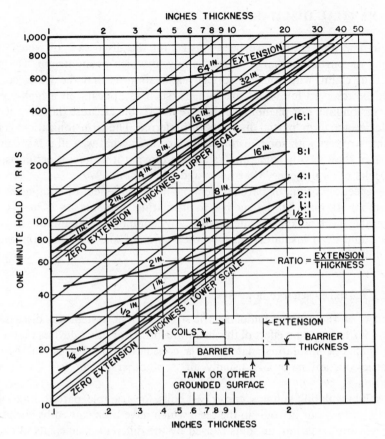

Figure 9.11. Creepage curves of solid insulation under oil.

breakdown of 60 kV on insulation 0.5 in. thick, whereas in oil only a 2.0-in. creepage distance is required.

The curves of Fig. 9.11 are for pressboard and Micarta. Some kinds of porcelain have less creepage strength than these; on the other hand, some grades of glass and polystyrene are much better, withstanding 150 kV for 1 minute with a 2.0-in. creepage path. As in all insulations sharp edges, points, and other stress concentrators should be avoided.

Only high grades of insulating liquids should be used. Tests on the quality of the insulation should be performed periodically and the insulation should be stored to keep out moisture and any other contaminants. If it is to be used again, it must be filtered before it is reused. Where very high voltages are used, as in X-ray apparatus, liquid filling is done under vacuum to remove entrapped air; the container must be sealed to prevent moisture from entering. Mica is not used in oil because the oil dissolves many flexible bonding materials.

9.10 PARTIAL DISCHARGE

"Much time and effort have been wasted on the discussion of which term, 'corona' or 'partial discharge,' should apply to which phenomenon. The most vociferous claim, with some historical justification, is that the word 'corona' should be reserved for visual phenomena, such as appears on a high-voltage transmission line. For phenomena that are not visible, because they are internal to a device, the term 'partial discharge' is preferred. The term 'ionization' is used by some workers with some justification. However, what we call it is much less important than having some understanding of what it is, how it performs, and the results of its presence" (Bartnikas and McMahon, 1979).

The effect of corona or partial discharge in a component or equipment may be twofold: It may be destructive to insulating elements or it may create undesirable electromagnetic interference. Any component, circuit, or structural element and even insulators may be the source of and may be damaged by partial discharge.

9.10.1 Partial Discharge in ac Systems

The effect on the distribution of electric stress in an ac system was discussed in Section 9.7.1. The main effect of this in a transformer or other devices that utilize more than one insulating material is a concentration of stress in the low-dielectric-constant materials, usually air or another gas in series with solid insulation.

In solid insulating systems, epoxy cast coils, for example, partial discharge is most likely to occur in voids in the resin which have been caused by shrinkage, improper impregnation, or splits caused by the different coefficients of expansion of the material used in the device.

In open coils, those which are not embedded and which operate in a gas, the corona inception voltage at a conductor edge opposite a ground plane varies as a function of the ratio of the thickness of the solid insulation to its dielectric constant as shown in Fig. 9.12. While the choice of barrier insulation is often limited, the high dielectric constant materials should be avoided so that the highest corona inception voltage may be achieved for a given spacing.

Liquid insulation systems can usually be stressed higher than gas systems. However, if gas is formed in the liquid by breakdown, overheating (boiling), or through any other means, partial discharges can occur in any bubbles that are formed.

9.10.2 Partial Discharge in dc Systems

The resistivity of the insulation controls the stress distribution in a dc system. When different materials are in series, the least stress will be across the most conductive materials; a liquid in series with most solid insulations will have very little stress on it because it is more conductive than the solid. A gas gap in series

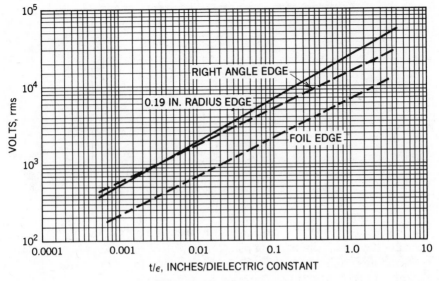

Figure 9.12. Edge corona.

with a solid, which is a site of partial discharge, will have little stress, and discharges will not occur until the charge leaks through the solid material.

9.10.3   Partial Discharge in Combined ac and dc Systems

In a rectifier circuit and others with combined ac and dc voltages, the ac crest-to-crest voltage determines the corona inception voltage. Unless the peak-to-peak ripple voltage is high enough to stress the gas gap to about twice its breakdown, continuous corona will not occur.

9.11   PARTIAL DISCHARGE DETECTION

Partial discharge can be detected in many different ways; often it can be heard or seen and if it is intense enough the ozone formed by the discharge can be smelled. Corona detectable by the senses is normally external, as on a power transmission line. However, the partial discharges of concern in transformers are most often internal to the device and cannot be heard, seen, or smelled. For this there are more sophisticated techniques for detecting and quantifying partial discharge; they fall into two main areas.

The first of these is the NEMA method, which utilizes a radio noise meter to measure the radio influence voltage generated by partial discharges in the transformer. The noise meter is calibrated in microvolts. This method is used primarily as a quality check on transformers; it cannot be used to isolate any specific noise source in a transformer.

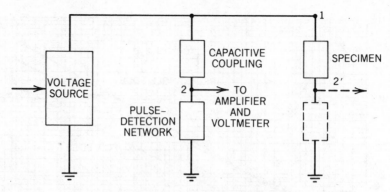

Figure 9.13. Partial discharge detection network.

The second method includes the transformer as a part of the test circuit and measures the partial discharge in picocoulombs. The detection of the discharges is usually accomplished with an *RC* power separation filter. Figure 9.13 is a diagram of a typical detection network. The voltage source may be an applied ac voltage, an applied dc voltage, an induced ac voltage, or an applied dc voltage superimposed on an induced ac voltage. The coupling network will be different for the applied ac and the dc detection networks.

Since it is the pulse energy in the partial discharges that is detrimental to the insulation, it is necessary not only to detect the discharges but also to measure the pulse amplitude and its repetition rate. The visual display of a corona discharge pattern on an oscilloscope can be used as a go/no-go acceptance test, but it provides little information as to the pulse energy in the discharges. The visual display can be very helpful in efforts to locate any sources of corona in complex high-voltage transformers.

The oscilloscope is of little use in detecting discharges in dc systems because the discharges occur randomly and are of such short duration that it is difficult to judge their amplitude and even more difficult to photograph them.

The electronic pulse counters developed for nuclear pulse spectroscopy permits the measurement of the cumulative count of discharges within predetermined regions of charge amplitude. The regions of charge amplitude are measured in picocoulombs. The pulse-height analyzer has eliminated much of the guesswork involved in the interpretation of partial discharge measurements.

9.12 HIGH-VOLTAGE TRANSFORMER DESIGN

The requirements for a high-voltage power transformer are:

Input: 200 V rms, three phase, 380 to 420 Hz, delta connected
Outputs:
   Secondary 1: 200 V rms, L-L, 90 VA, wye connected

Secondary 2: 13,910 V rms, 965 VA, operating into a full-wave rectifier, wye connected

Efficiency: 96%

Maximum weight: 6.5 lb

Conductive cooling, heat sink at 71°C ambient

Environment: SF_6 at 1.6 atm

No continuous corona greater than 10 pC

The steps that are followed in designing a high-voltage transformer are the same as those shown in Fig. 1.1. The first step in the design is the selection of the core to be used. Because minimum weight is a goal, a three-phase tape-wound, cut E core made of 0.004-in.-thick Supermendur was selected. The basic core dimensions are established for the load VA, and then the core window dimensions are increased to provide clearance for the high voltage. The core used in this design has the following dimensions:

Strip width 1.25 in.	Window height 1.25 in.
Strip build 0.38 in.	Window width 4.50 in.
Net core area 0.422 in.^{2}	Core weight 2.52 lb.

Primary turns per phase are figured from equation 3.4, where dimensions are in inches and B is in gauss. Primary turns are then

$$\frac{3.49 \times 200 \times 10^6}{380 \times 0.422 \times 19780} = 220$$

The primary line current at full load will be 2.27 A per phase; this will require AWG No. 20 wire with polyester enamel. The turns per phase for secondary 1 are 134. The S1 line current at full load is 0.265 A, which would require AWG No. 29 wire. To reduce coil build S1 is wound with two strands of AWG No. 32 bifilar. The turns per phase for secondary 2 are 9299 with a tap at 3767 turns. S2 is wound with AWG No. 36 enameled magnet wire.

The transformer was designed to be conductively cooled, and therefore the windings were placed on an aluminum coil form, split so that there would not be a shorted turn. The windings were placed on the coil form concentrically, the primary first followed by S1 and then S2, the high-voltage winding.

In Section 9.6.2 it was noted that solid insulations have higher intrinsic dielectric strengths than insulating gases but that this property could not always be exploited because the solid insulating materials often crack or develop voids. Because of this, it was decided to use an open construction with the SF_6 environment as part of the transformer insulation system; Kapton film was selected to be the sheet insulation between layers and windings.

The high-voltage secondary windings were designed to provide minimum electrical stress between the three high-voltage phases, between winding layers and between the low-voltage and high-voltage windings.

By using a wye connection on the secondary the voltage induced in each phase is only 58% of the line-to-line voltage, as compared to a delta, in which the phase voltage is the full line-to-line voltage. This permits a grading of the insulation at the neutral connection, usually the start of the winding.

If each secondary were wound with the minimum number of layers, 16 in this design, there would be more than 1000 V between the ends of adjacent layers or 500 V if the layers were wound back. In either case the amount of insulation required to provide a safe margin over the corona inception voltage, including stress concentrations at the end turns, resulted in a coil buildup which would not fit into the core window. By dividing the winding space into three sections the number of layers was more than tripled, to 59 layers in this case, with the maximum voltage between two layers reduced to 275 V. Using 0.005 in. of Kapton film the average stress is 55 V/mil. Assuming that the end turns approximate two cylinders, the d/r ratio from Fig. 9.5 is 2, giving a utilization factor of 0.65. This results in a peak stress at the end turn of 85 V/mil, about 35% of the stress required to initiate partial discharge as determined in Fig. 9.4(b).

The major insulation between the low-voltage and the high-voltage winding may be graded because the voltage across the coil is not constant. The insulation between the coil section at the neutral end of the winding and the low-voltage windings is 0.10 in. of Kapton and 0.15 in. of Kapton at the two higher voltage sections. This produces a peak stress of 150 V (rms)/mil on the Kapton film, which is satisfactory.

Figure 9.14. Three-phase winding arrangement to reduce electric stress.

To reduce electric stress between the adjacent phases the three coil sections on the middle phase are reversed with respect to those in the two outer phases. In this arrangement the maximum voltage between coils is the line-to-tap voltage, 8675 V, rather than the line-to-line voltage, 13,910 V. This allows less spacing between the coils, 0.10 in. rather than about 0.20 in. The three-section winding configuration is shown in Fig. 9.14.

Partial discharge is measured with the induced voltage at 110% of nominal and with the high-voltage winding floating at 4400 V dc. There should be no continuous discharges greater than 10 pC.

Except for the method of selecting the core and the manner in which the windings and coils are insulated and spacings are determined, the high-voltage design follows that of Fig. 3.9.

10 PULSE AND VIDEO TRANSFORMERS

In the previous chapters, all of the discussions were based on the basic waveform being sinusoidal. This was true whether the transformer was designed for one frequency or for a number of frequencies. In the next chapters, the waveforms being considered are nonsinusoidal. In Chapter 6 the class D amplifier was discussed, in which the voltage waveforms were square waves and the current waves were sine waves. In the following chapters neither the voltage or current waves are sine waves. The most common form of nonsinusoidal wave is the pulse or square wave. The difference between a pulse and a square wave is one of relative time. In a square wave, the wave is at its maximum and minimum voltages for equal times. A pulse train has unequal times at maxima and minima. The time ratio for pulses is usually greater than 10 to 1; that is, the pulse is off more than 10 times as long as it is on. Such pulses therefore have a high peak power and a low average power.

10.1 SQUARE WAVES

Square waves differ from sine waves in that the leading and trailing edges of the square wave are short in time compared to the flat top. Such pulses have many uses in television, radar, computers, and communication equipment.

A sine wave can be expressed as

$$F(t) = A \sin 2\pi f t \qquad (10.1)$$

where A is the peak amplitude and t is the time from 0 to $1/f$.

A square wave can be expressed as

$$F(t) = \sum_1^\infty b_n \sin 2\pi nt \qquad (10.2)$$

where $n = 1, 3, 5, \ldots$. The square wave is therefore the sum of a number of harmonically related sine waves, the lowest frequency of which is the reciprocal of the period of the square wave.

The frequency response of the transformer determines the fidelity with which the square wave or pulse is reproduced. It can be seen from equations 10.1 and 10.2 that a pulse transformer is equivalent to a broad-band transformer. The characteristics of a pulse transformer are expressed in the time domain rather than in the frequency domain.

A perfect square wave would rise from the minimum voltage to the maximum voltage in zero time, remain exactly at the maximum voltage for the specified time, and then return to the minimum voltage in zero time. Such pulses are only possible mathematically and are used to simplify the development of the design equations. Pulses and square waves can be generated in a number of ways. The method of pulse generation will not be discussed unless it is integrally related to the design of the pulse transformer. Practical pulses and square waves have finite rise and fall times. These rise and fall times may be very short but they are still finite. For example, rise and fall times of digital logic may be less than 10 ns. Transformers are not often used to couple these pulses.

Figure 10.1(*a*) shows a theoretical square wave. Figure 10.1(*b*) shows a pulse train. In the square wave, the time the voltage is at a maximum is equal to the time the voltage is at a minimum. For a pulse train, the interval between pulses greatly exceeds the pulse duration. The pulse duration is referred to as the pulse width. The frequency at which the pulses occur is called the pulse repetition rate (PRR) or pulse repetition frequency (PRF). It is expressed as pulses per second (pps). Common pulse widths lie between 0.05 and 100 μs. The intervals between pulses is normally between 10 and 10,000 times the pulse width. These values are representative only and in many cases may be exceeded. The general wave shape of Fig. 10.1(*b*), with short pulse duration compared to the interval between

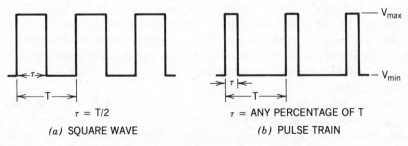

Figure 10.1. Ideal (*a*) square wave and (*b*) pulse train.

pulses, is the main subject of this chapter. The ratio of peak to average voltage or current may be very high. The rms values appreciably exceed the average values in such cases.

There are two ways in which the response of any circuit to a pulse or square wave can be found. The first of these consists of resolving the pulse into a large number of sine waves of different frequencies, finding the response of the circuit to each frequency and summing up these results to obtain the total response. This type of analysis can be formulated by a Fourier integral, but for most circuits the formulation is easier than the solution. An approximation to this method is the arbitrary omission of frequency components having negligible amplitude and calculating the frequency response to the relevant frequencies. This approximation has two subjective criteria: the number of frequencies to be retained, and the evaluation of the frequency components for which the circuit has poor response.

A second method, which was used originally to develop the equations and design curves used in this chapter, consists of finding the transient circuit response to the discontinuities at the leading and trailing edges of the square wave. It is possible to reduce the transformer to a circuit amenable to transient analysis without making any more assumptions than would be necessary for practical design work with the Fourier method. The transient method has the advantage of giving the total response directly and can be plotted as a set of curves which are of great convenience to the designer. The major assumption is that one transient disappears before another one begins. This assumption is justified if the transformer is to faithfully reproduce the input pulse.

A third method, which has been used to check the equations in this chapter and to develop the equations in Chapter 12, is similar to the method used in developing the equations in this chapter, but has far more flexibility than the solution of differential equations by classical methods. An equivalent circuit for the transformer and associated circuitry is developed. Then the differential equations for the various loops (or nodes) are written. By using Laplace transforms, the equations are converted to ordinary algebraic equations which can then be solved for the loop currents or node voltages. From these currents, expressions for the voltages at various parts of the circuit can be written. The advantages of solving the equations this way are many. First, a circuit with as many elements as required to give an accurate solution can be used. Second, by using equivalent circuits of some complexity, voltages internal to the transformer can be calculated. This can be extremely important in designing high-voltage transformers where the windings are split between two legs of the core. Third, the algebraic equations lend themselves to using computer programs to solve the equations. This reduces the work necessary to develop the designs for complex circuits.

Analyses are given of the influence of iron-core transformer and inductor characteristics on pulse waveforms. In all of these analyses the transformer or inductor is reduced to an equivalent circuit. This circuit changes for different waveforms, portions of a waveform, and modes of operation.

Initial conditions, resulting equations, and plots of the equations for design convenience are given in each case. Equations may be verified by the methods of operational calculus.

10.2   TRANSFORMER-COUPLED PULSE AMPLIFIERS

The analysis given in this section is for a square or flat-topped pulse impressed on the transformer from some source. Such a pulse is shown in Fig. 10.2. A generalized circuit for such an amplifier is given in Fig. 10.3. At least this is the circuit that applies to the leading edge of the pulse shown in Fig. 10.2 as rising from zero to some steady value E in zero time. The transformer OCL can be considered as presenting infinite impedance to this change and is omitted in Fig. 10.4. Transformer leakage inductance has an appreciable influence. It is shown as inductance L_s in Fig. 10.4. Resistor R_1 of Fig. 10.4 represents the source impedance. Transformer winding resistances are usually negligible compared to the source impedance. Winding capacitances are shown as C_1 and C_2 for the primary and secondary windings, respectively. The transformer load resistance, or the load impedance into which the amplifier works is shown as R_2. All these values are referred to the same side of the transformer. Since there are two capacitance terms C_1 and C_2, it follows that one or the other of these becomes predominant if the transformer turns ratio deviates from unity. Turns ratio and therefore voltage ratio affect these capacitances as discussed in Chapter 5. For a

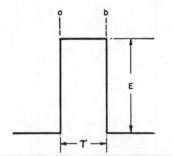

Figure 10.2. Flat-topped pulse.

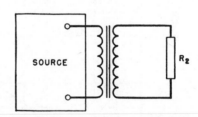

Figure 10.3. Transformer coupling.

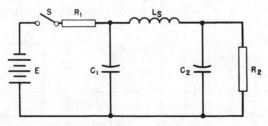

Figure 10.4. Equivalent circuit.

step-up transformer, C_1 may be neglected. For a step-down transformer, C_2 may be neglected. The first part of the discussion will be confined to the step-up case.

10.3   LEADING-EDGE RESPONSE

The step-up transformer is illustrated in Fig. 10.5. When the front of the pulse in Fig. 10.2 is impressed on the transformer, it is simulated by closing of the switch

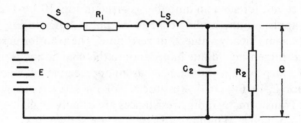

Figure 10.5. Circuit for step-up transformer.

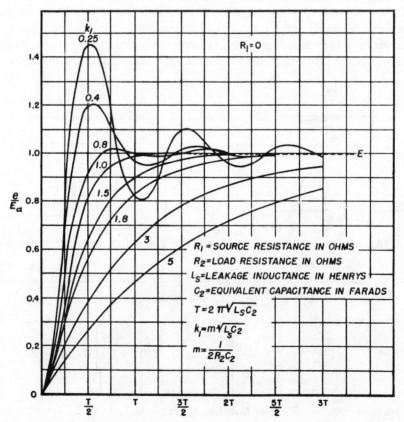

Figure 10.6. Influence of transformer constants on front edge of pulse ($R_1 = 0$). For equivalent circuit, see Fig. 10.5.

S. At this instant, voltage e across R_2 is zero, and the current from the source is also zero. These are the two initial conditions for equation 10.3, which expresses the rate of rise of voltage e from zero to its final steady state value $E_a = ER_2/(R_1 + R_2)$:

$$e = \frac{ER_2}{R_1 + R_2}\left[1 + \frac{m_2\epsilon^{m_1t}}{m_1 - m_2} - \left(1 + \frac{m_2}{m_1 - m_2}\right)\epsilon^{m_2t}\right] \qquad (10.3)$$

where $m_1, m_2 = -m(1 \pm \sqrt{1 - 1/k_1})$.

Figure 10.6 shows the rate of rise of the transformed pulse for $R_1 = 0$ and Fig. 10.7 for $R_1 = R_2$. In modulators using only part of the energy stored in a capacitor to form the pulse (hard tube modulators or the equivalent), the source impedance is comparatively small. For this type of modulator, the curves for $R_1 = 0$ are used. Line-type modulators are usually designed for $R_1 = R_2$.

The scale for the abscissas for these curves is not time but a percentage of the time constant T of the transformer. The equation for the time constant is given in Figs. 10.6 and 10.7. It is a function of leakage inductance and of capacitance C_2. The rate of voltage rise is governed by another factor, k_1, which is a measure

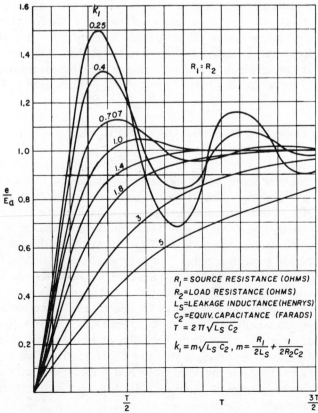

Figure 10.7. Influence of transformer constants on front edge of pulse ($R_1 = R_2$).

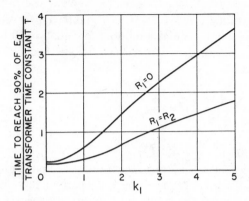

Figure 10.8. Time required to reach 90% of final voltage.

of the circuit damping factor. The relation of k_1 to the various transformer constants is given directly in Figs. 10.6 and 10.7. The greater the transformer leakage inductance and distributed capacitance, the slower the rate of voltage rise. This rate of rise is affected by R_1 and R_2 since these values are important in determining the damping factor k_1. If a slight amount of oscillation can be tolerated, the voltage rises faster than if no oscillation is present. If the circuit is damped very little, the oscillation may reach twice the steady-state voltage E_a. Usually, these high peaks are objectionable both from the operation of the circuit and from voltage breakdown considerations. The values for k_1 given in the figures are those which bracket the most acceptable practical values. The time required for the pulse voltage to reach 90% of E_a is given in Fig. 10.8.

10.4   RESPONSE AT THE TOP OF THE PULSE

Once the pulse top is reached, E_a is no longer dependent on the leakage inductance and distributed capacitance. It is now dependent on the OCL for the maintainance of the peak value. If the pulse stayed indefinitely at the value E_a, an infinite OCL would be required to maintain it. To keep the pulse at the value E_a for any period of time would also require an infinite inductance. Of course, this is not practical. Since the value of OCL is always finite, there will always be droop on the top of such a pulse. The equivalent circuit during this time is shown in Fig. 10.9. Here the inductance L is the OCL of the transformer, and R_1 and R_2 remain the same as before. Since the rate of change of voltage during this period is relatively small, capacitances C_1 and C_2 can be neglected. Leakage inductance is also usually small compared to the OCL. This, too, can usually be neglected during the pulse top. At the beginning of the pulse, the voltage e across R_2 is assumed to be at the steady-state value E_a. This is true if the voltage rise is rapid. Curves for the top of the pulse are shown in Fig. 10.10. Several curves are given. They represent several types of pulse sources ranging from a voltage

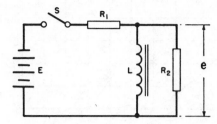

Figure 10.9. Circuit for top of pulse.

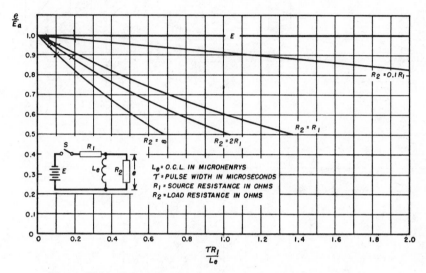

Figure 10.10. Droop at top of pulse transformer output voltage.

source where R_2 is one-tenth of R_1 to a current source where the load impedance is infinite and the source impedance is zero. In the latter curve, the voltage e has the source voltage as its initial value. All the curves are exponential, with a common point at 0,1. Abscissas are a product of time τ and R_1/L_e, time τ being the duration of the pulse between points a and b in Fig. 10.2. The greater the inductance L_e, the less the deviation from a constant voltage during the pulse.

10.5 TRAILING-EDGE RESPONSE

At time b in Fig. 10.2, it is assumed that switch S in Fig. 10.9 is opened. The equivalent circuit now reverts to that shown in Fig. 10.11, in which L_e is the OCL and C_D is total capacitance referred to the primary. Figure 10.11 illustrates the decay of the pulse voltage after time b (Fig. 10.2). The equation for the decay is

$$e = \frac{E_a}{m_1 - m_2}[(m_1 + 2\Delta m)\epsilon^{m_1 t} - (m_2 + 2\Delta m)\epsilon^{m_2 t}] \tag{10.4}$$

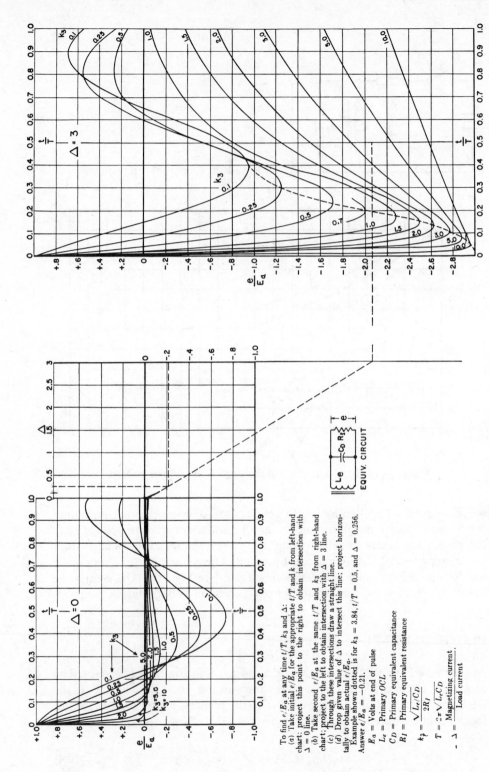

To find e/E_a at any time t/T, k_3 and Δ:

(a) Take initial e/E_a for the appropriate t/T and k from left-hand chart; project this point to the right to obtain intersection with $\Delta = 0$ line.

(b) Take second e/E_a at the same t/T and k_3 from right-hand chart; project to the left to obtain intersection with $\Delta = 3$ line.

(c) Through these intersections draw a straight line.

(d) Drop given value of Δ to intersect this line; project horizontally to obtain actual e/E_a.

Example shown dotted is for $k_3 = 3.84$, $t/T = 0.5$, and $\Delta = 0.256$. Answer $e/E_a = -0.21$.

E_a = Volts at end of pulse

L_e = Primary OCL

C_D = Primary equivalent capacitance

R_1 = Primary equivalent resistance

$$k_3^2 = \frac{\sqrt{L_e/C_D}}{2R_1}$$

$$T = 2\pi\sqrt{L_e C_D}$$

$$-\lambda = \frac{\text{Magnetizing current}}{\text{Load current}}$$

Figure 10.11. Interpolation chart for pulse transformer backswing.

294

where m_1, $m_2 = m(1 \pm \sqrt{1 - 1/k_3})$, $m = R_I C_D/2$, and the other terms are defined in Fig. 10.11. Ratio k_3 on these curves has a different meaning than k_1 in Fig. 10.6. The time constant T is also longer, but with low capacitance k_3 is higher and the pulse decay for higher values of k_3 is shorter. If the capacitance can be kept low, the slope of the trailing edge can be kept within reasonable limits. Accurate knowledge of the transformer and load capacitances is therefore important. Measurement and evaluation of transformer capacitance should be made as in Chapters 5 and 7.

If the transformer has appreciable magnetizing current, the shape of the trailing edge is changed. The greater the magnetizing current, the greater the negative voltage backswing. The ordinates at the left of Fig. 10.11 are given in terms of the voltage E_a at time a. They are plotted as if there is no droop on the top of the pulse. These curves apply when there is droop if the voltage E_a is multiplied by the ratio of the voltage at the end of the pulse to the voltage at the beginning.

The magnetizing current at the end of the pulse is

$$i_M = (E_a/mL)(1 - \epsilon^{-m\tau}) \tag{10.5}$$

where $m = R_1 R_2/(R_1 + R_2)L$ (see Fig. 10.9)

τ = pulse duration, s

L = primary OCL, H

Magnetizing current can be expressed as a fraction Δ of the primary load current $I (\Delta = i_M/I)$. For any R_1/R_2 ratio, $\Delta = [(R_1 + R_2)/R_1]$ times the ratio of the voltage at b (Fig. 10.2) to the voltage at point a, or

$$E'/E_a = 1 - R_1\Delta/(R_1 + R_2) \tag{10.6}$$

where E_a is the voltage at point a (Fig. 10.2), and E' is the voltage at point b.

This equation gives the multiplier for finding the actual voltage at the beginning of the pulse decay found from Fig. 10.11. With increasing values of Δ the backswing is increased. This is particularly true for the highly damped circuits corresponding to values of $k_3 \geqslant 1$. For lower values of k_3, the exciting current has less influence on the oscillatory backswings. These large backswings afford poor reproduction of the original pulse shape. Occasionally, transformers are designed for large backswings for specific applications. Some of these are discussed in Section 11.2.

Equation 10.4 is plotted at the left of Fig. 10.11 for $\Delta = 0$ and to the right for $\Delta = 3$. Instructions for finding the backswing in terms of E_a by interpolation for $0 \ll 3$, and for given values of k_3 and t/T are given below Fig. 10.11. This chart eliminates much of the labor of solving equation 10.4 for a given set of circuit conditions. Elements L_e, C_D, and R_1 in Fig. 10.11 sometimes include circuit components in addition to the transformer values. For linear resistance loads, the terms are interchangeable with L and R_2 of Fig. 10.9, and with C_2 of Fig. 10.5. All values must be referred to the same winding.

In transformers with oscillatory constants the backswing becomes positive again on the first oscillation. In some applications this would appear as an indication of an undesirable pulse. The conditions for no oscillations are all included in the real values of the equivalent-circuit angular frequency, that is, by the inequality

$$1/4R_I^2 C_D^2 > 1/L_e C_D$$

or

$$\sqrt{L_e/C_D} > 2R_1 \tag{10.7}$$

Terms are defined in Fig. 10.11.

The quantity L_e/C_D is considered the open-circuit impedance of the transformer due to the similar use of the expression for the open-circuit impedance of a transmission line. Its value must be twice the load resistance to prevent oscillations after the trailing edge. This requires low distributed capacitance.

The negative backswing may prove objectionable in some applications. Most real loads can be damaged if the negative backswing is excessive. The amount that is excessive must be determined by the actual application. Conditions for avoiding backswing are those in Fig. 10.11 for $k = 5$ and $\Delta = 0$. These require good core material, low exciting current, low distributed capacitance, and a relatively linear transformer load.

10.6 TOTAL RESPONSE

By means of the curves we can now construct the pulse shape delivered to a linear resistive load R_2. A transformer with the following properties is required to deliver a flat top pulse of 15-μs duration.

Primary leakage inductance (secondary short circuited) = 1.89×10^{-4} H

Primary open-circuit inductance   = 0.1 H

Primary/secondary turns ratio N_p/N_s = 1:3

Source resistance R_1 = 800 Ω

Load resistance (primary equivalent) R_2 = 5000 Ω

Primary effective capacitance C_2 = 448 pF

From the expressions given in Fig. 10.6,

$$m = 2.34 \times 10^6$$

$$T = 1.8 \, \mu s$$

$$k_1 = 0.68$$

The leading edge of the wave follows a curve between those marked $k_1 = 0.4$ and $k_1 = 0.8$ in Fig. 10.6. Value E_a is reached on $0.5T$ or $0.85\,\mu s$. An overshoot of about 10% occurs in $1.2\,\mu s$.

The top of the wave slopes down to a voltage determined by the product $R_1/L_e = 0.12$ by a curve that falls between those for $R_2 = \infty$ and $R_2 = 2R_1$ in Fig. 10.10. Voltage E' at b is $0.9E_a$.

The trailing edge is found from Fig. 10.11. Here

$$T = 42.2 \times 10^{-6}$$

$$k_3 = \frac{\sqrt{0.1/448 \times 10^{-12}}}{2 \times 5000} = 1.5$$

$$\Delta = \frac{5800}{800} \times 0.09 = 0.65$$

Load voltage reaches zero in $0.05T$ or $2.11\,\mu s$. The maximum backswing has an amplitude of 33% E' at $0.2T$ or $8.44\,\mu s$ after point b. The pulse delivered to the load E_2 is shown in Fig. 10.12 in terms of E instead of E_a.

So far it has been assumed that the pulse source is disconnected at the end of the pulse. In some applications the source remains connected. This would result if switch S (Fig. 10.9) were left closed and the voltage source E were short circuited. Under these conditions the leakage inductance remains in the circuit and an additional transient occurs. The transient has a shape similar to the one in Fig. 10.7 but is inverted and superposed on the backswing voltage due to OCL. In the example just given, this superposed oscillation has an amplitude of

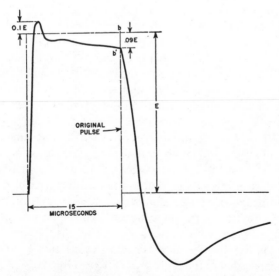

Figure 10.12. Output voltage of pulse transformer.

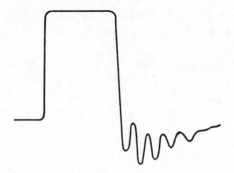

Figure 10.13. Oscillogram of voltage pulse.

10% of E, with a total result similar to the oscillogram of Fig. 10.13. Superposed backswing oscillations are discussed in more detail in Section 11.3. Because of the distributed nature of leakage inductance and capacitance, higher-frequency superposed oscillations may sometimes occur even when the load is disconnected at the end of the pulse. By their nature the conditions for oscillations are difficult to state with certainty. If oscillations appear on the leading edge, they are likely to appear on the trailing edge superposed on the voltage backswing.

10.7   STEP-DOWN TRANSFORMERS

The circuits of Figs. 10.5 and 10.9 for step-up transformers are essentially the same as those for wide-band transformers in Fig. 5.5(e) and (c). Low-frequency response corresponds to the top of the pulse and high-frequency response to the leading edge. In step-down transformers the pulse top is unchanged. The front edge corresponds to Fig. 5.11. Step-down transformer analysis shows that the form of the equation is similar to that for step-up transformers, except that the damping factor for the sine-wave term is greater by the quantity $R_2/L_s\beta$. Also, the decrement has the resistances R_1 and R_2 in the two terms reversed with respect to the corresponding terms for the step-up transformer. Except for this, the leading-edge curves are little different in shape from those of step-up transformers. Where $R_1 = R_2$ the curves are virtually the same as in Fig. 10.7. Constant-current amplifiers can be represented by the circuit of Fig. 10.14. Here I is the current entering the primary winding from the switching device. It is constant over most of the voltage range. The transformer is usually step down for reasons of impedance mentioned in Section 5.9. If the rise in the load current is expressed as a fraction of the final current, and the decrement is changed to $R_2/2L_s$, leading-edge response of these transformers is the same as in Fig. 10.6. It is reproduced in Fig. 10.15 with this change in constants. Flatness of the pulse top is approximately that of the curve $R_2 = 0.1R_1$ in Fig. 10.10. The trailing edge is the same as in Fig. 10.11.

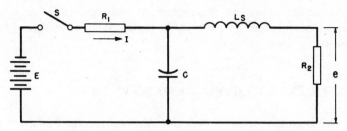

Figure 10.14. Step-down transformer equivalent circuit.

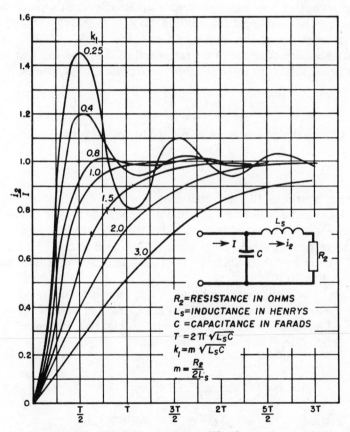

R_2=RESISTANCE IN OHMS
L_s=INDUCTANCE IN HENRYS
C =CAPACITANCE IN FARADS
$T = 2\pi\sqrt{L_sC}$
$k_1 = m\sqrt{L_sC}$
$m = \dfrac{R_2}{2L_s}$

Figure 10.15. Constant-current amplifier front-edge response.

Many practical cases are represented by the figures. If the transformer constants are outside the curve values, the pertinent equation should be plotted to obtain the response.

10.8   FREQUENCY RESPONSE AND WAVE SHAPE

It has been stated that there is a relationship between the frequency and time domains. The two concepts will now be correlated. The matter of phase shift enters into the concept. The relative phase of the different frequency components affects the wave shape. It is sometimes convenient to know whether a transformer whose frequency response is known can deliver a given wave shape. Starting with the low-frequency response, assume equal source and load impedances. Then the upper curve of Fig. 5.8 applies. This curve shows 90% of maximum response at the frequency for which $X_n/R_1 = 1$. How does this frequency response compare to the reciprocal of the pulse width at the end of which there is a 10% droop in the top of the pulse? X_N/R_1 can be written

$$2\pi f L/R_1 = 1 \quad \text{or} \quad f = R_1/2\pi L \tag{10.8}$$

Similarly, from the proper curve of Fig. 10.10, for 10% droop,

$$\tau R_1/L = 0.2 \tag{10.9}$$

Combining equations 10.8 and 10.9 gives $f = 0.0318(1/\tau)$, or the transformer should not be more than 1 dB down at a frequency about 1/30 of the reciprocal of the pulse width. For example, if a maximum of 10% droop is desired for a 2-μs pulse, the response should be not more than -1 dB at $0.0318 \times 0.5 \times 10^6 = 16$ kHz. Maximum phase shift is 27° (from Fig. 6.3), but this is taken into account in Fig. 10.10.

Similarly, the leading-edge rise time can be related to the transformer high-frequency response. This is found in Fig. 5.9 for the case of $R_1 = R_2$. The corresponding leading-edge curves are found in Fig. 10.7. Parameter k_1 of these curves is related to B in Fig. 5.8 as follows.

$$k_1 = m\sqrt{LC} = \frac{R_1\sqrt{LC}}{2L} + \frac{\sqrt{LC}}{2R_1 C} \quad \text{(for } R_1 = R_2)$$

$$B = X_C/R_1 = X_L/R_1 \quad \text{at frequency } f_r$$

$$= 2\pi f_r L/R_1 = L/R_1\sqrt{LC}$$

$$k_1 = 1/2B + B/2 \tag{10.10}$$

From equation 10.10 we can prepare Table 10.1.

If a transformer has a frequency response according to the curve for $B = 0.5$,

**TABLE 10.1   Parameters for
Frequency Response and
Wave Shape**

B	k_1
1.0	1.0
0.8, 1.25	1.025
0.67, 1.5	1.08
0.5, 2	1.25
0.25, 4	2.125

2 in Fig. 5.9, its leading edge will rise somewhere between the curves for $k_1 = 1$ and $k_1 = 1.4$ in Fig. 10.7.

Transformer OCL, leakage inductance, and effective capacitance must be known to make the comparison. These quantities must be known if the frequency response is given by Fig. 5.8 or 5.9, or the wave shape by Figs. 10.7 and 10.10. If conditions other than $R_1 = R_2$ exist, another set of response curves can be used. The corresponding relations can be found in the manner outlined.

Pulse transformer windings are similar to those in high-frequency transformers described in Section 7.1. Resonant frequency f_r is determined largely by leakage inductance and winding to winding capacitance. With pulse operation, the fast-rising voltage at the leading edge of the pulse impressed on the transformer may excite partial resonances in sections of the coil, or even turn to turn. If these resonances cause large oscillations in the output wave of the transformer, means must be taken to change these resonances so that they are outside the band of frequencies that could be excited by the rising voltage.

10.9   CORE MATERIAL

In Chapter 7 it was shown that effective core permeability in laminated cores decreases with increasing frequency. This decrease also occurs with short pulse widths. When a pulse is first applied to the transformer, there is initially very little penetration of the flux into the core laminations because of the field set up by the eddy currents. Only a fraction of the core is effective at this time. The apparent permeability is less than it is later in the pulse after the flux density becomes more uniform in the core.

A typical *B-H* curve for pulse transformers is shown in Fig. 10.16. Flux density builds up in the core in the direction shown by the arrows. For a typical loop such as *obcd*, the slope of the loop rises gradually to the end of the pulse *b*, which corresponds to point *b'* in Fig. 10.12. Since magnetizing current starts decreasing at this point, *H* also starts decreasing. Current in the windings does not decay to zero immediately but persists due to winding capacitance, and

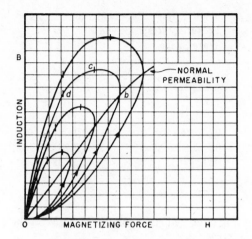

Figure 10.16. Pulse *B-H* loop.

sufficient time elapses for the permeability to increase. Flux density *B* may also increase for a short time after point *b*. The trailing edge of the pulse voltage soon reaches zero. This corresponds to point *c* in the loop. At some time later, the maximum backswing amplitude is reached. This corresponds to point *d* in the loop. At this time the slope or permeability is several times as great as it is at point *b*.

For any number of pulses of varying amplitudes but of the same pulse width, there are corresponding loops having respective amplitudes *c*. A curve drawn through point *b* of each loop is called the normal permeability curve. This is ordinarily given as the permeability curve for the material. The permeability μ for a short pulse width is less than the 60 Hz or dc permeability for the same material. Values of pulse permeability for 2-mil grain-oriented silicon steel are given in Fig. 10.17. The permeability values include the irreducible small gap that exists in type C cores. The cores on which the measurements were made had a ratio of gap to core length $l_g/l_c \approx 0.0003$, but the data are not critically dependent on this ratio. The effect of penetration time is clear.

Flux densities attained in pulse transformers may be low for small transformers where very little source power is available, or they may be high in high-power units. This is true whether the pulse width is a few nanoseconds or hundreds of microseconds.

The nickel-iron alloys in general have lower saturation flux densities than those of the silicon steels, but higher permeabilities below saturation. Depending on the flux density chosen, the increase in permeability with the use of nickel-iron alloy may vary from nothing to greater than 300%. This increase holds also for long time pulses, during which the permeability may approach the 60-Hz value.

To overcome the net dc pulse magnetization which is in the same direction throughout each pulse, an air gap may be inserted in the core to prevent it from

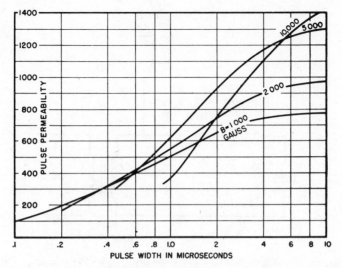

Figure 10.17. Effective permeability versus pulse width.

returning only to the residual magnetism value at the end of each pulse. This would limit the useful flux density range to the difference between the maximum flux density and residual B_r (see Fig. 10.18). This gap increases the effective length of the magnetic path and reduces OCL from the value it has with symmetrical magnetization. The reduction is less with materials of low permeability. To maintain the advantage of high permeability in nickel-iron alloys, the

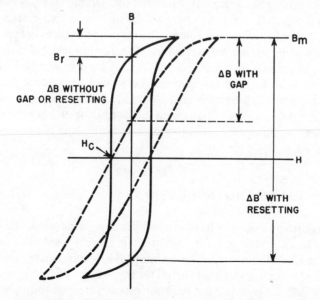

Figure 10.18. Flux density range in pulse transformer cores.

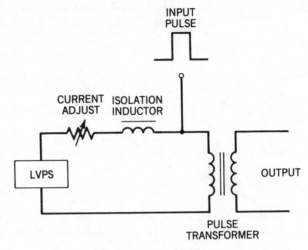

Figure 10.19. Pulse transformer reset circuit.

core is "reset." This is done by arranging the circuit so that sufficient negative current flows through the windings during the pulse "off" time to overcome coercive force H_c and set the flux density to the negative value of residual magnetism. Then nearly twice the previous maximum flux density ($\Delta B'$ in Fig. 10.18) is available for the pulse. Where resetting is possible, it is advantageous to use nickel-iron cores, although silicon iron or Supermendur cores have been used. Where resetting is not practicable, grain-oriented silicon steel or Supermendur is preferable. Figure 10.19 shows one circuit that has been used for resetting in order to improve the performance of a pulse transformer by reducing the number of turns required. The reduction in turns decreased leakage inductance and improved pulse rise time. Where very fast pulse rise times are required, a reset, high-permeability nickel-iron core in toroidal form can be used. The toroidal form of the core can be wound to produce extremely low leakage inductance (theoretically zero). With the transformer time constant low, the high-permeability core appears to be a resistor to the pulse rise time. Transformer rise times for reasonable powers in the tens of nanoseconds have been achieved.

10.10   WINDINGS AND INSULATION

Pulse transformers generally have single-layer concentric windings. In low- and medium-voltage windings, solid insulation is usually used. For high-voltage windings, solid insulation may be used but provisions are often made to support the windings in such a way that the major insulation is liquid or gas. For high load impedance, a single section each for primary and secondary may be used (Fig. 7.9). This type of winding has the lowest effective capacitance. For low load impedance, more interleaving is used to reduce the leakage inductance. To reduce capacitance to an absolute minimum, pie-section coaxial windings may be used. In these the coil capacitances are kept low by the use of universal windings. Interwinding capacitances are low because of the unity dielectric constant of the air between the sections. Such coils are more difficult to wind, require more space, and have higher leakage inductance. Such windings are used only where simpler types of windings will not meet the requirements of the transformer.

Coil sections can be wound with the same polarity as in Fig. 7.9, or with one winding reversed. Effective capacitance between *P* and *S* is given below for three turns ratios. Capacitance is based on 100-pF measurable capacitance.

Turns Ratio, N_1/N_2	Effective Capacitance Referred to Primary	
	Same Polarity	Reversed Polarity
1:5	533	1200
1:1	0	133
5:1	21	48

From this it can be seen that the polarity exemplified in Fig. 7.9 is preferable for reducing capacitance. The percentage difference is greatest for turns ratios near unity and decreases as the ratio increases.

Attention to insulation has so far centered around capacitance. The insulation must also withstand the voltage stress to which it is subjected. It can be graded to reduce the space required. Low-frequency practice is adequate for both insulation thickness and end-turn clearances. The subject of high-voltage insulation is discussed more fully in Chapter 9.

Small size is achieved by the use of high-quality insulation. The voltage strength and corona resistance of the insulation can be the primary factors in the size of a high-voltage transformer. High-quality insulation is relatively expensive, but the savings in core material costs usually more than offset the additional cost of insulation. This is particularly true if nickel-iron or Supermendur cores are used.

To utilize space as efficiently as possible, core-type construction is often used. Low capacitance between high-voltage coils is possible in such designs. The secondary windings are often split into unsymmetrical sections. There is greater leakage inductance in unsymmetrical windings, but there is less capacitance when the highest-voltage section of the winding has the shorter length. Lower capacitance is obtained with two coils than with a shell-type transformer of the same interleaving. The high-voltage windings in core-type transformers are usually the outer windings. Terminals or leads are best located in the coil or directly over the windings to maintain the margins. Insulating barriers may be located at the ends of the windings to reduce creepage. In liquid- or gas-filled high-voltage transformers, the mounting for the outer coils is preferably located at the ends of the coil form outside the high-voltage fields. Figure 10.20 shows example of all of the design techniques mentioned above. This arrangement permits the fullest utilization of the high electric strength of the insulating medium by increasing the creepage path breakdown to a value far above the breakdown strength of the liquid or gas.

Autotransformers, when isolation between primary and secondary is not needed, afford an opportunity for space saving. There are fewer total turns and therefore less winding space is needed. There is less leakage inductance, but not necessarily less distributed capacitance. The relative values of capacitance always depend on the voltage gradients.

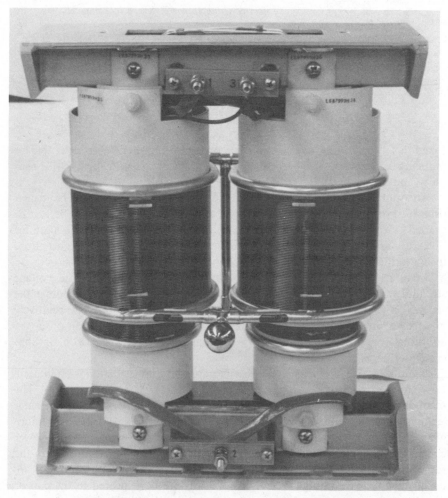

Figure 10.20. High-voltage, high-power pulse transformer.

Initial distribution of voltage at the leading edge of the pulse is not uniform because of the turn-to-turn and winding capacitance. In a single-layer coil the total turn-to-turn capacitance is small compared to the winding to ground capacitance. The turn-to-turn capacitances add in series. The winding to ground capacitances add in parallel. When a fast-rising voltage pulse is applied across the winding, current flows from the first few turns to ground. This leaves less voltage and current for the remaining turns. Initially, most of the pulse voltage appears across the first few turns.

After a short time, some of the current flows into the remaining turns inductively. Before long the capacitive voltage disappears. All of the current flows through all of the turns. The voltage becomes uniformly distributed throughout the winding. This condition applies during most of the top of the pulse. Between the initial and final current distribution, oscillations due to leakage inductance and winding capacitance may be present. There can be many different frequencies present. They may appear at the terminal of the transformer or they may only be present internal to the transformer, particularly when there is a discontinuity such as at a place where the coil is split into sections. These oscillations may cause the high turn-to-turn initial voltage stress to distribute to other turns in the winding.

Winding capacitance to ground is evenly distributed along the winding of a single-layer coil, and so is the turn-to-turn capacitance. If a rectangular pulse E is applied to one end of such a winding and the other end is grounded, the maximum initial voltage gradient is

$$(\alpha E/N)\coth \alpha$$

where N = number of turns in winding
$$\alpha = \sqrt{C_g/C_w}$$
C_g = capacitance of winding to ground
C_w = capacitance across winding

For the development of this expression, see BEAIRA (1941).

Practical values of α are large, and $\coth \alpha$ approaches unity. Then

$$\text{maximum gradient} \approx \alpha E/N \qquad (10.11)$$

or the maximum initial voltage per turn is approximately α times the final or average volts per turn.

If the other end of the winding is open instead of grounded, equation 10.11 still governs. This means that the maximum gradient is independent of load. If there is a winding N_1 between the pulsed windings N_2 and ground, α depends on C_{1-2} and C_1 in series. The initial voltage in winding N_1 is

$$E_1 = \frac{EC_{1-2}}{C_{1-2} + C_1} \qquad (10.12)$$

where E_1 = initial voltage in winding N_1
       E = pulse voltage applied across N_2
C_{1-2} = capacitance between N_1 and N_2
    C_1 = capacitance between N_1 and core

Thus the initial voltage in winding N_1 is independent of the transformer turns ratio. It is higher than the voltage which would appear in N_2 if N_1 were pulsed, because then current would flow from N_1 to ground without any intervening winding. If winding N_1 is the low-voltage winding, applying pulses to it stresses turn insulation less than if N_2 is pulsed.

Reinforcing the end turns of a pulsed winding to better withstand the pulse voltages is of doubtful value, because the additional insulation increases α and the initial gradient in the end turns. Increasing the insulation throughout the winding is more beneficial, for although α is increased, the remaining turns can withstand the oscillations better as the inductance becomes more effective. Decreasing the winding-to-core capacitance is better, for then α decreases and the initial voltage gradient is more uniform.

10.11   EFFICIENCY

Circuit efficiency should be distinguished from transformer efficiency. Magnetization current represents a loss in efficiency in the transformer, but the energy may be returned to the circuit after the pulse. Circuit efficiency may be estimated by comparing the area of the actual wave shape across the load to that impressed on the primary of the transformer. It includes the loss in the input impedance R_1 (Fig. 10.9). Except for this loss, the circuit and transformer efficiency are the same when the source is cut off at the end of the pulse. It is important in testing for losses to use the proper circuit.

Core loss can be expressed in watt-seconds per pound per pulse. The core loss in small transformers can be measured by using a calorimeter. The transformer is placed in the calorimeter, and the necessary connections are made through feed-through insulators. Dielectric loss is included in such a measurement. It is appreciable only in high-voltage transformers using solid insulation between windings. It may be separated from the iron loss by first measuring the loss of the complete transformer and then repeating the measurement with the high-voltage winding removed. At 6 kG and 2-μs pulse width, the loss for 2-mil grain-oriented silicon steel is approximately 6000 W/lb or 0.01 W-s/lb per pulse. For square pulses, core loss varies (1) as B^2 or E^2 for constant pulse width, and (2) as pulse width for constant voltage and duty τf, where τ is the pulse width and f is the pulse repetition rate. Dielectric loss is independent of pulse width and varies (1) as the repetition rate for constant voltage, and (2) as E^2 for a constant repetition rate.

Copper loss is often negligible because of the comparatively few turns required for a given rating. This is true if the wire size is chosen for a circular mil area comparable to the area which would be used for the same rms current in a power transformer. If the windings are used to carry other currents, such as filament current or reset current, the copper loss may be appreciable. This loss is not included in the output/input loss, but must be included when calculating the temperature rise of the transformer.

Efficiencies over 90% are common in pulse transformers, and with high-permeability materials, efficiencies over 95% have been obtained. These figures are for high-power pulse transformers. As with power transformers, maximum efficiencies occur when the losses are equally distributed among the iron, copper, and dielectric. This equality is very seldom achievable in modern fast-rise-pulse transformers.

10.12   NONLINEAR LOADS

The role played by leakage inductance and distributed capacitance in determining the pulse shape was discussed in Sections 10.3 and 10.5. It has been shown that the first effect is a more or less gradual slope on the leading edge of the pulse. The second effect consists of the oscillations superposed on the voltage backswing after the pulse has decayed to zero voltage. Consider the additional influence of nonlinear loads on the leading edge of the pulse.

Figure 10.6 is based on the following assumptions:

1. Load and source impedances are linear.
2. Leakage inductance is "lumped."
3. Distributed capacitance is "lumped."

Assumptions 2 and 3 are approximately justified. Pulses effectively cause the windings to operate beyond natural resonance. This is similar to the higher-frequency operation of coils described in Section 7.10. The distributed nature of the capacitance and leakage inductance, as well as partial coil resonance, may cause superposed oscillations that require correction. The general outline of the pulse shape is still caused by the low-frequency leakage inductance and capacitance.

Assumption (1) may be a source of serious error. Most load impedances for pulse transformers are nonlinear. Examples of nonlinear loads are traveling-wave tubes, klystrons, crossed-field amplifiers, magnetrons, and bipolar transistor bases driven by pulse transformers. In a nonlinear load with a constant voltage applied, the problem is chiefly one of current pulse shape. An example of this is given assuming a highly nonlinear load such as a magnetron or crossed-field amplifier. First assume that no current flows into this load for such a time as it takes to reach steady voltage E. During this first interval, the transformer is therefore oscillatory. After voltage E is reached, the current rises rapidly at first and then more slowly, as determined by the new load R_2. The sudden application of this load at voltage E damps out the oscillations that would exist without this load and furnishes two conditions for finding the initial current. A rigorous solution of this problem by analytical methods is complicated and time consuming. If we know the characteristics of the load and the OCL, leakage inductance, and distributed capacitance, the differential loop equations can be solved readily on a digital computer (see Carter and Rambo, 1967).

The problem can be simplified by making the assumption that the voltage E is constant. When the pulse voltage reaches E, capacitance C_2 no longer draws current. At the instant t_r (Fig. 10.21) when voltage E is first reached, the current in L_s which was drawn by capacitance C_2 flows immediately into the load. Since the voltage was rising rapidly at instant t_r, the energy that would have resulted in the first positive voltage half-cycle (shown shaded in Fig. 10.21) must be dissipated in the load. The remaining oscillations are also damped. Prior to the time t_r, all current through L_s flowed into C_2. The value of this current is $C_2 de/dt$. We may therefore find the slope of the appropriate leading-edge voltage curve and multiply it by the transformer capacitance to obtain the initial current. An unloaded transformer leading edge means a small k_1 in Figs. 10.6 and 10.7. The leading-edge slope at voltage E is given in Fig. 10.22. The ordinates of this curve are $(T\, de/E)/dt$, with E corresponding to the E_a of Fig. 10.6. Ordinates of Fig. 10.22 are multiplied by $C_2 E/T$ to find the initial current.

Few nonlinear loads have absolutely zero current up to the time the voltage E is reached. The foregoing solutions are thus approximate. In spite of this, the

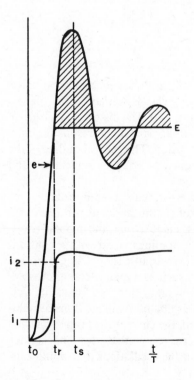

Figure 10.21. Voltage and current waveshapes for magnetron or CFA loads.

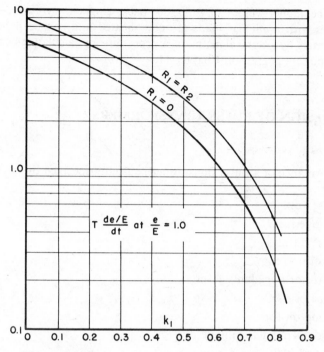

Figure 10.22. Front-edge slope of pulse transformers.

procedure gives fair accuracy. A summary of the procedure:

1. Find the initial capacitive current as just outlined.
2. Estimate the current at which the load e-i curve departs from a straight line (i_1 in Fig. 10.21).
3. Add currents 1 and 2. This gives i_2 (Fig. 10.21) as the total initial current.

Pulse current continues to rise beyond the value i_2 if the initial current value is less than the final operating current corresponding to voltage E. It will droop if the initial current is higher than the load current at voltage E. To obtain constant current over the greater part of the pulse width, i_2 should equal the load current at voltage E. When this equality does not exist, the rate of rise or droop is determined by transformer leakage inductance, source impedance, and load impedance. Magnetrons depend on the rate of rise of voltage to oscillate in the proper mode. The initial current may drop to nearly zero before the main current pulse starts. When there is negligible initial current i_2, the condition for a good current pulse is $E/i_2 \approx \sqrt{L_s/C_2}$, where L_s is the leakage inductance.

At the end of the pulse, when the voltage source is reduced to zero (point b, Fig. 10.2), the circuit reverts to that shown in Fig. 10.11, but the transformer loses all the load except its own losses. Since by this time it has drawn exciting current, the higher values of Δ in the backswing curves apply. Backswing amplitudes with nonlinear loads are complicated and can be predicted only approximately by graphical methods. A procedure for calculating backswing for line-type pulsers is given in Chapter 11.

10.13   DESIGN OF PULSE TRANSFORMERS

10.13.1   Requirements

The performance of a pulse transformer is usually specified by the following:

1. Pulse voltage
2. Voltage ratio
3. Pulse duration
4. Repetition rate
5. Power or impedance level
6. Slope of rise
7. Droop on top
8. Amount of backswing permissible
9. Type of load

Design data for ensuring that these requirements are met are provided in the preceding sections in several sets of curves. The steps for applying these curves to a design will now be demonstrated.

10.13.2   Start of Design

The first step in a design is choosing the core. It is helpful if some previous designs exists, but a first try at a core can be taken by following a method similar to the one used for selecting a core for a wide-band transformer (Section 5.10).

After choosing a core, the designer must next calculate the number of turns in the various windings. In pulse transformers operating at high voltages, the limiting factor is usually the operating flux density, but it may be the required voltage clearances in lower-average-power medium-voltage transformers with short pulse widths. For unidirectional pulses, the number of turns may be derived from

$$e = \frac{N\,d\phi}{dt} \times 10^{-8} = NA_c \frac{dB}{dt} \times 10^{-8}$$

$$\int e\,dt = NA_c \int dB \times 10^{-8} \tag{10.13}$$

For a square wave $e = E$ and

$$E = NA_cB \times 10^{-8}$$

or

$$N = \frac{E\tau \times 10^8}{6.45BA_c} \qquad (10.14)$$

where E = pulse voltage
τ = pulse duration, s
B = allowable flux density, G
A_c = core section, in.^{2}
N = number of turns

In many designs, the amount of droop or the backswing that can be tolerated at the end of the pulse determines the number of turns, because of their relation to the OCL of the transformer. After the turns are determined, appropriate winding interleaving should be estimated and the leakage inductance and capacitance calculated.

With the leakage inductance and winding capacitance estimated, the rise time for linear loads can be found from Fig. 10.6 or 10.7. Similarly, from the OCL and distributed capacitance, the shapes of the top of the pulse and trailing edge are found from Figs. 10.10 and 10.11. If performance from these curves is satisfactory and the coil fits the core, the design is completed.

10.13.3 Final Calculations

Preliminary calculations may show too much slope on the leading edge of the pulse, too slow a fall time, too much backswing, or that the transformer may not meet some of the other requirements. This is not an unusual situation on the first try. If the leading-edge response is too slow, it is necessary to make adjustments. Two damping factors $R_1/2L_s$ and $1/2R_2C_2$ contribute to the leading-edge rise time. Preliminary calculations will show which is preponderant. Sometimes it is possible to increase leakage inductance or capacitance without increasing time constant T greatly.

If the leading-edge response is still too fast or too slow, it will probably be necessary to use a different core. The smallest core that will provide sufficient space for the necessary insulation and windings is the best core to use. Small core area A_c may require too many turns to fit the core.

If the calculated leading-edge slope is approximately correct, it may be adjusted by the following means:

1. Change number of turns.
2. Reduce core size.
3. Change interleaving.
4. Increase insulation thickness.
5. Reduce insulation dielectric constant.

High capacitance is a common cause of poor performance. One or more of items 2 to 5 may often be changed to reduce capacitance. It is sometimes possible to rearrange the circuit in such a way that the performance of the transformer is enhanced. One illustration of this is the termination of a transmission line. Line termination resistance may be placed on either the primary or secondary side. If it is placed on the primary side, the leading edge of the pulse is improved. Figure 10.7 does not show the improvement since it was plotted for Fig. 10.5. For resistance on the primary side, the damping factor reduces to the single term

$$\alpha = \frac{R_1 R_2}{2(R_1 + R_2)L_s} \tag{10.15}$$

Improvement of the trailing edge usually follows improvement to the leading edge.

The leading- and trailing-edge performance of transformers operating into nonlinear loads may be quite different from that calculated from Figs. 10.6 and 10.11. A nonlinear load may look like a very high impedance at the beginning of the voltage rise. This can cause the voltage to rise more rapidly than predicted. If the load is nonlinear, it is best to set up a computer program to see what the transformer rise times and fall times are with the nonlinear load.

Core permeability is important because fewer turns are required on a high-permeability core to obtain the required OCL. If a large gap is used, the permeability becomes less important. If a high-permeability core is used, a reset winding should be used rather than a core gap. Permeability at the beginning of the trailing edge (point b', Fig. 10.12) is important for two reasons: the amount of droop at this point depends on the OCL. For a given droop the turns on the core are fixed. Second, the normal permeability data apply to b'. Flux density is chosen for two reasons: It should be as high as possible for small size, but not so high as to result in excessive magnetizing current and backswing voltage.

Example. Assume the following performance requirements:

Pulse voltage ratio = 11,700 : 117,000 V

Pulse duration = 10 μs at 70% voltage points

Pulse repetition rate = 250 pps

Impedance ratio = 15 : 1500 Ω (linear)

To rise to 90% of final voltage in 0.500 μs or less

Droop not to exceed 5% in 10 μs

Backswing amplitude not to exceed 20% of pulse voltage

15-Ω source

Minimum size

Major insulation 30 psig of sulfur hexafluoride

The calculated performance parameters are:

Peak power = 9.6 MW
Secondary rms current = 4.2 A
Primary rms current = 420 A

A reset core will be used to reduce size. The reset circuit will be as shown in Fig. 10.19. The final design has the following characteristics:

Primary turns = 20
Secondary turns = 200
Core: 2-mil silicon steel with 1-mil gap per leg
Core area = 3.52 in.^{2}
Core length = 37.42 in.
Core weight 39.0 lb, window 4.62 in. × 11.0 in.
Primary leakage inductance = 7 μH
Effective primary capacitance = 3000 pF
Flux density = 25,800 G

We are required to achieve 5% droop from $R_1 = R_2$ in Fig. 10.10. The required total primary inductance is 1.5 mH. If we assume that the isolation inductor has a minimum inductance of 12 mH, the primary OCL must be 1.77 mH. There is no load loss equivalent to 139 Ω (referred to primary):

$$\text{flux density} = \frac{11,700 \times 10 \times 10^2}{6.45 \times 3.51 \times 20} = 25,840 \text{ G}$$

At 12,900 G and 10 μs, $\mu_\Delta = 1800$:

$$\text{primary OCL} = \frac{3.2 \times 400 \times 3.51 \times 10^{-8}}{0.002 + 37.42/1800} = 1.97 \text{ mH}$$

This paralleled by 10 mH in the reset isolation inductor gives an effective primary inductance of 1.65 mH.

The front-edge response is determined as follows:

$$m = \frac{15 \times 10^6}{2 \times 7.5} + \frac{10^6}{0.03 \times 2.9} = 1.218 \times 10^7$$

$$T = 2 \times \pi \times \sqrt{7.5 \times 0.0029 \times 10^{-6}} = 0.911 \, \mu s$$

$$k_1 = 12.18 \times \sqrt{7.5 \times 0.0029} = 1.796$$

To rise from 10% to 90% of voltage on $k_1 = 1.8$ requires $0.55 \times T = 0.501 \, \mu s$. The top slope is

$$\frac{\tau R_1}{L_e} = \frac{10 \times 15 \times 10^{-6}}{1.65 \times 10^{-3}} = 0.091$$

and from the curve $R_1 = R_2$ in Fig. 10.10 the top droop = 4.5%
    The magnetizing current is

$$(R_1 + R_2)/R_1 \times 4.5\% \text{ or } 9\% \text{ of the load current}$$

For backswing,

$$m = \frac{10^6}{0.03 \times 3} = 1.11 \times 10^7$$

$$T = 2\pi \times \sqrt{1.65 \times 3} \times 10^{-6} = 14 \, \mu s$$

$$k_3 = \frac{1.65/3 \times 10^3}{30} = 24.7$$

This is much larger than the largest value on the curve. If we use $k_3 = 10$, it can be seen that the fall time is less than $0.50 \, \mu s$ and the backswing would be about 7%. The larger k_3 would give a larger backswing, but it would still be less than the 20% specified.
    If the load were nonlinear, the voltage would rise at a different rate:

$$m = \frac{15 \times 10^6}{2 \times 7} + \frac{10^6}{0.178 \times 3} = 3.3 \times 10^6$$

$$k_1 = 3.3\sqrt{7 \times 0.003} = 0.478$$

The voltage would then begin to rise to 90% final voltage in $0.15 \times T$ or $0.137 \, \mu s$. The actual rise time will be between the two calculated values. The backswing under these conditions will be as calculated in Chapter 11.
    A section through the transformer is shown in Fig. 10.23. The primary winding is triple-enameled copper strap. The secondaries are wound with AWG No. 16 triple-enameled wire. The average voltage stress is 585 V per turn.
    The core loss is about 2500 W. Data are not available for accurately calculating the core loss, but it can be estimated by extending the core-loss curves for the equivalent ac frequency and flux density. Multiplying this number by the duty cycle will give a good starting point for calculating core loss. Actual losses are best determined by measurement. Figure 10.20 is a photograph of the transformer.

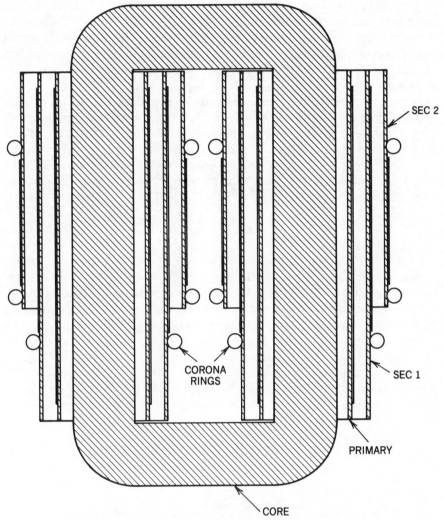

Figure 10.23. Section of pulse transformer of Fig. 10.20.

10.14 TESTING

Tests for open circuits and short circuits, inductance, turns ratio, and dc resistance are made on pulse transformers in the same manner as in other transformers. The equipment used must be suitable for the low values of inductance encountered, but otherwise no special precautions are necessary. The dc resistance is usually low, so that test equipment for very low resistance is necessary.

Various methods have been used to check effective pulse OCL. These may involve substitution of known inductances or the current buildup or decay may be measured. The OCL can be calculated from the time constant of the transformer inductance and an external known resistance. When such measurements are attempted under pulse conditions, there is usually a certain amount of error due to reflections, incidental capacitance, and the like. A method involving the measurement of pulse permeability and calculation by the OCL equation is given here.

If the air gap and pulse permeability are known, the OCL for a given core area and number of turns can be calculated. If the gap used is purposely made large to reduce saturation, proper allowance for it can be made in equation 5.13. If the gap is the minimum obtainable, it is necessarily included in the permeability measurement. This is often done in taking pulse permeability data, as it was in the data of Fig. 10.17. With this definition of permeability equation 5.13 reduces to

$$OCL = (3.2N^2 A\mu \times 10^{-8})/l_c \qquad (10.16)$$

Equation 10.16 is valid only when $l_c \gg l_g$.

B-H data for a pulse transformer are taken by means of a circuit similar to that of Fig. 10.24. The primary circuit gives a horizontal deflection on the oscilloscope proportional to I and therefore H for a given core. The current can be measured either by placing a small resistor in series with the primary or a calibrated current probe. If a resistor is used, it should be low enough in ohmic value not to influence the magnetizing current waveform appreciably. If the voltage drop across a high-resistance load R_2 (≈ 50 times normal pulse load) is almost the entire secondary voltage e_2, then voltage e_c applied to the vertical plates is the time integral of e_2 and is therefore proportional to flux density at any instant (see equation 10.13).

Properly terminated and compensated probes must be used to obtain accurate measurements. Distributed capacitance of the winding, shown dashed,

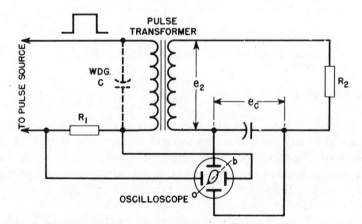

Figure 10.24. *B-H* test for pulse transformers.

should be minimized, as it introduces extraneous current into the measurement of H. One way to minimize this capacitance is to omit the high-voltage winding and make the measurement, using the low-voltage winding only. The pulse source must have enough power to drive the core to the required flux density.

With a calibrated oscilloscope it is possible to determine the slope of the dashed line ob in Fig. 10.24, drawn between the origin and the end of the pulse, and representing effective permeability μ at instant b (Fig. 10.2). This value of μ can be inserted in equation 10.16 to find OCL. Cores failing to meet the OCL should first be examined for excessive air gap.

Effective values of leakage inductance and capacitance are difficult to measure. The calculations of capacitance and leakage inductance are based on the assumption of "lumped" values, the validity of which can be checked by observing the oscillations in an unloaded transformer when pulse voltage is applied. The frequency and amplitude of these oscillations should agree with those calculated from the leakage inductance and effective capacitance. The pulse source should be chosen for the squareness of its output pulse. Because of the light load, the transformer will usually be oscillatory and produce a secondary pulse shape of the kind shown in Fig. 10.25. In this figure the dot-dash line is that of the impressed pulse and the solid curve is the resulting transformer output voltage.

The first check of leakage L_s and C_2 is made by finding the time constant T from

$$T = 2\pi\sqrt{L_s C_2}$$

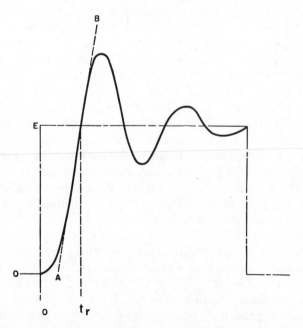

Figure 10.25. Finding transformer constants from pulse shape.

This time constant can be related to the time interval t_0-t_r in Fig. 10.25 by consulting Fig. 10.6. Equations in this figure can be used for finding values of parameter k_1 using L_s, C_2, source resistance R_1, and load resistance R_2. This load resistance will be that corresponding to transformer losses only; hence $R_2 \gg R_1$ for a pulse source with plenty of power, and

$$k_1 = \sqrt{L_s/C_2}/2R_2$$

With this value of k_1, the overshoot of the first voltage oscillation over the flat top value E may be found from Fig. 10.6 and may be compared with that observed in the test. When the load is resistive or when the voltage pulse is the criterion of pulse shape, these are the only checks that need to be made on leakage inductance and distributed capacitance.

When the load is a magnetron, triode, blocking oscillator, grid circuit, or other nonlinear load, the shape of the current pulse is important. Ordinarily, the current will not build up appreciably before time t_r in Fig. 10.25. The shape of this current pulse and sometimes the operation of the load are determined to a large extent by slope AB of the no-load voltage at time t_r. This time is the instant when the first oscillation crosses the horizontal line E in Fig. 10.25. As indicated in Fig. 10.21, there is a relationship between this slope and the parameter k_1. If the slope AB is confirmed, the correct current pulse shape is also assured.

Insulation can be tested in one of two ways, depending on whether the insulation and margins are the same throughout the winding or whether the insulation is graded to suit the voltage. In the former case an equivalent 60-Hz peak voltage, applied from winding to ground at the regular 60-Hz insulation level, is sufficient. But if the winding is graded, this cannot be done because the voltage must be applied across the winding and there is not sufficient OCL to support low-frequency induced voltage; hence a pulse voltage of greater-than-normal magnitude must be applied across the winding. Adequate margins support a voltage of the order of twice normal without insulation failure.

Such pulse testing also stresses the windings as in regular operation, including the nonuniform distribution of voltage gradient throughout the winding. The higher-voltage pulse test is done at a shorter pulse width so that the core does not saturate. If the core does saturate, the voltage backswing may exceed the pulse voltage of normal polarity and thus subject the insulation to an excessive test voltage. This backswing may be purposely used to obtain higher voltage than the equipment can provide, but it must be carefully controlled. Corona tests are sometimes used in place of insulation tests. Where the insulation is not graded, this can be done by using a 60-Hz voltage in a standard corona test set.

Often the function of the transformer is to invert the pulse for each stage; that is, the transformer changes it from a negative pulse at the output of one stage to a positive pulse at the input of the next. Polarity is therefore important and should be checked during the turns ratio test. If the transformer fails to deliver the proper shape of pulse, it may be deficient in one of the properties for which tests are mentioned above.

11 PULSE CIRCUITS

Pulses of energy are used for many applications. The energy required in these pulses may be as small as that required to clock a computer (microjoules) to the energy required by some of the high-energy lasers (megajoules). Low-energy pulses are usually low voltage and can be formed directly at the voltage and power level required. These pulses are usually directly coupled into the load and use no transformers. At somewhat higher powers, these pulses may be generated by circuits that use a transformer in the circuit which generates the pulses. It is also necessary in some circuits to couple pulses from one impedance level to another.

In the high-power pulse circuits used in radar and lasers, the pulse can very seldom be generated at the voltage level at which it will be used. High-power microwave tubes may require voltages well in excess of 100 kV. Unless these tubes are made with grids so that the collector current can be turned on and off by pulsing the grid, the full cathode voltage and current must be pulsed. Circuits to pulse this power can rarely operate at voltages above 25 kV, so the voltage must be stepped up to be used.

Even when the power levels are low, transformer coupling is often used to isolate low-voltage from high-voltage circuitry. One type of pulse circuit that is usually transformer coupled is the circuit for generating very fast rising trigger pulses. This type of circuit has many applications in modern equipment.

In Chapter 10 the basic requirements for designing transformers for pulse use were discussed. The following sections discuss some of the various applications for pulse transformers. One application for pulse transformers is in the various types of power inverter circuits that are in use today. The discussion of the design and application of these transformers is left for Chapters 12 and 13.

11.1   BLOCKING OSCILLATORS

Blocking oscillators are often used to generate pulses at certain repetition rates. The pulse may be used to drive a pulse amplifier as a trigger pulse or it may be used in other applications where the pulse shape is not highly critical. Powers of a few watts average or tens of watts peak may be generated.

A typical blocking oscillator circuit is shown in Fig. 11.1. The input current is usually comparable in magnitude to output current. Input and output winding turns are approximately equal. The operation of a bipolar transistor in a blocking oscillator is described in the following paragraphs. The operation of vacuum tubes or FETs is similar.

If the base is only slightly positive, the transistor draws collector current and because of the large number of base turns the transformer drives the base positive, increases the collector current, and starts a regenerative action. During this period, the base draws more current, charging the bias capacitor to a voltage depending on the base current flowing into the bias resistor-capacitor circuit. The negative collector voltage swing is determined by collector saturation, so that large positive swings of base current result in virtually constant collector voltage. This continues for a length of time determined by the constants of the transformer, after which the regenerative action is reversed. Because of lowered collector voltage swing, the collector circuit can no longer drive the low impedance reflected from the base, and the charge accumulated on the bias capacitor becomes great enough to decrease collector current rapidly in a degenerative action. Collector current soon cuts off, and then the collector voltage overshoots to a high positive value and the base voltage to a high

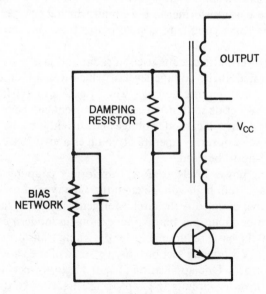

Figure 11.1. Blocking oscillator.

negative value. The base voltage decays slowly because of the discharge of the bias capacitor through the base bias resistor. The next pulse occurs when the negative base voltage decreases sufficiently so that regenerative action starts again. Hence the repetition rate depends on the R and C in the base circuit. Either the negative or the positive pulse voltage may be utilized. Instantaneous voltages and currents are shown in Fig. 11.2 for a load that operates only on the positive pulse. The general shapes of these currents and voltages approximate those in a practical oscillator, except for superposed ripples and oscillations, which often occur.

The negative pulse has a much squarer wave shape than the positive pulse, and consequently it is used where better wave shape is required. No matter how hard the base is driven, collector impedance cannot be lowered below a certain value, so a limit to the negative amplitude is formed. There is no such limit to the positive pulse, and this characteristic may be used for a voltage multiplier.

If the transformer has low OCL, the leakage inductance may be high enough to perform like an air-core transformer. That is, there are optimum values of coupling for maximum power transfer, base drive, and negative pulse shape, but they are not critical.

The leading-edge slope of the negative pulse is determined by leakage inductance and capacitance as in Fig. 10.6 with two exceptions: The pulse is negative and the load is nonlinear; hence there are no oscillations on the inverted top. The slope of this portion of the pulse can be computed from Fig. 10.10, provided that transistor and load impedances are accurately known. The positive pulse can be found from Fig. 10.11 if these curves are inverted.

Pulse width, shape, and amplitude are also affected by the ratio of base turns to collector turns in the transformer. The voltage rise is steeper as this ratio is greater. Since the base capacitance increases as the square of the base turns, a turns ratio greater than unity is seldom used. The exact ratio for close control

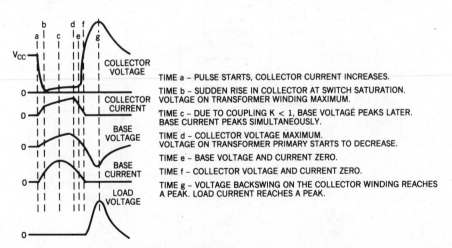

Figure 11.2. Blocking oscillator voltages and currents.

depends on device data which may not be available and must be determined experimentally. The situation parallels that of the class C low-Q oscillators mentioned in Chapter 6.

The circuit of Fig. 11.1 is called a free-running blocking oscillator. When it is desired to synchronize or otherwise control the pulse repetition rate, an external "trigger" pulse is applied to the blocking oscillator base or emitter.

11.2   LINE-TYPE PULSERS

Figure 11.3 is the schematic diagram of a line-type pulser or modulator. This pulser uses dc resonant charging with a hold-off diode to apply the charge to the pulse-forming network. The operation of the pulser is as follows.

During the charging period of each cycle, the diode permits direct current to flow through the charging reactor to the pulse-forming network and through the primary of the pulse transformer to ground. The rate of charging is determined by the inductance of the charging reactor and pulse-forming network capacitance. The inductances in the pulse-forming network and the leakage inductance of the pulse transformer are so small that they are negligible during this period.

The pulse transformer load can be one of many types of devices, each with a different volt-ampere characteristic. Magnetrons and crossed-field amplifiers (CFAs) exhibit a volt-ampere characteristic which is similar to that of a biased diode. This characteristic is shown in Fig. 11.4. Klystrons and traveling-wave tubes (TWTs) exhibit a volt-ampere characteristic which can be expressed

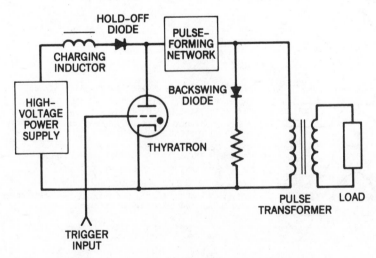

Figure 11.3. Simplified schematic of line-type pulser.

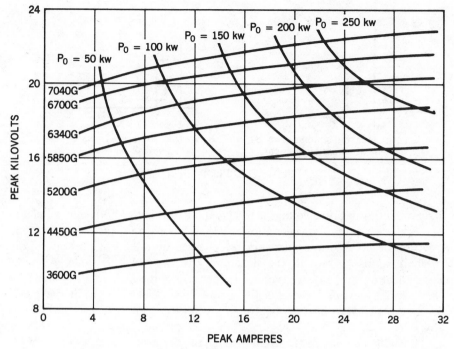

Figure 11.4. Typical magnetron volt-ampere characteristics.

analytically by equation 11.1.

$$I = kE^{3/2} \tag{11.1}$$

where k is the microperveance of the tube, a characteristic of the tube design. Klystrons and TWTs are operated with high negative potential on the cathode. The secondary of the transformer may be wound as either a single or a double winding. If the rms current required by the tube filament is of the same order of magnitude as the rms pulse current, a double winding can be considered and a low-voltage filament transformer can be used. If the filament current would increase the pulse transformer losses and therefore the size, it may be better to use a single secondary and a high-voltage isolation transformer to supply tube filament voltage. Since any capacitance from primary to secondary in the filament transformer is across the secondary of the pulse transformer, it must be included when calculating the rise and fall times of the pulse transformer. The combined reactance of the pulse-forming network capacitance and the charging reactor inductance at their resonant frequency is high compared to the circuit equivalent series resistance. Therefore, the pulse-forming network charges up to a voltage of approximately $2E$, where E is the dc supply voltage. A negligibly small voltage appears on the pulse transformer and load during the charging

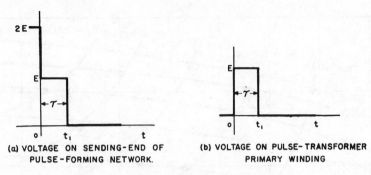

Figure 11.5. Pulser voltages.

interval. If this negative voltage is too great for the load, a diode may be placed across the winding to provide a lower-impedance path for the reverse current.

After the pulse-forming network has charged up fully, it is prevented from discharging back through the dc source by the hold-off diode. At some subsequent instant, a trigger voltage on the grid of the hydrogen thyratron (or other switch device) causes the thyratron to conduct and permit the pulse-forming network to discharge through the very low internal resistance of the thyratron.

The sudden discharge of current through the thyratron causes a voltage wave to start down the pulse-forming network as in Fig. 11.5. This voltage is an inverted step function with a value $(2E - E) = E$. Initial voltage $2E$ on this network is divided equally between the network and the pulse transformer primary, and produces pulse voltage E, of duration τ. The pulse width τ is the length of time that the pulse takes to travel down the pulse-forming network (PFN) and back. After time t_1 the circuit is ready to charge slowly again through the charging reactor. Output current and voltage rise at the pulse leading edge in general accordance with the explanation of Section 10.12, but sometimes a "despiking" network $R'C'$ is included to reduce current oscillations. In this pulser the load equivalent impedance R_M (referred to the pulse transformer primary) is equal to Z_N, the pulse-forming network impedance.

If the load circuit does not conduct current during pulser operation, voltage applied to the pulse transformer primary is doubled. This may cause insulation failure if the open circuit continues. For this reason, spark gaps are often provided on pulse transformers in line-type pulsers.

11.3   FALSE ECHOES AFTER MAIN PULSE

Trailing-edge oscillations are of two general types: (1) a long-term or low-frequency oscillation (see Fig. 11.6) dependent on capacitance C_D and pulse transformer open-circuit inductance L_e, and (2) a superposed high-frequency oscillation dependent on capacitance C_D and L'_N ($=$PFN inductance L_N plus

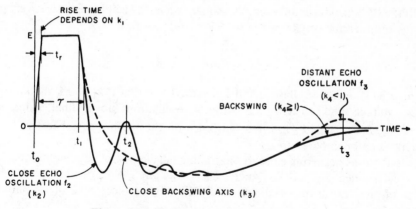

Figure 11.6. Oscillations on pulse backswing.

pulse transformer leakage inductance L_S). If these oscillations exceed zero in the positive or main pulse direction, false "echoes" of two kinds may occur: close echoes, adjacent to the main pulse, and distant echoes, which appear later at a comparatively longer time interval. Either of these cause the output tube to generate a spurious RF signal. The distant echo corresponds to oscillations (1), and the close echo superposed on this long-term backswing to oscillation (2). These are represented in Fig. 11.6 as oscillations f_3 and f_2, respectively. With proper attention to the circuit constants, it is possible to eliminate both types of echoes.

The low-frequency backswing oscillation or axis is affected by C_N only while it is still in the circuit. When the conduction of the switch (thyratron or other device) ceases, C_N is cut out of the circuit. This happens soon after the trailing edge becomes negative. Once this happens, the presence or absence of distant echoes is determined only by C_D in combination with L_e and R_e. Moreover, if $C_N \gg C_D$, close echoes are also determined by C_D and L'_N, whether or not the switch device is conducting during the close echo interval.

Both low- and high-frequency oscillation amplitudes depend on the amount of resistance in the circuit. At instant t_1 this resistance is R_M, the output tube impedance E/i_2 (Fig. 10.21), in comparison with which the transformer core loss equivalent resistance R_e is negligibly high. After the output tube ceases conducting, only R_e remains. During the trailing-edge interval, the circuit resistance varies from R_M to R_e. Resistances R_M and R_e may be replaced by their geometric mean R_I during the trailing-edge interval and the part of the backswing immediately following. This applies to both low- and high-frequency backswing oscillations during the interval $t_2 - t_1$ (Fig. 11.6). The low-frequency or long-term axis of backswing may then be found from values of parameter k_3 determined by L_e, C_D, and R_I, as indicated in Fig. 10.11. If the oscillations are damped out, the impedance ratio still determines wave shape. This ratio is designated k_1, k_2, k_3, or k_4, depending on the portion of the pulse as indicated in Fig. 11.6.

If the PFN produces an essentially square wave, the leading-edge wave shape at the output tube is determined by the impedance ratio

$$k_1 = R_M/2\sqrt{L_s C_D}$$

It may be shown (see Glascoe and Lebacqz, 1948) that if $k_1 = 0.5$, the output tube voltage and current rise without oscillations to a final value at $t_r = 1.6\sqrt{L_s C_D}$. This k_1 has little influence on the trailing edge because L_s is usually small compared to L_e.

Oscillations occurring close to the trailing edge of the pulse are of frequency f_2 [determined by $L'_N C'_D$, where $C'_D = C_D C_N/(C_D + C_N)$, and L'_N is the sum of the transformer leakage and PFN inductance] and of amplitude determined by

$$k_2 = 1/2R_I\sqrt{L_N/C'_D}$$

This amplitude is superposed on the backswing as an axis, which, if oscillatory, has frequency f_3 determined by $L_e C'_D$. Since one purpose of good pulser design is the elimination of false echoes, the backswing axis considered here is always nonoscillatory. Assuming that the switch device is nonconducting for most of the backswing interval, the condition for nonoscillatory backswing is

For the close part of the backswing:

$$\sqrt{L_e/C'_D} \geqq 2R_I$$

For the distant part:

$$\sqrt{JL_e/C_D} \geqq 2R_e$$

where L_e is the OCL at time t_1 and J is the ratio of low-frequency core permeability to pulse permeability. If

$$k_3 = 1/2R_I\sqrt{L_e/C'_D}$$

and

$$k_4 = 1/2R_e\sqrt{JL_e/C_D}$$

then

$$k_3 > k_4$$

because generally

$$R_e/R_I > \sqrt{JC'_D/C_D}$$

So if the pulser is designed to prevent distant echoes, $k_4 \geqslant 1$ and k_3 is several times the value of k_4. Time intervals influenced by these impedance ratios are illustrated in Fig. 11.6. In general, for a good pulse, the close part of the backswing axis follows a nonoscillatory pattern of relatively high impedance ratio, such as those shown on Fig. 10.11 for $k_3 > 1$.

This general effect of pulse transformer magnetizing current is to depress the backswing axis. Magnetizing current does not affect the criterion for absence of false echoes $k_4 > 1$; hence a high ratio J is helpful in eliminating close echoes. The ratio Δ of magnetizing current to load current is much less for the close echo than for the distant echo because R_I is less than R_e. The close echo Δ is approximated by

$$\Delta = \tau R_I / L_e \tag{11.2}$$

To prevent close echo, the first positive peak of oscillation should be no greater than the negative backswing axis voltage at instant t_2 in Fig. 11.6. This equation leads to a transcendental relation between k_2, k_3, and Δ, which is plotted in Fig. 11.7. [This equation is developed in Lee (1954).] It will be noted that all values of k_2 in Fig. 11.7 are less than unity. Under the conditions here assumed, there is always a certain amount of high-frequency oscillation superposed on the backswing axis. If k_2 is greater than the value given by Fig. 11.7, there is no false close echo. To prevent false echoes it is best if k_2 exceeds the curve value substantially, since the equivalent resistance R_I is an approximation. The various impedance ratios are tabulated in Table 11.1.

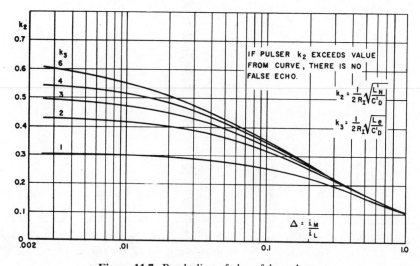

Figure 11.7. Borderline of close false echoes.

TABLE 11.1   Pulser Impedance Ratios

Part of Pulse Affected	Value of Load Resistance	Impedance Ratio Defined	Condition for Good Pulse Shape
Front edge	R_M	$k_1 = \dfrac{R_M}{2\sqrt{L_s/C_D}}$	$k_1 = 0.5$ for min. t_r, with flat-top current pulse
Close echo	$R_I = \sqrt{R_M R_e}$	$k_2 = \dfrac{1}{2R_I}\sqrt{\dfrac{L'_N}{C'_D}}$	$k_2 \geqq$ value in Fig. 255 for no close echo
Backswing axis close to pulse	$R_I = \sqrt{R_M R_e}$	$k_3 = \dfrac{1}{2R_I}\sqrt{\dfrac{L_e}{C_D}}$	$k_3 > k_4$ by definition
Distant echo	R_e	$k_4 = \dfrac{1}{2R_e}\sqrt{\dfrac{L_e J}{C_D}}$	$k_4 \geqq 1$ for no distant echo

Example. Suppose that the transformer of Section 10.13 were used in a line-type pulser with $L_n = 150\,\mu H$, $C_n = 0.667\,\mu F$, $R_I = \sqrt{15 \times 139} = 46\,\Omega$, and $J = 2.0$. Using equation 11.2 gives us

$$\Delta = \frac{46 \times 10 \times 10^{-6}}{1.65 \times 10^{-3}} = 0.278$$

$$C'_D = \frac{0.667 \times 3000 \times 10^{-18}}{(0.667 + 0.003) \times 10^{-6}} = 2987\,\text{pF}$$

$$L'_N = 150 \times 10^{-6} + 7 \times 10^{-6} = 157\,\mu H$$

Following are the impedance ratios:

$$k_2 = \frac{1}{2 \times 46}\sqrt{\frac{150 \times 10^{-6}}{2987 \times 10^{-12}}} = 2.44$$

$$k_3 = \frac{1}{2 \times 46}\sqrt{\frac{1.65 \times 10^{-3}}{3000 \times 10^{-3}}} = 8.06$$

$$k_4 = \frac{1}{2 \times 139}\sqrt{\frac{2 \times 1.65 \times 10^{-3}}{3000 \times 10^{-12}}} = 3.77$$

From ratio k_4 we can see that there are no distant false echoes present. In Fig. 11.7 minimum k_2 for close false echoes is 0.22. Since k_2 calculated above is 2.44 there are no close false echoes. According to Fig. 10.11, the transformer has 10% backswing. This transformer is satisfactory for its application.

11.4   DC CHARGING INDUCTORS

In the pulser of Fig. 11.3, an inductor is used to charge the pulse-forming network to nearly double the voltage from the dc power supply. There are several modes of charging available through an inductor. The three methods usually considered are dc resonant charging with a hold-off diode, dc resonant charging without a hold-off diode, and linear charging.

Inductor voltages and currents for the three methods are shown in Fig. 11.8. These voltages and current are all based on an infinite Q. Reactor Q's should be as high as possible to minimize losses.

The design of inductors for resonant charging without a hold-off diode and linear charging is very critical. The inductance value must be held to fairly tight tolerances and must be linear or else the voltage on the PFN may vary from pulse to pulse, causing undesirable performance of the pulser.

Dc resonant charging with a hold-off diode is the most commonly used method of PFN charging. (Inverter charging is also used. This is discussed in Chapter 13.) The inductance value is not highly critical, but the value of inductance needs to be such that charging time is less than one interpulse period (1/PRF; resonant frequency > PRF/2). Most inductors are designed so that the maximum charging time, when the inductance value is maximum, would be 0.85/PRF. If the minimum value of inductance is not controlled, the rms current through the winding can cause overheating of the reactor.

The average current through the inductor is given by the equation

$$I_{av} = f_r C_N E_N \tag{11.3}$$

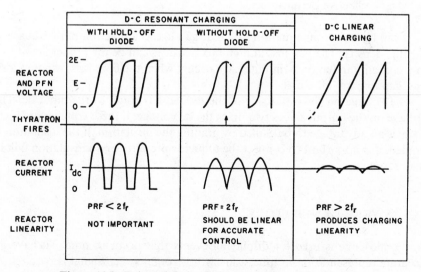

Figure 11.8. Pulser charging inductor voltage and currents.

The rms current is

$$I_{\text{rms}} = \frac{1.11 I_{\text{av}}}{\sqrt{f_r T}} \tag{11.4}$$

This determines the losses in the winding of the charging inductor.

The number of turns is given by

$$N = \frac{2.47 \times 10^6 E_N f_r T}{f_r A_c B_m} \tag{11.5}$$

where A_c is the core area in square inches and B_m is the peak flux density in gauss. To keep the inductor small, the flux density should be kept to a maximum.

The current in the inductor has a dc component. The core will saturate unless there is a gap to prevent it. The gap is given by

$$l_g = \frac{3.2 N^2 A_c \times 10^{-8}}{L} - \frac{l_c}{\mu} \tag{11.6}$$

The gap loss is

$$W_g = 10^{-11} d l_g \mu_e f_r B_m^2 \sqrt{T/2} \tag{11.7}$$

where d = lamination width, in.

l_g = gap length, in.

μ_e = effective permeability

It can be seen from equation 11.7 that gap loss is a function of the square of the flux density. The gap loss at high flux densities can be the predominant loss in a charging inductor. Minimum size occurs when the core loss, winding loss, and gap loss are all equal.

The voltage on the input side of the inductor is equal to the input dc. The voltage on the output side is two times the dc voltage. If these voltages are high, there is an advantage to be gained by grading the insulation. If the insulation is graded, it is important to connect the inductor properly or an insulation failure may result.

11.5 SWEEP GENERATORS

An oscilloscope is used for displaying events that occur in time. To have an accurate representation in time, a linear horizontal sweep signal is required. If the sweep rate is nonuniform, distortion of the displayed waveform results. In magnetic deflection a uniform sweep is produced by a current that varies linearly with time during the sweep interval. A transformer for magnetic deflection is used in the circuit of Fig. 11.9. Pulses of sweep duration τ_s applied to the input of a constant-current amplifying device causes the output current to increase linearly throughout the pulse. If the transformer had no losses, a pulse of constant amplitude E_a would cause current to increase linearly with time until the pulse ended, in accordance with equation 11.8:

$$E_a = -L(di/dt) \tag{11.8}$$

Losses in a practical transformer are equivalent to a resistance in series with L, and the rise in current is exponential. If the losses are small, current rise may be confined to the part of the exponential curve that is nearly linear. The transformer load is usually the deflection coil of a picture tube. If this coil also has sufficiently low loss, the deflection coil current has the same waveform as the transformer primary, and a linear sweep results. Feedback is often used to obtain better linearity of the input voltage.

At the end of the pulse, current i does not stop flowing immediately, because of the transformer and deflection coil inductance. A large-amplitude voltage backswing results, corresponding to large values of Δ in Fig. 10.11. Values of L, C, and R_e are such that the backswing is oscillatory with a period $2\tau_r$ which is small compared to sweep duration τ_s. During τ_r, the first or negative half of the backswing oscillation, the oscilloscope beam is "blanked" or cut off quickly, to allow the beam to retrace to the extreme left, ready for the next sweep period. For a bright picture, the retrace period τ_r should be short compared to τ_s, so that scanning occurs during a large percentage of the time. In television receivers, the sweep frequency is 15,750 Hz and the backswing frequency approximately 77 kHz. Thus the retrace time is about 10% of the sweep period. Positive voltage backswing is used in starting the next sweep trace. This will be described later. Magnetic deflection sweep transformers are made with low-loss cores. Manganese-zinc ferrites are the most widely used core material. Design of sweep transformers is closely integrated with the sweep circuit.

Electrostatic deflection is accomplished by application of sawtooth voltages to the horizontal plates of an oscilloscope. Such a voltage is shown in Fig. 11.10.

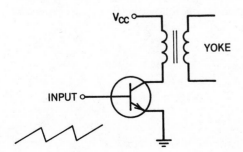

Figure 11.9. Sweep generator.

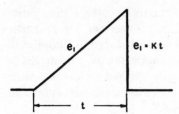

Figure 11.10. Linear sawtooth wave.

Sawtooth voltages may be formed in several ways. If the pulse requires amplification before being applied to the CRT deflection plates, a sweep amplifier is necessary. Here again linearity is important. The spot is moved at a uniform rate across the screen and quickly returned to repeat the trace. In such a circuit the load on the transformer can be regarded as negligible. Assume a linearly increasing voltage as shown in Fig. 11.10 to be applied to the equivalent circuit of Fig. 11.11.

$$e_1 = Kt$$

Then

$$Lpe_1 = LK$$

where $p = d(\cdot)/dt$, and the voltage across the transformer primary is

$$e = \frac{Lpe_1}{Lp + R_1} = \frac{LK}{R_1}(1 - \epsilon^{-R_1\tau/L}) \tag{11.9}$$

The slope of e is

$$de/dt = K\epsilon^{-R_1t/L} \tag{11.10}$$

This voltage e has the same slope as the applied voltage times an exponential term that is determined by the resistance R_1 of the amplifier, the OCL of the transformer, and the time between the beginning and the end of the linear sweep. Under the conditions assumed, the value of the exponential for any interval of time can be taken from the curve marked $R_2 = \infty$ in Fig. 10.10. For example, suppose that the sweep lasts for $500\ \mu s$, that the sweep amplifier output impedance is $800\ \Omega$, and that the transformer inductance is 10 H. The abscissa of

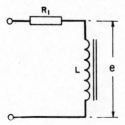

Figure 11.11. Sawtooth transformer circuit.

Fig. 10.10 is 0.04. Since the slope of this exponential curve equals its ordinate, at the end of the interval the slope of the voltage applied to the plates of the oscilloscope will be 96% of the slope that it had at the beginning of the interval.

Assume that at the end of the time interval t (Fig. 11.10) the amplifier is cut off. Then the sweep circuit transformer reverts to that of Fig. 10.11, in which C_p has the same meaning as before but R_I includes only the losses of the transformer, which were neglected in the analysis for linearity of sweep. The voltage does not immediately disappear but follows the curves of Fig. 10.11 very closely, as in magnetic deflection. Backswing voltage may be kept from affecting the screen by suitable spacing of the applied waveforms or biasing the cathode-ray grid.

Vertical sweep transformers are used in television receivers to displace the horizontal sweep lines at a 60-Hz rate, in order to produce a picture. The vertical displacement is fairly linear, retrace rapid, and a sawtooth waveform is necessary here also. Because of the relatively slow vertical displacement, yoke inductance is negligible, so that vertical sweep amplifiers effectively operate into resistive loads during trace periods. The transformers present no particularly difficult problem beyond that of high OCL at low cost.

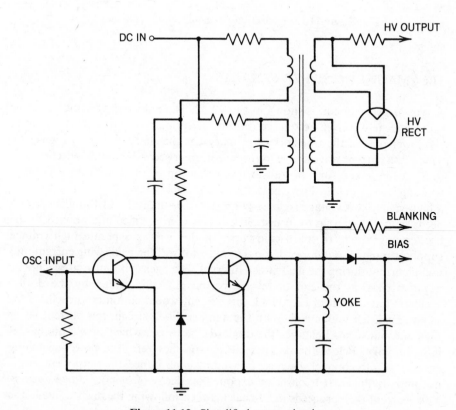

Figure 11.12. Simplified sweep circuit.

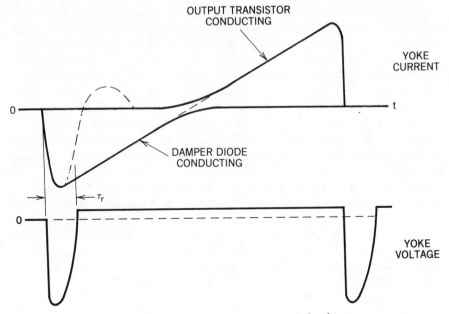

Figure 11.13. Deflection yoke current and voltage.

11.6 MAGNETIC SWEEP CIRCUITS

A typical sweep circuit is shown in Fig. 11.12. The pulse voltage applied to the input appears inverted across the transformer primary winding. The current in the lower part of the transformer primary has the shape shown in Fig. 11.13. This is the current wave shape in the transformer secondary and deflection coil (termed the yoke). An autotransformer extension of the transformer primary winding is used to transform the pulse voltage backswing shown in Fig. 11.13 to a high value. This voltage is actually much larger than Fig. 11.13 indicates and needs only 8:1 step up to furnish 30 kV. It is then rectified and applied to the accelerating anode of the oscilloscope. In this way, a separate high-voltage supply is avoided. A damper diode is used to convert the backswing current into useful current during the next sweep interval. The backswing current reaches its negative peak at the end of retrace period τ_r. As indicated by the dotted oscillation at the left of Fig. 11.13, this current would continue to oscillate for some time if left undamped. With the damper diode circuit, this current never oscillates but instead charges the diode RC network, which slowly discharges into the yoke. Before the damper current reaches zero, the power-switching device starts to conduct. Because of winding capacitance, the output current is not initially linear. It is offset by exponential decay of damper diode current. Yoke current then proceeds in a linear manner, following the dashed line in the transition from damper to switch device current, as in Fig. 11.13.

12 INVERTER TRANSFORMERS

Inverters are becoming more and more common. The design of inverter transformers is therefore important. A circuit that converts dc into ac is usually called an inverter. The term "converter" is usually applied to a circuit that converts ac into dc. Some inverters are used with an ac output, but the large majority of electronics inverters are used to convert dc to high-frequency ac which is then transformed to some higher or lower voltage and rectified to dc. Inverters may vary in usage, but certain features that are important to transformer design are present in all the circuits.

In this chapter we discuss the characteristics of various types of inverter transformers, techniques for deriving inverter transformer parameters, and then the procedure for designing the transformer for meeting the parameters. Some applications of inverters are discussed briefly in Chapter 13.

12.1 COMPARISON OF INVERTER TRANSFORMERS TO CONVENTIONAL PULSE TRANSFORMERS

In Chapter 10 the two broad groups of transformers were described. The first group, which includes power transformers, audio transformers, and RF transformers, operates in the frequency domain. The largest number of transformers in the world are in this class. The second group are those transformers operating in the time domain. Inverter transformers belong in this class. Inverter transformers then fall into the same class as pulse transformers. With this increased use of inverters in power supplies for electronic equipment, the number of transformers in this group may soon be predominant.

TABLE 12.1 Comparison of Pulse Transformers With Inverter Transformers

	Pulse	Inverter
PRF	$< 100\,\text{Hz}$ to $> 10\,\text{kHz}$	$10\,\text{kHz}$ to $1\,\text{MHz}$
Duty	$< 10\%$	Up to 50%
Load[a]	Fixed	Variable
Pulse width	Fixed[b]	Variable
	Nanoseconds to microseconds	Microseconds
Power		
Peak	Watts to megawatts	Watts to kilowatts
Average	Watts to megawatts	Watts to kilowatts
Pulse rise time	Fast ($< 0.5\,\mu\text{s}$)	Medium ($< 2\,\mu\text{s}$)

[a]The load on a pulse transformer may be highly nonlinear during the pulse but is usually relatively constant from pulse to pulse. An inverter load may not only be nonlinear, but may vary between pulses by one or more orders of magnitude.
[b]While the pulse width is usually fixed in a pulse transformer or it may have a few discrete pulse widths, an inverter transformer may vary continuously from pulse to pulse in such a manner as to create a dc imbalance even from a balanced source.

Inverter transformers are a special case of pulse transformers. Before we try to develop design parameters, it might be helpful to look at the similarities and differences between pulse transformers and inverter transformers. A comparison is shown in Table 12.1. The most significant difference is the fact that most pulse transformers work into a constant (if nonlinear) load, while most inverter transformers operate into a load which varies widely in impedance and may also be nonlinear. The significance of this will be seen when we start deriving the parameters for the design. The other differences that contribute to the design are the power level, duty cycle, and impedance levels. Inverter transformers are often step-down transformers. The load impedances are often a fraction of an ohm.

12.2 DEVELOPING DESIGN PARAMETERS

Considering inverter transformers as a class of pulse transformers suggests that the equations and curves used for designing pulse transformers may be useful in designing inverter transformers. This is true if the curves are used with caution. An examination of the operating states of a typical inverter can be used to illustrate the conditions that must be considered. Figure 12.1 is the equivalent circuit of one half of a voltage-fed inverter. The operation of the inverter is described as it cycles through one input pulse. At the beginning of the cycle, assume that the inverter has been running long enough to stabilize the operating conditions. The voltage at the output is E_1 at t_0. The current through L_3 is equal to the current immediately at the end of the preceding pulse. SW$_1$ is open, D$_1$ is not conducting and D$_2$ is conducting. In this condition no power is being

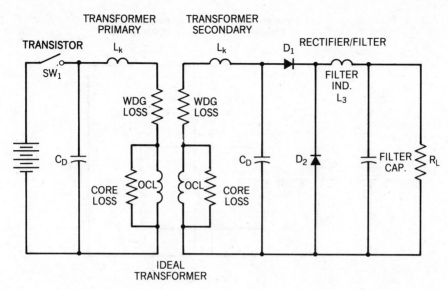

Figure 12.1. Equivalent circuit of one-half of a voltage-fed inverter.

supplied from the source. At time t_1, SW_1 is closed and current begins to flow in the primary loop. The voltage on the secondary starts to rise toward E_1 at a rate determined by L_s and C_d and the damping factor k_1. At time t_2, E_1 equals the output voltage. At this time, D_1 conducts and D_2 stops, causing current to now be drawn from the source. The voltage continues to rise, until at time t_3 it is equal to the input voltage. The voltage on the circuit remains at the input voltage until time t_4, at which time SW_1 opens. The voltage at L_1 in the secondary circuit starts to decrease at a rate determined by the OCL and C_d and the damping factor k_3. At time t_5, the voltage again is equal to the output voltage. D_1 stops and D_2 conducts, once again disconnecting the load from the source and making the current delivered to the load to be dependent on the energy stored in L_3 and the filter capacitor. The voltage on the secondary of the transformer decays toward zero. The damping factor changes to a value controlled by the core loss of the transformer. This may lead to a k_3 less than 1 with both backswings and superposed oscillations.

12.3   VOLTAGE RISE

The general operation of one-half of a voltage-fed inverter was described in Section 12.2. In a manner similar to that used for pulse transformers, the active parts of the equivalent circuit during the various segments of the pulse can be described. From the time t_1 to t_2, the equivalent circuit of the inverter transformer is shown in Fig. 12.2(a).

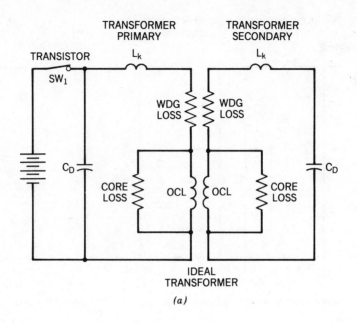

(a)

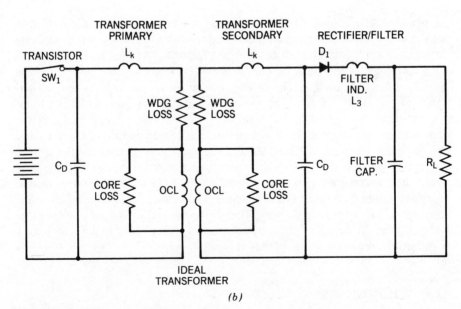

(b)

Figure 12.2. Equivalent circuit of an inverter transformer: (a) during voltage rise $E_{in} < E_{out}$; (b) during voltage rise $E_{in} > E_{out}$; (c) during voltage decay $E_{in} > E_{out}$; (d) during voltage decay $E_{in} < E_{out}$.

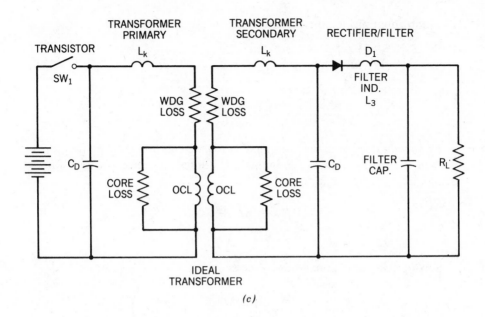

(c)

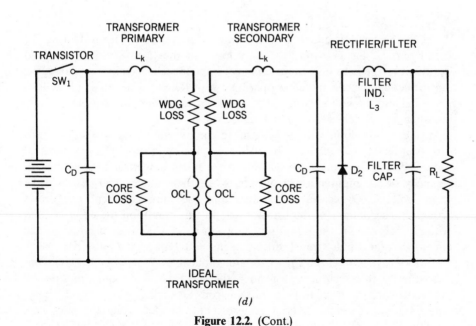

(d)

Figure 12.2. (Cont.)

The loop equations for the current in the two loops can be written

$$E = i_1 R_1 + L\frac{di_1}{dt} + \frac{1}{C}\int (i_1 - i_2)\, dt$$

$$0 = \frac{1}{C}\int (i_2 - i_1)\, dt + i_2 R_2$$

By solving these equations simultaneously we can find the equation for the voltage across R_2. Since the initial conditions are zero, the transformed equations will take the form

$$E/s = R_1 I_1 + Ls I_1 + 1/Cs(I_1 - I_2)$$

$$0 = R_2 I_2 + 1/Cs(I_2 - I_1)$$

Solving these equations leads to the equation

$$e = \frac{ER_2}{R_1 + R_2}\left(1 + \frac{m_2\epsilon^{m_1 t} - m_1\epsilon^{m_2 t}}{m_1 - m_2}\right) \tag{12.1}$$

where $m_1, m_2 = -m(1 \pm 1 - 1/k_1^2)$.

This is the same as equation 10.3 expressed in a different form. During the critical part of the rise time the only load on the transformer is the load equivalent to the core loss. Since this load impedance is usually very large, the transformer can be considered as operating with $R_1 = 0$. The value of k_1 will be very small and the voltage will rise in an oscillatory fashion. From time t_2 to t_3 the equivalent circuit is shown in Fig. 12.2(b).

At time t_2 the input voltage is equal to the output voltage. Energy is now being transferred to the load, so the value of k_1 increases rapidly toward its final value. The time constant of the output filter is much longer than the equivalent frequency of the input pulse. However, when the voltage at D_1 exceeds the output voltage, D_2 ceases to conduct and the output current now flows through the secondary of the transformer. In this condition the effective load on the transformer is the dc output impedance. This equivalent circuit is the same as that for a pulse transformer with $R_1 = 0$. During the pulse rise, therefore, the design curves in Fig. 10.6 are applicable. When the load current starts flowing through the transformer, the rise time slows and prevents the overshoot from the oscillatory condition from occurring. This type of operation will occur in most inverter transformers. To determine the rise time, equation 12.1 must be solved for two different values of R_2. The first of these values is equal to the equivalent core loss. The second value, after the voltage equals the output voltage, is equal to the load resistance. If a snubber is used, the value of R used in the snubber becomes the resistance value used for calculating the first part of the rise time.

12.4   TOP OF PULSE

Once the input voltage reaches the dc voltage (less the voltage drop of the switching device), the voltage across the primary of the transformer remains essentially constant until the switch is turned off. Even though the voltage remains constant, the current through the primary continues to rise during the time of the pulse. The amount of the current rise is according to the equation

$$i_m = (E_a/mL)(1 - \epsilon^{-m\tau}) \tag{10.5}$$

The amount of current rise that can be allowed determines the value of open-circuit inductance. This current rise is often chosen as a percentage of the load current. Unless great care is used in making the selection, the final results can be unacceptable and may in the ultimate case lead to the failure of the inverter switching transistors.

As an example of what could happen, consider the following situation. An inverter to operate with an output of 5 V dc at 100 A is to be designed. The maximum exciting current is chosen as 10% of the load current or 10 A. The inverter is to operate, however, down to an output current of 20 A. At this load the exciting current is 50% of the load current. This condition is illustrated in Fig. 12.3. This does not cause any problems during the top of the pulse. It would perhaps be thought that an exciting current of 10% of the minimum current would be better, but this can cause other problems with the design. The implications of the exciting current on the other transformer parameters are discussed in Section 12.7.

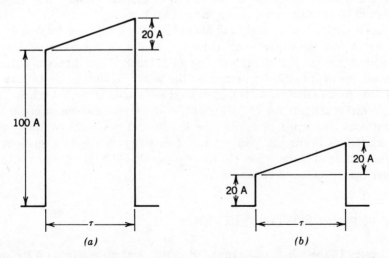

Figure 12.3. Typical inverter primary waveforms: (*a*) 100-A load; (*b*) 20-A load.

12.5 VOLTAGE DECAY

At the end of the predetermined time, the input switching device will be turned off. Figure 12.2(c) shows the equivalent circuit when this happens. When the input switch is opened, the current flowing in the primary open circuit inductance is equal to the load current plus the exciting current of the transformer. Since this current is flowing in an inductance, it cannot change direction immediately and so continues to flow. The voltage on the primary reverses. When the secondary voltage then becomes less than the output voltage, D_1 stops conducting and the secondary voltage also reverses. It is the ringing due to the stored energy in the transformer that contributes to the output voltage ringing of the power supply.

We can write the loop equations for the currents in the various loops of the circuit. By solving these simultaneous equations, we can arrive at an equation for the voltage decay across the primary of the transformer. The equation for the primary voltage decay is given in equation 12.2.

$$e = \frac{E_a}{m_1 - m_2}[(m_1 + 2\Delta m)\epsilon^{m_1 t} - (m_2 + 2\Delta m)\epsilon^{m_2 t}] \qquad (12.2)$$

If we compare this equation to equation 10.4, it can be seen that the equations 10.4 and 12.2 are the same. The only difference is in the relative value of R_2. In the case of the inverter transformer, this is supplied by the equivalent losses in the transformer. Figure 10.11 can be used to calculate the backswing produced by the transformer primary.

The factor Δ in equation 12.2 is the ratio of the exciting current to the load current. As the exciting current is increased with respect to the load current, the ringing amplitude increases. In the design of the transformer, it is therefore important to keep the open-circuit inductance high.

The important factors in the initial fall time of the pulse are the open-circuit inductance and the capacitance. However, as during the rise time the load impedance changes, so it changes during the fall time. When the voltage starts to decrease, the load is still connected. As the voltage continues to decrease, the load impedance changes to the equivalent resistance of the core loss. If the transformer is designed for a k_3 of about 10 with the load impedance equal to the output load, the value of k_3 may be less than 1 when the voltage on the secondary falls below the load voltage. This will give a large overshoot and ringing at the end of the pulse. The transformer should be designed to have a k_3 greater than 1 under all conditions.

12.6 SUPERPOSED OSCILLATIONS

In Chapter 11 the effects of leakage inductance and capacitance on the major backswing were discussed in detail. The same conditions prevail in an inverter transformer. Since the load is very nonlinear the conditions for the superposed ringing are potentially present in all inverter transformers.

If the switch has essentially a zero turn-on time, the rise time of the leading edge of the pulse is determined by the impedance ratio

$$k_1 = R_M/2\sqrt{L_s/C_D}$$

If the conditions for k_1 are such that the voltage rises without oscillation, k_1 has very little influence on the trailing edge of the pulse. Although this was usually true for a pulse transformer working with a fixed, nonlinear load, it may be difficult to achieve in all cases in an inverter transformer unless great care is used in choosing L_s and C_D.

Oscillations near the trailing edge of the pulse are of frequency f_2 determined by $L'_N C'_D$, where $C'_D = C_D C_c/(C_D + C_c)$ and L'_N is the sum of the leakage inductance and the stray wiring inductance. C_s is the stray capacitance of the circuit. The amplitude of the oscillations is determined by

$$k_2 = \frac{1}{2R_I}\sqrt{\frac{L_N}{C'_D}}$$

This amplitude is superposed on the backswing as an axis which may also be oscillatory, as discussed in the preceding paragraph. The switch is open for the total backswing interval. However, in a poorly designed transformer, the backswing may not be returned to zero before the switch is closed again. The conditions for nonoscillatory backswing are:

For the close part:

$$\sqrt{L_e/C'_D} \geqq 2R_I$$

For the distant part:

$$\sqrt{JL_c/C_D} \geqq 2R_e$$

where J is again the ratio of low-frequency core permeability to pulse permeability. Ferrite cores have essentially the same permeability under pulse conditions as they do under low-frequency conditions. At the lower inverter frequencies (longer pulse widths) where laminated core materials may be used, the pulses are long enough so that J is also approximately 1.

Then

$$k_3 = \frac{1}{2R_I}\sqrt{\frac{L_c}{C'_D}}$$

and

$$k_4 = \frac{1}{2R_e}\sqrt{\frac{JL_e}{C_D}}$$

Since the value of C'_D is two capacitances in series and one of these is C_D, C'_D is less than C_D and therefore $k_3 > k_4$. In an inverter circuit the time between pulses is usually short compared to the pulse width, so oscillations at more than five times the pulse width do not occur. It is therefore important to design for $k_3 > 1$ under all conditions.

12.7   SELECTING PROPER TRANSFORMER PARAMETERS

Taking all of the various operating conditions into consideration, guidelines for determining the open-circuit inductance, leakage inductance, and distributed capacitance can be developed. It can be seen that the value of open-circuit inductance is most important in determining the exciting current, and that this determines the conditions between the pulses. With a value of open-circuit inductance selected higher than the value needed for full load, a high value of k_3 can be maintained for the condition of no load.

The impedances for most inverter transformers are low, from a few ohms to fractions of an ohm. For example, a 5-V 200-A power supply has an impedance of 0.025 Ω. For this reason it is also important to keep the leakage inductance of the transformer low. The transformers are usually step down, so the curves in Fig. 10.15 apply to the leading edge of the pulse. Typical values of pulse rise time would be on the order of 250 ns. If a k_1 of 1 were selected, T would be 500 ns, so a value of $L_s C$ would be $0.5 \times 10^{-6}/2 = 0.0796 \times 10^{-6}$.

$$m = k_1 \sqrt{L_s C}$$

$$= 1.26 \times 10^7$$

For $R_2 = 0.025$,

$$L_s = R_2/2m$$

$L_s = 1.99$ nH referred to the secondary. For a typical transformer step down of 24.1, this would be 1.15 μH reflected to the primary. This is still a very low value to achieve.

It was mentioned in Section 12.3 that the k_1 value changes during the leading edge of the pulse. Since this is true, the rise time may be achievable with a much higher value of T. Figure 12.4 shows a leading edge where a k_1 value of 0.25 is assumed for 70% of the rise time. This is followed by a change of k_1 to 1.0. In this case the 90% point is reached in $0.375T$ instead of $0.5T$. Real conditions will often be better than this since a k_1 of 0.1 or less is possible.

With the value of L_s calculated in the preceding paragraph, and for $T = 0.5$ μs,

$$C_D = T^2/4$$

$$= 3.18 \ \mu\text{F}$$

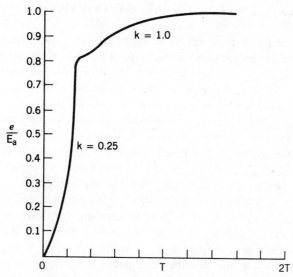

Figure 12.4. Leading edge of a pulse with two k_1 values.

This is the amount of distributed capacitance allowable on the secondary to achieve the desired rise time. This value would be difficult to design into a transformer. One ten-thousandth of this value would be high for most inverter transformers.

When the damping factors for the rise time, fall time, and superposed oscillations are considered, it is obvious that C_D wants to be as small as possible so that the values of R_I can be kept as large as possible. With ferrite-core materials it is impossible at the lower inverter frequencies to keep the capacitance low enough so that the oscillations are critically damped by the core loss.

To summarize the requirements for an inverter transformer:

1. The open-circuit inductance should be selected so that the magnetizing current at the lightest expected load does not cause excessive backswing.
2. The leakage inductance should be selected to be as low as reasonable to give reasonable values of k_1 and m in Fig. 10.15.
3. The distributed capacitance should be kept low to provide for critical damping for reasonably high values of R_I.

Meeting all of these conditions simultaneously may not be easy, particularly the condition of having high open-circuit inductance and low leakage inductance.

12.8 CONTROLLING UNDESIRABLE CONDITIONS

When all the design parameters from the preceding section are considered, a combination of design parameters that cannot be realized in the actual design may result. If this happens, there are still ways to develop design parameters that will result in a practical design and still give very good performance. One method of design is to select a value of snubber resistor that will give good performance and provide critical damping during the time when the transformer is normally unloaded. A simple RC snubber network is shown in Fig. 12.5. By selecting a value for the snubber and then selecting the design parameters by substituting the snubber value for the core-loss resistor value, the value of capacitance, open-circuit inductance, and leakage inductance can be chosen simultaneously to achieve practical values for transformer design.

The selection of a snubber value prior to knowing values for capacitance and inductance does not follow hard and fast rules. The usual limitation is the amount of power that is dissipated in the snubber resistor. This power can be minimized by keeping the distributed capacitance low. Since the snubber is active only during the rise and fall times, the amount of power dissipated during each pulse is $CV^2/2$, where C is the snubber capacitor. Since this occurs four times per cycle, the power in watts is $4 \times \mathrm{PRF} \times$ joules per pulse. If a value of R equal to the lightest expected load is chosen, it will usually result in a satisfactory transformer design. The value of C will be realizable. When the transformer is loaded to full load, the k factors will still be reasonable. After the transformer is designed as closely as possible to the design values, the snubber resistor should be adjusted to $R = 2\sqrt{L_e/C_D}$.

When the values of leakage inductance, open-circuit inductance and distributed capacitance do not give a reasonable value for a single resistor for damping both rise and fall time, a more complex snubber must be designed. An example of a snubber with two time constants are shown in Fig. 12.6. In this snubber, the voltage is applied to the capacitor through resistor R_1. The voltage remains across the capacitor until the switching device is turned off. At the end of the pulse the capacitor is discharged through resistor R_2. It is difficult to use this type of circuit across the primary of a transformer, due to the voltage reversal on the primary of the transformer when the two switching devices operate. This type of snubber can be used very well when the winding has a single polarity such as across center-tapped or split windings.

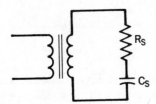

Figure 12.5. Simple RC snubber network.

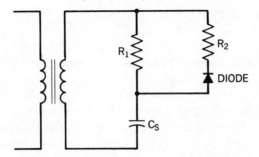

Figure 12.6. Multiple-time-constant snubber.

Snubbers cannot be expected to eliminate the ringing on a poorly designed transformer. If care is taken in selecting the transformer parameters, a snubber can be very effective in removing unwanted oscillations. Figure 12.7 shows the effect of placing a snubber across a well-designed transformer. Snubbers can be placed across the primary, secondary, or both. The best method is to place the snubber across both primary and secondary. To do this, calculate the value of R and C required for either the primary or the secondary. The value of R is doubled and the value of C halved. The values of R and C are then transformed by N^2, where N is the turns ratio.

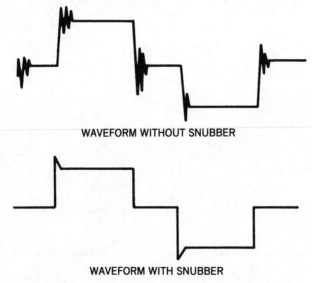

WAVEFORM WITHOUT SNUBBER

WAVEFORM WITH SNUBBER

Figure 12.7. Effect of snubber on transformer ringing.

12.9   CORE SELECTION

The size of the core to use for an inverter transformer can be determined in a manner similar to the manner of choosing a core for power transformers. There are factors that need to be considered which will make the choice of an inverter transformer core more complicated. The major factors to be considered are core material, inductance/capacitance ratios, and skin effect in the windings.

Many different core materials are used for inverter transformer cores. The most commonly used are ferrites, Permalloys, and 50% nickel-irons. Each of these materials has both advantages and disadvantages at the frequencies normally used for inverters. At first, ferrites would seem to have a distinct advantage over the other types of core materials since the losses in ferrites are very low. Most ferrites have two disadvantages. The first is that the saturation flux density is low when compared to other core materials. The second disadvantage is the relatively low working temperature of the ferrites. This may not be a problem when the ambient temperature is slightly above room temperature, but when designed for the high-temperature aircraft ambient, the applications of ferrites can be limited. Figures 12.8, 12.9, and 12.10 are the loss curves for a low-loss manganese-zinc ferrite, 1-mil Permalloy, and 1-mil 50% nickel-iron, respectively. For comparable flux densities the ferrite has the lower loss. Allowing the core loss to increase may allow the transformer to be made

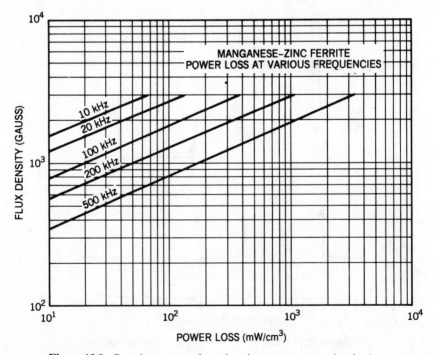

Figure 12.8. Core-loss curves for a low-loss manganese-zinc ferrite.

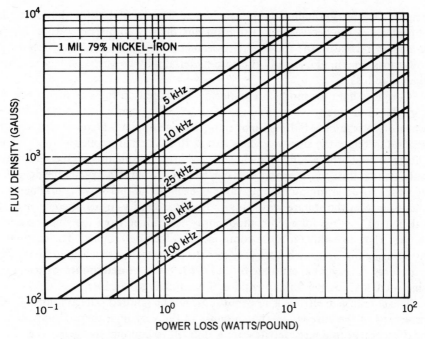

Figure 12.9. Core-loss curves for 1-mil 79% nickel-iron.

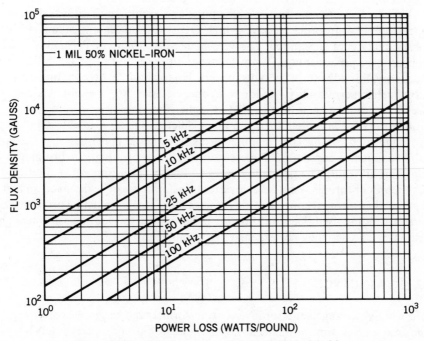

Figure 12.10. Core-loss curves for 1-mil 50% nickel-iron.

smaller on the Permalloy core. One other feature of ferrites that must be considered is the decrease of saturation flux density with temperature. As the frequency increases, the use of ferrites is increasingly advantageous because of lower loss. The final choice of the core material may be based on factors other than core loss.

Core-loss data for most core materials are based on the flux density in the core due to a sine wave of voltage. Most inverter transformers have a square wave of voltage or a bidirectional pulse voltage applied. Data do not exist in general to relate the sine-wave loss curves to the actual applied voltage. Most efforts to relate these conditions have been rather inconclusive. One set of tests indicate that the square-wave losses can be determined to a reasonable degree of accuracy by determining the rms value of each of the sine-wave components of significance in the square-wave input. The flux density for each component is calculated. The loss for each component is determined from the sine-wave loss curves. The rms value for these components is then determined. The calculated losses using this approach were very close to the core losses measured on silicon-iron cores. A very good estimate to the core loss can be determined by using the sine-wave loss for the fundamental and multiplying by 1.10.

The core loss in ferrites is low, but it cannot be ignored. The amount of heat generated in the core is low, but because of the very high thermal impedance, heat generated in a large ferrite core is difficult to conduct to the surface of the core, where it can be dissipated. The losses in the tape-wound cores, even if higher, are easier to remove and so the end result may be a smaller transformer.

The effect of OCL on the top of the pulse and exciting current was discussed in Section 12.4. In order to have a high OCL with minimum leakage a high-permeability core is required. The very high permeability of Permalloy would appear to be an advantage. With the apparent reduction of permeability with frequency, this advantage soon disappears. Up to about 50 kHz Permalloy does have higher permeability. At lower frequencies the 50% nickel-irons can be useful because they can be operated at higher flux densities.

If ferrite cores are used, there are many different shapes available. Cup cores are often used. Many core manufacturers have made core shapes which are modifications of cup cores. These cores have large openings for bringing out high current leads. The permalloy cores and 50% nickel-iron cores are used as cut C cores or toroids. Very good inverter transformers can be designed on high-permeability toroidal cores. Any dc component of current must be prevented or a reset winding used. Despite superior performance, toroidal transformers are seldom used since they are expensive to manufacture.

12.10 WINDINGS

The winding arrangements for inverter transformers can be simple, core, or shell, the same as in a conventional transformer. Ferrite cup cores permit only a core-type winding. Many times this type of winding is used even though the

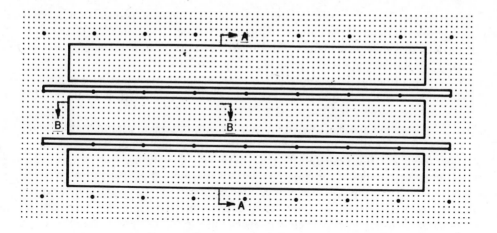

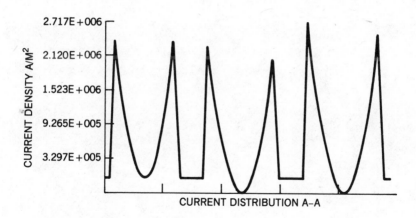

CURRENT DISTRIBUTION A–A

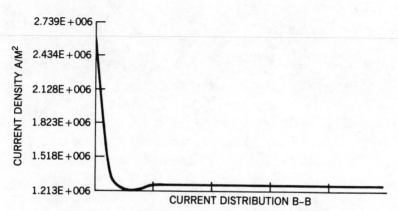

CURRENT DISTRIBUTION B–B

Figure 12.11. Skin effect in multiple-layer flat copper straps.

transformer design is not optimum, since this type of transformer is less expensive to build. If the current required is not too large, some interleaving of the windings may be achieved. If the inverter is a bridge or half-bridge, the core type of winding can usually be made to operate satisfactorily.

If the primary or secondary is center tapped, a shell-type winding is advantageous. If optimum performance is required, a shell-type winding should be used for inverter transformers.

The types of windings used are basically conventional in arrangement. Some of the winding techniques and insulation techniques require special consideration. All inverter transformers have very few turns. For low-voltage outputs the number of turns may be one, or if center tapped, one turn on each side of the center tap. These windings often must carry high currents. One way of winding this one turn is to use heavy copper sheet and form the turn. For a single-layer winding or for even two layers, this can be done and the skin effect is reduced over round wire. If more than three layers are used, the current in the

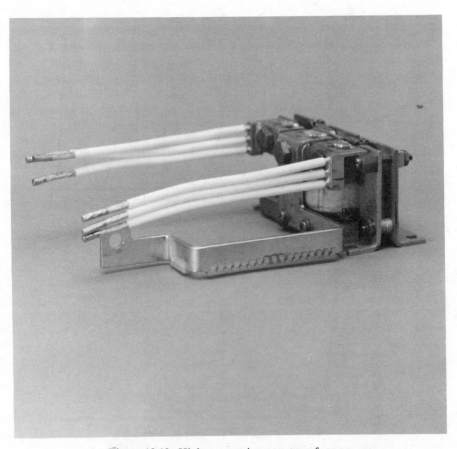

Figure 12.12. High-current inverter transformer.

inner layer or layers is forced to the outer edge by the proximity effect. In a round wire the proximity effect forces the current to the outer edges of the wire more uniformly. In a sheet this same effect is produced but is much more pronounced (see Fig. 12.11). The current is forced to the outside edges. The effective resistance may increase many times over the dc resistance. The winding losses in the center layer of the winding may cause hot spots which will lead to early transformer failure.

Skin effect is also a problem in other windings. If the transformer is small, large solid wire is difficult to wind. One solution to this problem is to wind several smaller wires in parallel. This type of winding is subject to the same kind of problem as in a multilayer sheet winding. Unless the windings are transposed as they are wound, the current will remain in the outer wires and not be conducted in the inner wires.

Litz wire can be useful in winding transformers at these frequencies and is used where maximum current density is required. The most usual winding is with solid magnet wire and flat copper strip in two or fewer layers. A transformer wound this way is shown in Fig. 12.12.

Under all conditions, the effect of the area reduction due to skin effect in the wires must be taken into effect when calculating the copper loss. This is not a simple problem, so the actual losses are estimated. If the losses are critical, they are measured under operating conditions.

12.11   HIGHER-FREQUENCY INVERTERS

Until recently, most power inverters have been limited in frequency to about 50 kHz. The limitation was basically caused by the limited current capabilities for very high speed switching devices. If high power were needed at high frequencies, several devices had to be connected in parallel. Since these devices differed widely in characteristics, it was not feasible to parallel more than two or three devices without selecting the devices for matching characteristics. At lower switching speeds more devices could be paralleled.

With the advent of power field-effect transistors, the operating frequencies for power inverters is increasing. Powers to several hundred watts are now achievable at frequencies up to 250 kHz, with some inverters being designed to 1 MHz or more. Transformers for these inverters have the same type of performance requirements as those at low frequencies. The problem lies in achieving the requirements. The reactive drop for a transformer operating at 250 kHz is 10 times as great as a transformer at 25 kHz with the same leakage inductance. Since the one-turn minimum limitation applies and the IR drop and copper loss are the same for the same current, the only way to reduce the reactive drop is to use the means available for reducing the leakage inductance of the one turn. Some of the ways to reduce the leakage are to increase the winding length b, decrease the thickness of the primary/secondary insulation, reduce the mean turn, reduce the winding build, and maximize interleaving.

Care must be taken when doing these things so as not to increase capacitance. If the capacitance is allowed to increase, the rise time of the pulse, which now must be tens of nanoseconds, will suffer. Since the current density in the conductors is the same as at lower frequencies, the core must be large enough for a wire with sufficient current-carrying capacity to fit.

Ferrite cores are used at these frequencies. Perhaps the best design for a transformer for high frequencies is as shown in Fig. 12.13. Windings arranged in this way have very low leakage inductance. The capacitance may be a little higher than other configurations but not enough higher to cause problems.

Dc must be kept from the windings to prevent transformer saturation. This was difficult with bipolar transistors, due to the difference in storage times. It would not appear to be a problem with FETs.

Example. Inverter Transformer. A transformer is needed for an inverter circuit to supply 28 V dc at 12.5 A. The inverter input is 208 V line to line, 60 Hz, three phase. This is rectified directly to produce 270 V dc. This voltage is chopped in a half-bridge circuit at 20 kHz to produce the ac voltage for step down. In the half-bridge circuit, the peak primary voltage in each direction is 135 V.

Dc input voltage = 270 V
Dc output voltage = 28 V
Inverter frequency = 20 kHz
Nominal duty = 0.667

Figure 12.13. High-frequency toroidal inverter transformer.

Transformer turns ratio $= 3:1$ step down

Core: 66-mm diameter, manganese-zinc cup

$b = 1.50$ in.

$a = 0.40$ in. max.

Mean turn $= 5.12$ in.

$$N_p = \frac{135 \times 10^8}{25.8 \times 20 \times 1.6 \times 1.11 \times 10^6} = 15 \text{ turns}$$

$$N_s = 15/3 = 5 \text{ turns per half}$$

The secondary current per half:

$$I_p = 12.5/0.667 = 18.74 \text{ A}$$

$$I_{rms} = 18.74\sqrt{.667} = 15.3 \text{ A}$$

The primary current:

$$I_p = 18.74/3 = 6.25 \text{ A}$$

Allowing for 80% efficiency, we have

$$I_p = 6.25/0.8 = 7.8 \text{ A}$$

$$I_{rms} = 6.37 \text{ A}$$

The transformer windings are Litz wire wound, as shown in Fig. 12.14. Each section has five turns and is wound with several wires in parallel.

Primary: four wires in parallel, each wire 48 strands of AWG No. 38, Litz

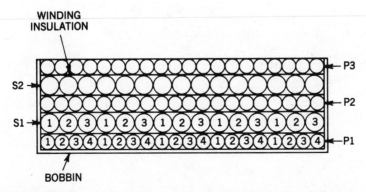

Figure 12.14. Cup core winding arrangement for low leakage inductance.

Secondary: three wires in parallel, each wire 112 strands of AWG No. 38, Litz

Winding insulation: 0.007-in. polyamide film (0.005 + 0.002)

$$\text{OCL} = \frac{3.2 \times (15)^2 \times 1.11 \times 24.9 \times 10^{-6}}{4.84} = 4.11 \text{ mH}$$

$$I_m = 21.6(1 - e^{-16.6 \times 10^{-3} \times 21.6/4.11 \times 10^{-3}})$$

$$= 1.8 \text{ A or } 23\% \text{ of the load current}$$

The leakage inductance referred to the primary

$$L_s = \frac{10.6 \times (15)^2 \times 5.12 \times (2 \times 3 \times 0.007 + 0.4)}{9 \times 1.5 \times 10^9}$$

$$= 0.4 \, \mu\text{H}$$

There is no layer-to-layer capacitance, so the capacitance will be winding-winding and winding-core.

The primary-to-secondary capacitance is

$$\frac{0.225 \times 5.12 \times 1.5 \times 3}{0.007} = 741 \text{ pF}$$

The voltage is uniformly distributed, so

$$C_e = C/3 = 247 \text{ pF}$$

There are four insulation spaces, so

$$C_{\text{tot}} = C_e \times 4 = 988 \text{ pF}$$

$$\text{winding to core} = 18 \text{ pF}$$

$$\text{total capacitance} = 1006 \text{ pF}$$

core loss at 1600 G and 20 kHz $= 90 \text{ mW/cm}^3 \times 88.3 \text{ cm}^3 = 7.95 \text{ W}$

$$Z = E^2/W = (135)^2/7.95 = 2293 \, \Omega$$

$$= 2\pi\sqrt{0.4 \times 1006 \times 10^{-18}} = 0.126 \, \mu\text{s}$$

$$\text{initial } m = \tfrac{1}{2} \times 2293 \times 1006 \times 10^{-12}$$

$$= 2.17 \times 10^5$$

$$k_1 = 2.17 \times 10^{-4}\sqrt{0.4 \times 1006} = 4.35 \times 10^{-3}$$

When $E_{in} > E_{out}$,

$$m = \tfrac{1}{2} \times 17.3 \times 1006 \times 10^{-12} = 2.87 \times 10^7$$

$$k = 2.87 \times 10^{-2}\sqrt{0.4 \times 1006} = 0.576$$

This will give about 20% overshoot on the leading edge of the pulse.
For the decay time,

$$R_I = \sqrt{17.3 \times 2293} = 199\,\Omega$$

$$k_3 = \sqrt{4.11 \times 10^{-3} \times 1006 \times 10^{-2}/398} = 5.08$$

$$T = 2 \times \pi \times \sqrt{4.11 \times 1006 \times 10^{-15}} = 12.8\,\mu s$$

For $k_3 = 5$, the fall time is $10.08 \times T = 1.0\,\mu s$. The backswing is 25% with $k_3 = 5$ and $\Delta = 0.23$. $k_2 = k_3$; therefore, there are minimum oscillations.

12.12   TESTING

Inverter transformers should be tested for the same parameters as power transformers. A turns ratio test is very important since the number of turns is small, often less than 50 turns. The open-circuit inductance must be checked. This test should be done at the flux density at which the transformer was designed to operate. A sine wave of voltage at 60 or 400 Hz with the voltage adjusted for the flux density is an adequate test for transformers with ferrite cores. Transformers with laminated cores are better tested at higher frequencies, where the decrease in apparent permeability due to skin effect in the core is similar to actual operating conditions.

DCR measurements are extremely important in inverter transformers. The DCR of both primary and secondary windings is low. A good DCR measurement is necessary for finding poor connections before high currents can cause overheating. These poor connections may have one of a number of causes, such as cold solder joints, broken strands in Litz wire, strands of Litz wire not soldered, or a bolt joint not tight enough. Any one of these can cause local heating and lead to transformer failure.

Leakage inductance is very difficult to measure because of the very low value. A measurement of leakage inductance under laboratory conditions on several samples can be used to establish an expected range. A routine test can be made by checking the ringing on the pulse as was done in Chapter 10. A polarity test is important to verify that the effective capacitance will be as calculated. Core loss, temperature rise, and permeability tests may be run as engineering tests to verify the design, but are not necessary to run as production tests. The final test on any design is how well it works in the circuit. Many times this is the best test to run, and then compare future transformers to the engineering model in production.

13 INVERTER CIRCUITS

Circuits that change dc to ac are called inverters. There are many types of inverter circuits in use. Only a few of the basic types are described in this chapter. The constraints imposed by the circuit on the inductive component design will be described. Inductive component design also has an effect on the circuit performance. It should be possible to extend the design techniques to any type of inverter inductive component from the types of inverter circuits described. The outputs of some inverters are rectified to produce dc; the outputs of others are used as ac. Both dc-to-dc and dc-to-ac circuits will be reviewed. Dc-to-ac inverters are used where the desired output is some ac frequency. The output of a dc-to-ac inverter may or may not be a sinewave. The dc-to-dc circuits are basically used for dc power supplies. The dc-to-dc circuit converts the input dc to ac for transformation.

The transistor properties that are important in inverter circuits will be considered. The application of thyristors to inverters will also be described. The design requirements of the transformers for many different types of inverters are similar.

13.1 TYPES OF INVERTER CIRCUITS

Inverter circuits are used at power levels of a few watts or less to thousands of kilowatts or more. Operating frequencies may range from 50 Hz to 1 MHz or more. Input voltages may be 5 V dc for low power or automotive applications to hundreds of kilovolts for electric transmission lines. The output may be ac or dc. There are also differences in the types of switching devices used. With these variables it would appear that the differences would outweigh the similarities. Table 13.1 is a list of some of the variables of inverter circuits given in general terms. Of the 16 types listed in Table 13.1, eight are immediately shown to be similar to other types. Of the eight remaining, there are only four basic types of operation.

The inverter circuits described in this chapter are only a few of the many variations that exist, but the ones described here are the basic types in common use. All the circuits described here use some means of regulating the output voltage. The dc-to-dc inverters regulate the output dc voltage. The dc-to-ac inverters regulate the rms voltage or average voltage of the output wave. One circuit that is not truly an inverter, but will be described, is the series switching regulator.

There are several types of switching devices used in inverter circuits, whether dc to dc or dc to ac. For power levels less than a kilowatt, bipolar transistors are commonly used. In Chapter 6 various classes of amplifiers were discussed. In classes A, B, and C the amplifying devices were not operated in the saturation mode. In the class D amplifier the amplifying devices were actually operated as switches, with the positive element being at saturation during the conduction. The input currents and output currents were both sinusoidal in shape.

TABLE 13.1   Inverter Transformer Classification[a]

Category	DC Bias	Rectified	Filtered	Interrupted
A	N	Y	Y	Y
B	N	Y	Y	N
C	N	N	Y	Y
D	N	N	Y	N
E[b]	N	N	N	Y
F[b]	N	N	N	N
G[c]	N	Y	N	Y
H[c]	N	Y	N	N
J	Y	Y	Y	Y
K	Y	Y	Y	N
L[d]	Y	Y	N	Y
M[d]	Y	Y	N	N
N[e]	Y	N	Y	Y
P[e]	Y	N	Y	N
R	Y	N	N	Y
S	Y	N	N	N

[a]Y, yes; N, no.
[b]Conventional pulse transformers.
[c]Operate in the same manner as E and F.
[d]Operate in the same manner as R and S.
[e]Possible usage but rarely used.

The basic transistor voltage-fed inverter, shown in a simplified schematic in Fig. 13.1, operates in a manner similar to a class D amplifier. The difference in operation is seen in the waveform from the main switching devices. The ideal voltage and current waveforms are shown in Fig. 13.2. The most common type of voltage-fed inverter adjusts the width of the pulse to compensate for variations in voltage. This type of regulation is called pulse-width modulation (PWM). The dead time between the end of the pulse in one direction and the start of the opposite pulse gives one transistor time to turn off before the other transistor turns on.

FETs can be used in place of the bipolar transistors. Circuit operation is similar to operation with bipolar transistors but the operating frequencies are usually higher.

There are many variations of the basic inverter circuit shown in Fig. 13.1. The current-fed inverter operates in a manner similar to the voltage-fed inverter except that the input current is limited. In this type of inverter, the voltage into the main switch is usually regulated.

A third basic type of inverter is the flyback inverter. In this inverter the switches permit the current to build linearly in the primary of the transformer until the desired amount of energy is stored. The switch is then turned off and the energy is transferred to the output. There are also variations of this type of inverter circuit.

The inverter circuits described have used devices for switches which turn off when the input drive is removed. Both bipolar transistors and FETs operate in this way. However, both types of devices have peak currents which are limited to a few times the average current. In addition to this limit, the absolute maximum current for fast-switching bipolar transistors is about 100 A; the peak current in FETs is even lower, which limits the maximum power that can be switched with

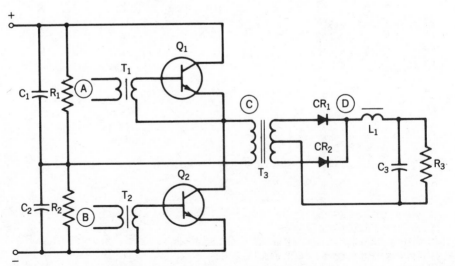

Figure 13.1. Transistor voltage-fed inverter.

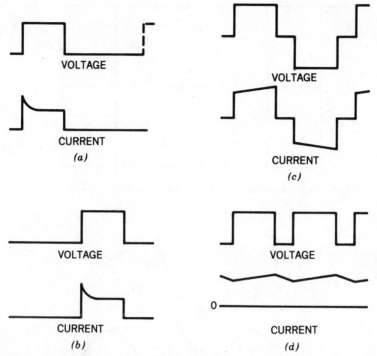

Figure 13.2. Ideal voltage and current waveforms for a voltage-fed inverter. (A, B, C, D indicate points in Fig. 13.1).

a single device. Bipolar transistors particularly decrease in gain toward the maximum collector current so that proportionately more drive is needed as the current increases.

The thyristor is a device that is turned on by an input signal, but once current flow starts it will continue conducting until the current flow direction is reversed. This type of device will not work in the circuits described in the previous paragraphs. To operate in an inverter, a thyristor must have some means of forcing the current to reverse. This is usually done by some type of resonant circuit. A simplified schematic of a type of inverter using thyristors is shown in Fig. 13.3. In this circuit, when SCR1 is switched on, the voltage is applied to the load through the series resonant circuit C_cL_c. The voltage shown in Fig. 13.4(a) starts a current buildup in a sinusoidal fashion [Fig. 13.4(c)], which follows the sine wave until the voltage reaches a peak at the time the current is zero. With no gate signal present on SCR1, the SCR will turn off. If SCR2 is turned on at this time, the sine wave of current will continue in the opposite direction. There are many variations of this basic circuit, some that use a parallel resonant circuit for commutation, others use a circuit that is resonant at some harmonic of the fundamental operating frequency. Each of these inverters has advantages and disadvantages. These circuits are uniquely designed to operate with thyristors, but at lower powers can be used to advantage with bipolar transistors or FETs.

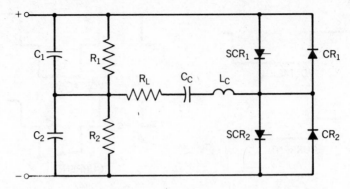

Figure 13.3. Simplified schematic of a series resonant inverter.

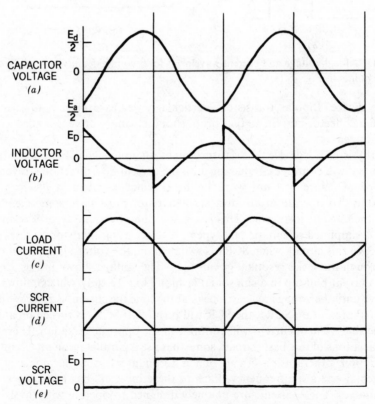

Figure 13.4. Ideal waveforms in an SRI thyristor inverter.

364

13.2 TRANSISTOR OPERATION

Inverters came into common use with the fast-switching high-voltage power transistor. Before the high-voltage transistor became common, a few inverters were used for special applications, such as traction drives, motor controls, and converting low-voltage dc to ac in order to effect a level change. These were mostly applications where special vacuum or gas tubes could be used. These tubes were replaced by thyristors or low-voltage transistors.

When inexpensive transistors with ratings up to 400 V and switching speeds of 1 μs or less became available, a new product line came into common usage, the off-line inverter power supply. These power supplies could be made smaller and are more efficient than the linear regulator type of power supply, particularly for the 5-V power supplies needed to operate digital integrated circuits. These power supplies convert the line voltage directly to dc. This dc is chopped at some high frequency (usually above the audio-frequency range), transformed in level through a high-frequency transformer, and then rectified to produce dc at the desired level.

The transistors are driven by a low-voltage driver (Fig. 13.5). When the transistor turns on the voltage on, the collector goes to the saturation voltage. The transistor remains at the saturation voltage as long as there is enough base drive to maintain the current. When the base drive is removed, the transistor returns to the off-state.

During the voltage rise as shown in Fig. 13.6(a), the current is also changing,

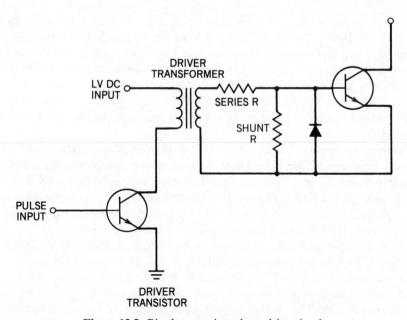

Figure 13.5. Bipolar transistor base drive circuit.

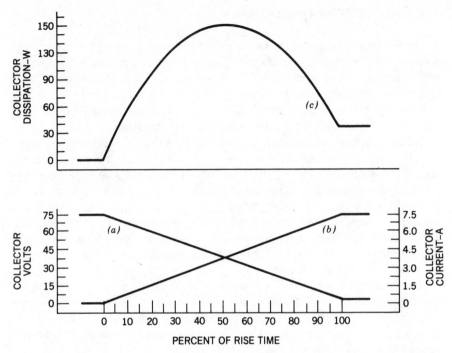

Figure 13.6. (*a*) Turn-on voltage of a switching transistor; (*b*) current; (*c*) power.

as shown in Fig. 13.6(*b*). The power dissipation at this time is as shown in Fig. 13.6(*c*). The power dissipated during this time is very high. If it is too high, the transistor can be destroyed; therefore, it is important to keep the rise and fall times as short as possible. The transistor safe operating area (SOA) rating determines how much current the transistor can conduct at a given voltage. A typical SOA curve is shown in Fig. 13.7. If the current being conducted is high and the time change of voltage and current is too slow, the transistor is subject to damage.

Conditions similar to turn-on occur at turn-off. Again large amounts of power are dissipated for a short time. A typical power dissipation curve during a pulse is shown in Fig. 13.8. The power dissipated during the rise and fall times is very high, but the time is so short that the average power is not excessive. Even though the average power is not above the power rating of the transistor, the device may be destroyed if the rise and fall times are too long.

For example, a bipolar transistor is operating at $V_{cc} = 225$ V. The peak current is 10 A and the saturation voltage is 2 V. The rise and fall times of the transistor are 250 ns each. The transistor is on for 16.5 μs at the base of the pulse. The total period is 50 μs. Assume that the voltage rise and fall times are linear, as shown in Fig. 13.8. The average current is 5 A. The power during the rise and fall times is 725 W each. The power dissipated during the time when the transistor is

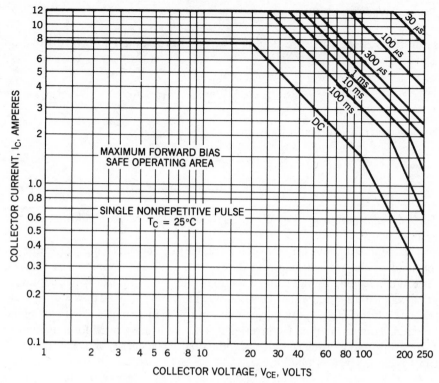

Figure 13.7. Power transistor SOA curve.

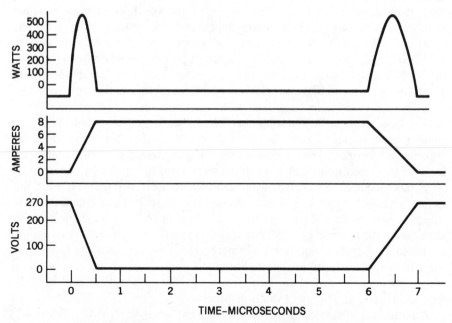

Figure 13.8. Pulse power dissipation in a bipolar transistor.

saturated is 20 W. These are all peak powers. For rise or fall the total dissipation is

$$P_{\text{rise}} = P_{\text{pk}} \times t_{\text{rise}}/T$$
$$= 725 \times 0.25/50 = 3.625 \text{ W}$$

During the transistor saturation time

$$P_{\text{av}} = P_{\text{pk}} \times t_{\text{on}}/T$$
$$= 20 \times 16/50 = 6.4 \text{ W}$$

The total dissipation is 13.65 W with 56% of the loss occurring during the rise and fall times.

The argument above is based on the voltage and current changing together. If the load in the transformer is inductive, this will not happen. The current, instead of decaying to half value when the voltage reaches half value, continues at full value for a longer time. This not only increases the peak power during the fall time but this higher peak power is concentrated in a much smaller area of the junction. Transistors used for inverter service are usually rated for a maximum current through a specified value of inductance. The energy that the transistor can handle is given in millijoules.

When a transistor is driven to saturation, excessive carriers are placed in the collector region. When the base drive is removed, these carriers must be dissipated before the transistor starts to turn off. The time it takes for these carriers to be dissipated is called the storage time. This time is dependent on a number of factors which are circuit dependent. The storage time is usually expressed as a maximum. Proper circuit design can be used to reduce the storage time. In Fig. 13.2 the voltage was shown with a period of time between the turn-off of one transistor and the turn-on of the other. This time is a small amount greater than the maximum storage time of the transistors. If this time is shorter than the storage time of a transistor, both transistors are on at one time, causing a very large current flow during that time. This high current will probably destroy the transistors. This dead time limits the maximum duty at which the inverter can operate. It also determines the minimum pulse width, since the inverter cannot switch a pulse shorter than the transistor storage time.

Voltage-fed inverters are usually ac coupled into the transformer. If this is not done, the transformer design must allow for a dc current unbalance due to unequal storage times. This is usually an air gap in the core magnetic path. In cut C cores the residual gap and flux density may absorb this unbalance, but in ferrite cores a specific gap is usually necessary. This increases the number of turns required to maintain the required OCL. The increase in the number of turns may increase the leakage inductance, with the possibility of exceeding the second breakdown energy of the transistors.

A more recent development than the bipolar transistor is the power FET. There are many variations in the types of FETs but they all work by using a voltage to control the output current. In general, these devices individually do not have the power-handling capability of bipolar transistors. The rise and fall times of FETs are much faster than bipolar transistors and they have no storage time. With higher-power FETs the operating frequencies of many inverters are being extended into the megahertz region.

Bipolar transistors have large and varied storage times. If a number of bipolar transistors are in parallel, the one with the longest storage time will carry all of the current in the last few nanoseconds of the turn-off. This could lead to transistor failure. Bipolar transistors usually require matching or complicated circuitry to force current sharing.

FETs, with no storage time, can be easily paralleled. A few parallel devices can easily equal the power of all but the largest bipolar devices. The FETs can be safely used in voltage-fed inverters without the problems seen with bipolar transistors.

13.3 THYRISTOR OPERATION

If two transistors are connected as shown in Fig. 13.9, a signal applied to the base of Q_1 will start current flowing through Q_2. However, when the signal is removed from Q_1, Q_2 continues to conduct until the current through it tries to

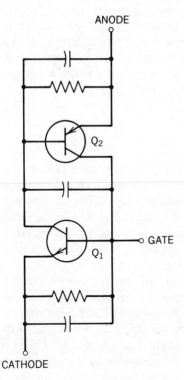

Figure 13.9. Two-transistor thyristor model.

reverse, at which time current flow stops. This is exactly the same action that occurs in a thyristor. The two-transistor connection in Fig. 13.9 is often used as a computer model for a thyristor.

Thyristors are used in inverters called forced commutated inverters. They are usually used for high-power, low-frequency applications. Thyristor turn-on power requirements are almost independent of the main current rating. High di/dt ratings are required for using thyristors in pulse circuits. This limitation is not too serious in inverters where the currents are sinusoidal.

A thyristor requires time for the junction to turn on. If the current rises too high or too rapidly, the thyristor may be destroyed. The di/dt rating of the thyristor is an expression of this factor. Some power thyristors have di/dt ratings as low as 25 A/μs, where others are as high as 1000 A/μs. Thyristors for inverters usually have di/dt ratings of 100 to 200 A/μs.

A second factor to be considered is the dv/dt rating. If a fast-rising voltage is applied to the anode of a thyristor, it can turn on by itself through internal capacitive coupling to the gate. This is usually a gate signal barely sufficient to turn the thyristor on and may result in damage to the device. Fast-rising voltages may occur in thyristor inverters when one thyristor is turning on when another is turning off. It is customary in inverter design to bypass the thyristor with an RC circuit whose time constant is longer than the critical dv/dt for the thyristor.

Fast-switching transistors turn off in 1 μs or less. FETs turn off in nanoseconds. A fast-recovery thyristor will recover its blocking properties in 10 μs. At high powers the turn-off time may be 25 μs or longer. This turn-off time limits the frequency at which a thyristor inverter can operate. Thyristor inverters up to 20 kW have been operated in the 10- to 20-kHz region. Higher-power inverters usually operate at lower frequencies. Most thyristor inverters are used for dc-to-ac conversion, although a few are used for other applications.

13.4 DC-TO-DC INVERTERS

By far the most common type of inverter for power supplies is the pulse-width modulated (PWM) inverter, using bipolar transistors or FETs as switching devices. When a unidirectional pulse train is applied to an LC filter, the output of the filter is a dc voltage which is proportional to the average value of the pulse train. If the pulse width is adjusted to compensate for variations in pulse amplitudes, a constant dc voltage output is obtained. Most switching power supplies use this technique for voltage regulation.

The PWM-controlled inverters fall into two groups, the current-fed inverter and the voltage-fed inverter. The voltage-fed inverter is the most common since it is the least expensive to build. A schematic of a voltage-fed half-bridge inverter is shown in Fig. 13.1. The voltage-fed inverter power supply shown in Fig. 13.10 used the transformer whose design example was given in Chapter 12. The two most common problems with this type of circuit are too much storage time in

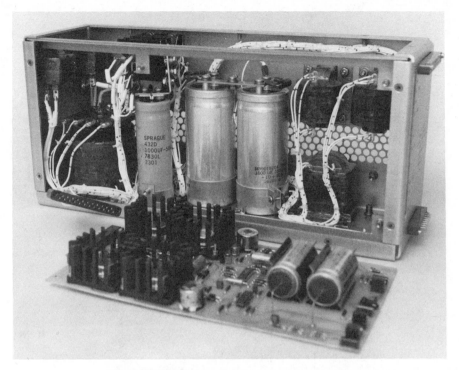

Figure 13.10. Voltage-fed inverter power supply.

the switch transistors and not enough capability to withstand the peak current during transistor turn-off. Both of these conditions may be potential problems in a particular application, but the proper choice of inductive component parameters can eliminate the problems.

With the voltage-fed inverter, if one half of the bridge stays on when the other half turns on, a current limited only by the dc resistance of the circuit can flow through the transistors. This current will probably destroy one or both of the transistors.

To overcome this problem, current-fed inverters are used. Figure 13.11 is a simplified schematic of a current-fed inverter. In the current-fed inverter an inductance is placed in the circuit between the switching transistors and any energy storage capacitors. The transistors drive the power transformer with a square wave. There is crossover between the time the first transistor switches off and the second transistor is on, but the inductor in the input prevents the current from rising significantly during this time.

Regulation of this type of inverter is usually achieved by controlling the voltage into the main switch with a series regulator. A general circuit for series regulator is shown in Fig. 13.12. The dc output voltage is sensed and compared to a reference voltage V_{ref}. The output voltage is controlled by the series element Q_1. The voltage across Q_1 must be great enough to maintain control at the

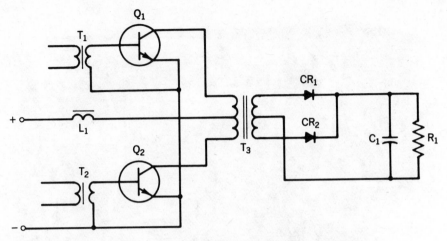

Figure 13.11. Simplified schematic of a current-fed inverter.

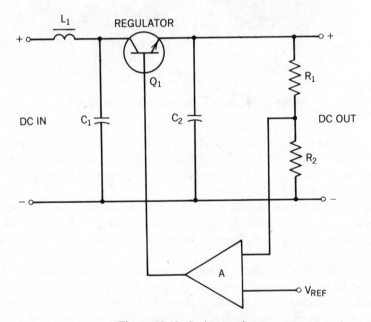

Figure 13.12. Series regulator.

lowest input voltage and at the lowest point of the ripple voltage. If the current is a few amperes, the series element must dissipate many watts.

To prevent this high loss a series switching regulator is used (Fig. 13.13). In the switching regulator, the series element is turned on to saturation. When the output voltage reaches a preset value, the transistor is switched off.

Current flow through the load is supplied during the off-time by the inductor

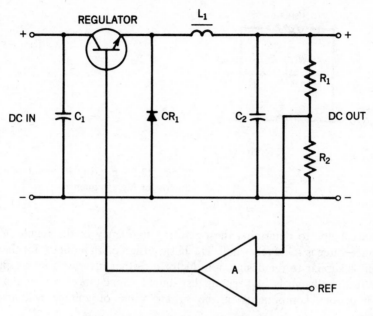

Figure 13.13. Switching series regulator.

L_1 and the capacitor C_2. The transistor is turned on by a pulse. When the voltage at the input of the inductor rises, the current through the inductor supplies energy to the output. When the voltage on the capacitor reaches a preset level, the switch is turned off. The choice of the inductor value is determined by how fast energy is transferred to the output. The value of L is determined by

$$L = \frac{V_{\text{out}}(V_{\text{in}} - V_{\text{out}})}{\Delta I_L \times f \times V_{\text{in}}} \tag{13.1}$$

where ΔI_L is the change of current through the inductor. The power loss in the transistor is now the peak current times the saturated voltage averaged over the total period. The turn-on time of the series switch is often synchronized with the turn-on time of the main switching transistors. The main switching transistors are used strictly as switches and do not function in the regulation of the output voltage.

The design of the inductor is determined from these waveforms. The equivalent circuit of the inductor is shown in Fig. 13.14. The sudden application of voltage to this equivalent circuit can have large instantaneous currents since the inductance is shunted by the distributed capacitance. It is therefore necessary to design the inductor with low distributed capacitance. The inductor is also subjected to relatively large voltage swings. At most inverter frequencies the ac flux density is still low enough to keep the core loss low. Since the value of

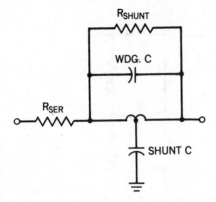

Figure 13.14. Equivalent circuit of an inductor for high frequencies.

inductance helps to determine the operating frequency of the regulator, the inductance value is relatively small. The basic design of an inductor for the type of circuit is similar to the design of inductor for other inverter power supplies. The design challenge is in keeping distributed capacitance low so that the higher-frequency components in the square wave of voltage will also be attenuated.

In the voltage-fed inverter each transistor is turned off before the other transistor is turned on. This means that twice during each cycle the transformer disconnected from both the input and output circuits. The effect of this turn-off is to produce ringing between the pulses. In the current-fed inverter, the transistor on-times overlap each other. The transformer is therefore loaded at all times. There are still spikes produced when the transistors turn off, but the energy involved in these spikes is less. The load change is from full load to one-half load back to full load, so the energy is one-fourth as such as in the voltage-fed inverter. The input regulator still produces spikes. All of these spikes can be capacitatively or inductively coupled into the output unless great care is taken in laying out the circuit.

A third type of transistor circuit which is finding usage is the flyback inverter. This is a higher-power modification of the flyback circuit used for horizontal sweep and high-voltage generation in CRT circuits (Chapter 11). In the flyback inverter, the transistor switches are turned on and current flows through the primary of the inverter transformer. When a predetermined current is reached, the transistors are turned off. The energy in the transformer primary transfers to the output. The waveforms for this type of inverter are shown in Fig. 13.15 along with a simplified schematic. The flyback inverter requires some very careful transformer design to prevent ringing from appearing on the output waveform. Unless great care is taken in design, the ringing can approach the same amplitude as the main pulse (Section 11.3).

A flyback inverter can operate in either a continuous or a discontinuous mode. In the continuous mode more energy is stored in the transformer than is

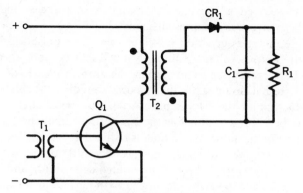

Figure 13.15. Flyback circuit.

delivered to the load. The amount of energy put into the transformer during each on cycle is equal to the energy delivered to the load. In the discontinuous mode, the total energy stored in the transformer is delivered to the load and some time is allowed to elapse before the recharge is started. The energy delivered to the load in the discontinuous mode is the peak stored energy times the duty cycle. Continuous-mode operation makes less perturbation on the power line than operation in the discontinuous mode. Some flyback inverters are designed to work in both modes, depending on the amount of power output required.

The energy transferred to the output is $LI^2/2$ joules for each pulse. A square wave of voltage is applied to the primary of the transformer, but the current is a linear ramp, as described in Chapter 11. The voltage is regulated by controlling the amount of energy stored in the primary of the transformer. Since the inductance is fixed, the current through it is controlled to regulate the output voltage. The transformer must be designed so that the time required to reach the desired current is longer than the time it takes to transfer the energy into the output capacitor. This imposes a severe limitation on the amount of leakage inductance. A high-permeability core is useful in this type of design. If the losses in the transformer are too great, much of the energy stored in the primary can be lost before it is transferred to the output.

A recent modification of the flyback inverter is the feedforward inverter. Schematically, it looks the same as the flyback inverter. In the feedforward inverter energy is transferred to the output while the transistors are on. The output current is continuous. The input current is ramped in a manner similar to the flyback inverter. Energy is transferred to the load in both the on and off cycles. The transformer design is similar to a flyback transformer.

All of the types of inverters mentioned above can use any switch that can be turned off by a control element. Most of these inverters were developed to use bipolar transistors. Many of these inverters are now using FETs. As switching devices are developed which operate at higher frequencies, the transformer design for these inverters will present many challenges.

All of these inverters generate spikes of very short pulse widths which can appear on the output voltage. Although much can be done to remove these spikes, the best prevention is to design the transformer to minimize spike generation. External circuitry can then be used to reduce the spikes that are left.

Filter inductors for these types of inverters also need careful design or the inductor may be self-resonant below the major ripple frequency. If this happens the effectiveness of the inductor as a filter is reduced and at frequencies where the harmonics of the ripple are large, may produce little or no attenuation. For example, an inductor with 100 pF of distributed capacitance and inductance of 55 mH is resonant at 120 kHz, which is the sixth harmonic of the output ripple of a 20-kHz inverter. At frequencies above 120 kHz the shunt capacitance of the inductor becomes the predominant factor. The impedance decreases, until at some higher frequency the impedance of the inductor is less than the inductive impedance of the filter capacitor. Any frequency at which there is a resonance will probably be amplified. It is possible to have ripple spikes on the output of the power supply which exceed the amplitude of spikes at the transformer. Inductors designed for minimum capacitance are required to filter inverter power supplies.

Often the noise generated by an inverter cannot readily be removed from the output by a conventional *LC* filter. Some of this noise is common mode; that is, it appears equally on both the go and return leads and is in phase on these leads. To filter this noise a common-mode inductor is used. A common-mode inductor has two or more closely coupled windings of equal turns on the same core. The voltage impressed on one winding generates a flux in the core. The windings are so connected that the flux generated by the voltage on the second winding is out of phase with the flux generated by the first winding. Since the fluxes cancel by transformer action, there is no coupling of the noise to the output. When a common-mode inductor is used in a dc path, the windings are arranged so that the dc flux also cancels. Since there is no net dc flux, a very high permeability core with no gap can be used. Common-mode inductors can usually be made very small, with a few turns of wire large enough to carry the current. Very high current outputs may use a core around the wires.

Common-mode inductors may also be used on ac power lines or signal lines to prevent unwanted noise coupling. When used to decouple signal lines, the signal may be affected by the leakage inductance. Therefore, very few turns very closely coupled are used.

For ac power lines, the common-mode noise cancels but the fundamental component of the ac current does not cancel. The inductor must be designed for the ac flux density. A common-mode inductor for ac lines may have many turns to keep the flux density below saturation at the fundamental frequency. The effectiveness at higher frequencies is usually affected by the leakage inductance and distributed capacitance.

13.5   DC-TO-AC INVERTERS

Transistor circuits have found great use as inverters, particularly in dc-to-dc inverters. There are many applications for converting dc to sine-wave ac. Switching devices used in inverters must handle large peak currents. Transistors are limited in peak current capability. As the collector reaches saturation the base current must be increased to the point where the base current is 20% or more of the collector current. This has limited the power in transistor inverters to about 1 kW. (There are some transistor inverters that go to 20 kW at 400 Hz.)

Most of the inverters that convert dc to ac sine waves have high output power. These inverters are used for changing frequencies, providing an uninterruptible power source for computers or to produce ac to step up from some source, such as a thermoelectric generator or fuel cell.

Higher-power inverters usually use thyristors as switches. Thyristors, once turned on, do not turn off until the current through them reverses and stays reversed for the recovery time of the thyristor. The types of inverters discussed in Section 13.4 were turned off at the peak of the current by turning off the switching devices. A thyristor must have current reversals to turn off. Since the dc input current is unidirectional, some means of forcing this current reversal must be provided. This is usually done with a *LC* resonant circuit.

There are two basic types of inverters using resonant circuits, the series commutated and parallel commutated. In the series commutated inverter the resonant circuit is in series with the load. The parallel commutated inverter has the resonant circuit in parallel with the load. Many variations of these types of inverters have been developed. These are all variations of the two basic types. The most commonly used of these two types of forced commutated inverters is the series resonant inverter (SRI).

Series resonant inverters have many uses. They are used to generate ac power from batteries for uninterruptible power systems, change frequencies (60 to 400 Hz or 400 to 60 Hz), radar transmitter pulser charging circuits, and other applications. The powers handled can be from a few kilowatts for small specialized power sources to many megawatts, as in the case of inverters for changing extra high voltage dc to ac for normal 60-Hz power transmission.

The design criteria of the inductive components is not significantly different than for any of the applications discussed previously. Figure 13.3 is a simplified schematic of an SRI. There are two inductive components present in this type inverter, the transformer and the commutating inductor. In low-frequency applications, the design of the transformer is similar to the design of a transformer for power applications. The leakage inductance is more important than it is for a power transformer, since the leakage inductance acts in series with the commutating inductance. If the leakage inductance is too high, the resonant frequency may be too low. If a harmonic of the fundamental frequency is used for commutation, the leakage inductance becomes even more important. The SRI can readily be connected to give multiphase output for ac circuits.

13.6 SPECIAL CIRCUITS

In Chapter 11 dc resonant charging of pulse networks was discussed. Another way of charging PFNs or capacitors is through a series resonant inverter (Fig. 13.16). In this method pulses of constant energy are used to charge the PFN capacitor. The amount of energy to be stored in the PFN is calculated as

$$E = CV^2/2$$

The energy per pulse is merely the total energy divided by the number of pulses. If the number of pulses is reasonably large, the total energy can be regulated by turning off the inverter when the desired energy is reached. The energy stored will be within $1/N$ of the desired energy. This is an important consideration in modern radar, where the performance of the system may be determined by the transmitter stability from pulse to pulse. The pulse-to-pulse transmitter stability is in turn determined by how accurately the pulse-to-pulse voltage on the final power amplifier tube can be controlled.

An inverter for charging a PFN may be operated at frequencies between 5 and 20 kHz. The top operating frequency is determined by the turn-off times of thyristors which will handle the current required. The output is rectified. The resonant frequency is determined by the energy storage capacitor in the inverter and the PFN capacitance in series, transformed to either the primary or secondary level. The commutating inductor is usually placed in the primary of the transformer, where the voltage is lower. If the leakage inductance is too great, the energy transfer will not take place in the desired time and the PFN will

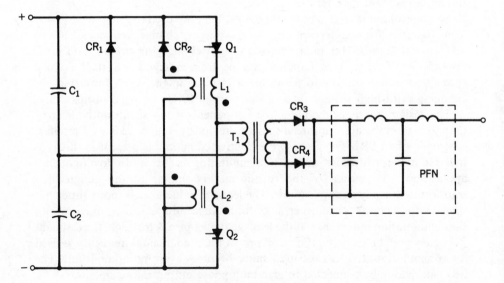

Figure 13.16. PFN charging.

not charge to the desired energy. The inductance of the commutating inductor may be decreased to compensate for the leakage inductance if the value of leakage inductance is not too great.

If the leakage inductance could be controlled, the commutating inductor could be eliminated. Controlling the leakage for setting the inverter frequency is risking an expensive transformer for the saving of one component. It is best to try to make the leakage inductance small compared to the commutating inductor value. In this way random changes in leakage inductance from transformer to transformer have an insignificant effect on inverter operation.

14 OTHER MAGNETIC DEVICES

The *Standard Dictionary of Electrical and Electronics Terms* (IEEE, 1977) defines an "electronic transformer" as "any transformer intended for use in a circuit or system utilizing electron or solid state devices...." In the previous chapters the emphasis has been placed on what could be identified as power transformers and inductors, those that operate in circuits at a fixed low frequency or over a relatively narrow frequency band, and those transformers that must operate actually or effectively in circuits over some frequency range: wide-band, pulse, and inverter transformers.

There are many other magnetic devices which, based on the definition above, could be classified as electronic transformers. This chapter includes a brief discussion of some of these other magnetic devices which have unique applications in electronic circuits.

14.1 THYRATRON TRANSFORMERS

Anode transformers used for supplying thyratrons resemble rectifier anode transformers but generally have higher rms current for a given direct current in the load, and are more subject to voltage surges. With resistive loads, anode current has the same wave shape as the shaded portion of the anode voltage in Fig. 14.1. The relation of peak, rms, and average currents is shown in Fig. 14.2 as a function of firing angle θ for single-phase full-wave circuits. Voltage reduction as a function of θ is shown in Fig. 8.13. If a transformer is designed for operation with zero firing angle, maximum current flows in any given load; the transformer is then capable of carrying the current with greater firing angle, so long as the

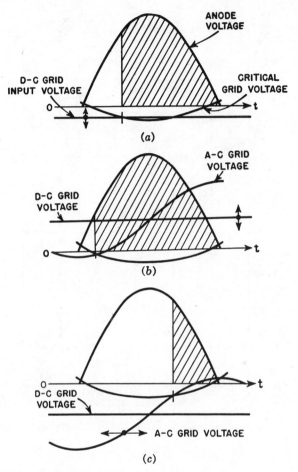

Figure 14.1. Grid control of thyratron with (a) variable dc grid voltage, (b) variable dc grid voltage with superposed ac grid voltage, and (c) fixed dc grid voltage with superposed ac grid voltage of variable phase position.

load impedance remains the same. If the load impedance is changed with $\theta > 0$ to keep the load current as high as possible, the limiting value may be found from Fig. 14.2. The average load current which may flow without overheating the transformer decreases as θ increases.

High-voltage surges occur when capacitance input filters are used with grid-controlled rectifiers. To a degree these surges are likely to occur even when the load is nominally resistive, because of incidental capacitance in the transformer, wiring, and other components. If the load is a radio-frequency generator, the RF bypass capacitor adds to this effect. In Fig. 14.3(a) the total amount of external capacitance is designated C_1. A half-wave anode transformer is shown for simplicity, but each half of a full-wave transformer, or each phase of a three-

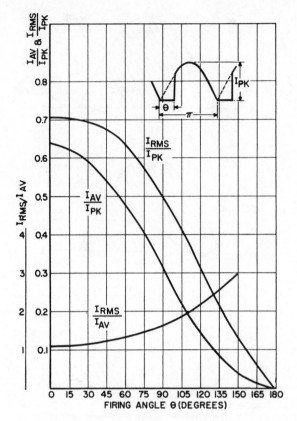

Figure 14.2. Single-phase thyratron currents.

phase transformer, behaves similarly. When the thyratron firing angle is greater than zero, a steep voltage wavefront occurs at the instant of firing t_θ, [Fig. 14.3(b)], as follows:

Normal voltage induced at point A in the secondary winding is e_1 volts above ground, just prior to t_θ. As soon as the thyratron fires, the external wiring and circuit capacitance C_1 momentarily forms an effective short circuit from A to ground. A large surge current flows into this short circuit, but initially this current cannot be drawn from the primary because of the inevitable inductance of the windings. The initial current is therefore supplied by the secondary winding capacitance. Since point A is momentarily short-circuited, a surge voltage, equal and opposite to e_1, is developed in the secondary winding. This voltage surge appears across the turns or layers of winding nearest to A. Unless precautions are taken in the design of the anode transformer the voltage may be high enough to damage the winding insulation.

As is shown in Chapter 10, with steep wavefronts in single-layer windings initial voltage distributes most equally between turns when ratio $\alpha = \sqrt{C_g/C_w}$ is

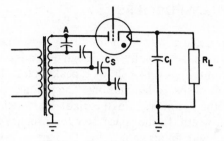

(a) SCHEMATIC CONNECTIONS FOR EACH PHASE

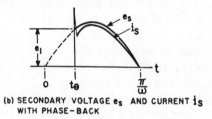

(b) SECONDARY VOLTAGE e_s AND CURRENT i_s
WITH PHASE-BACK

Figure 14.3. Thyratron plate transformer operation.

small, C_g being the capacitance of the winding to ground and C_w the series capacitance across the winding. If the secondary of Fig. 14.3 were a single-layer winding of n turns. C_w would be C_s/n. In multilayer coils, ratio α is not so readily defined. In general, small effective layer-to-layer capacitance means small effective C_g in relation to C_w, small α, and more linear initial distribution of voltage. Many layers are better than few layers in keeping capacitance C_g small. In the limit, a one-turn-per-layer coil would have small α and good initial voltage distribution. In practice this extreme is not necessary to avoid layer insulation breakdown. It is usually sufficient to split the secondary into part coils, like S_1 and S_2 in Fig. 3.13. This reduces C_g to a quarter of the corresponding capacitance of full-width coils. Ratio α is reduced, and voltage distribution improved.

Even with part coils there is some nonlinearity of voltage distribution, especially in the top layer. This nonlinearity may be minimized by providing a static shield over the top layer and connecting it to point A (Fig. 14.3). The momentary voltage described above appears within the winding, and unless there are taps it may not be observable. If a surge suppressor circuit (usually a capacitor-resistor network across the secondary) is used, it does not appreciably diminish the internal winding voltage surge, but such a surge suppressor may be necessary to damp out oscillations in the external circuit due to firing of the thyratrons.

14.2 PEAKING TRANSFORMERS

To produce a steep peaked waveform for firing thyratron tubes, sometimes special transformers are used. Usually, the design depends on the nonlinearity of the magnetizing current. Figure 14.4 shows a peaking transformer in which the magnetic core is made of special laminations. The primary is wound on the full-width left leg, and the secondary on the right leg which is made of a few laminations of small width. In the space between primary and secondary is a laminated shunt path with an air gap. In Fig. 14.5 are shown the core fluxes ϕ_m and ϕ_s, linking the primary and secondary coils, respectively. At low inductions, the same flux links both coils. As the flux rises from zero in each cycle, at first all the flux links the secondary coil, but because of the smaller cross section of the right leg it saturates at the value ϕ_s and the main flux ϕ_m flows through the shunt path. Thus there is a long interval in each cycle during which the flux change is substantially zero, and no voltage is induced in the secondary coil. During the short period θ_s, a voltage is induced in the secondary coil which has a very peaked waveform. This happens twice in each cycle.

Because of the shorter time for the change in ϕ_s, $d\phi/dt$ would remain nearly constant over the angle θ_s if there were no leakage flux, and for $1:1$ turns ratio there would be approximately equal volts in the primary and secondary coils. The secondary flux change takes place over a much shorter period of time, and the flux rises to only a fraction of its maximum value ϕ_m. Therefore, less core area is needed in the secondary leg to obtain the desired voltage e_s. This leads to

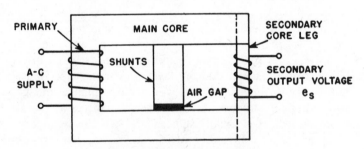

Figure 14.4. Peaking transformer.

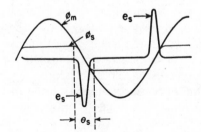

Figure 14.5. Fluxes and secondary voltages in peaking transformer.

the following approximate ratio:

$$\frac{A_s}{A_p} = \frac{\theta_s}{\pi} \sin \frac{\theta_s}{2} \tag{14.1}$$

where A_s = core area in the secondary leg
A_p = core area in the primary leg

Peaked secondary voltage may be made steeper by the use of nickel-iron laminations in the secondary leg, because these alloys have sharp saturation. The air gap in the center leg prevents it from shunting all the primary flux, which would reduce the secondary voltage to zero. This air gap should be no more than 5 to 10% of the window height, to keep leakage flux from threading through the secondary coil and giving a less peaked waveform. With this length of air gap and a total window length of twice the window height, the secondary turns are, for a 1 : 1 voltage ratio,

$$N_s \approx 2N_p \tag{14.2}$$

where N_p is the number of primary turns.

Transformers may be made to peak by the use of special circuits instead of special cores. Voltage waveforms like Fig. 14.5 are obtainable if the primary winding is operated at a voltage exceeding saturation but is connected in series with a large linear inductance or other high impedance. Grain-oriented core material, with its rectangular hysteresis loop, is well suited to peaking transformers. When primary and secondary windings are wound over the same magnetic path, the same volts per turn are induced in both windings, and equation 14.2 no longer applies. A peaking inductor circuit is shown in Fig. 8.24.

14.3 CURRENT-LIMITING TRANSFORMERS

Filaments of large vacuum tubes sometimes must be protected against the high initial current they draw at rated filament voltage. This is done by reducing the starting voltage automatically through the use of a current-limiting transformer, with magnetic shunts between primary and secondary windings. The shunts carry very little flux at no load; as the load increases, the secondary ampere-turns force more of the flux into the shunts until at current I_{sc} [Fig. 14.6(b)], the output voltage is zero. The same principle is used to limit current in transformers for oil-burner ignition, precipitrons, and neon or other gas-filled tube signs.

Cross-sectional area through each shunt path is the same as that of the upper or lower leg of the shell laminations; then flux in the shunts does not exceed that in the core, shunt iron loss is not abnormal, and secondary voltage is sinusoidal. At short-circuit current I_{sc}, half the total flux flows through each set of shunts.

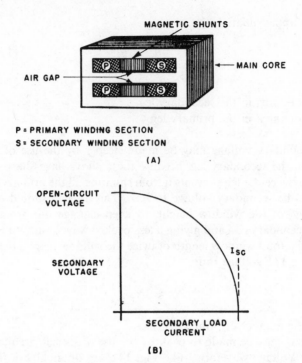

Figure 14.6. (*A*) Current-limiting transformer; (*B*) output voltage versus current curve.

The air-gap length in each shunt path can be found from equation 3.6:

$$l_g = 0.6NI_{sc}/B_m \quad \text{in.}$$

where N = secondary turns
B_m = allowable induction in the shunts, G

The constant 0.6 is generally too small because of the flux fringing around the gap. The increase of gap made necessary by fringing may be found from Fig. 3.24. If the shunts are too short, the transformer does not limit the current properly. It is best to have slightly less air gap than necessary, and find by trial the right length of shunt. Fringing flux heats the coils and core somewhat more than in an ordinary transformer. If the secondary current is heavy, coils are wound pancake fashion and connected in parallel; they may have to be cross-connected for the coils to divide the load equally.

If the ordinate for open-circuit voltage and abscissa for short-circuit current in Fig. 14.6(*b*) are equal, the curve is a quarter-circle for a perfect transformer because the secondary current at short-circuit is all reactive. With core, shunt, and winding losses the curve for an actual transformer falls some 10 to 15% less than the quarter-circle at currents 0.5 to 0.75 times I_{sc}.

14.4   AUTOTRANSFORMERS

An autotransformer has a single winding which is tapped as shown in Fig. 14.7 to provide a fraction of the primary voltage across the secondary load. The connections may be reversed so that a step-up voltage is obtained. The regulation, leakage inductance, and size of an autotransformer for a given rating are all less than for a two-winding transformer handling the same power. Where the voltage difference is slight, the gain is large. Where the voltage difference is great, there is not much advantage in using an autotransformer, nor can it be used where isolation of the two circuits is required.

Autotransformers are used in electronic applications chiefly for the adjustment of line voltage, either to change it or to keep it constant. Examples are the reduction of plate voltage for tuning an amplifier and the maintenance of constant filament voltage. Taps may be chosen by means of a tap switch to adjust the load voltage. The load voltage may be adjusted to within half the voltage increment between taps.

If the voltage is adjusted while load remains connected, bad switching arcs occur, either from breaking the circuit or from short-circuiting turns. To provide for adjustment under load conditions, a resistor may be momentarily connected in the circuit as the tap switch bridges from one tap to the next, and current is limited to full-load value. In large power tap changers, an inductor replaces the resistor to avoid heating and losses.

The VA rating of an autotransformer depends on the ratio of input to output voltage. In Fig. 14.7 the output current $I_2 = I_1 + I_3$. Let p = percent tap/100 $= E_2/E_1$. Neglecting losses, $I_2 = I_1/p$ and $I_3 = (1/p - 1)I_1$. Then

volt-amperes (in the upper portion) $= (1 - p)E_1I_1$

volt-amperes (in the lower portion) $= pE_1I_3 = (1 - p)E_1I_1$

which satisfies equality of volt-amperes in each section. For ratio p close to unity, the VA rating and hence size for a given output can be made very small; for small values of p the size is not much less than that of a two-winding transformer, but the autotransformer has much less regulation. Its effective

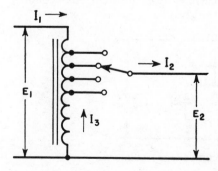

Figure 14.7. Autotransformer voltages and currents.

winding reactance and resistance decrease as $(1 - p)^2$; that is, for a given unit,

$$\frac{X \text{ (or } R) \text{ as autotransformer}}{X \text{ (or } R) \text{ as two-winding transformer}} = (1 - p)^2 \qquad (14.3)$$

Appreciably less regulation is obtained in an autotransformer, even when size is not reduced much, because the right-hand term in equation 14.3 is squared.

When the power for electronic equipment is supplied by a 230-V line, but auxiliary items such as relays and small motors are used at 115 V, a convenient way of obtaining the latter voltage is to center-tap the primary of a large plate transformer, and use it as a $2:1$ step-down autotransformer. The larger primary winding copper requires little extra space, and an additional transformer is thereby saved.

To improve the closeness of voltage control, a variable autotransformer has been developed in which the moving tap is a carbon brush which slides over exposed turns of the winding. Brush resistance prevents excessive transition current and permits smooth voltage control; yet it offers little additional series resistance to the load. The same idea can be applied to two-winding transformers for secondary voltage adjustment.

When autotransformers are used on three-phase supply lines, they may be connected the same as two-winding transformers in star, delta, open-delta, or Scott connections. The last two connections are less subject to objectionably high regulation in autotransformers and, if they supply three-phase anode transformers, cause no serious primary voltage unbalance for voltage ratio p close to unity.

14.5 STATIC VOLTAGE REGULATORS

Automatic regulators of various kinds have been devised for keeping comparatively small amounts of power at a constant voltage. Figure 14.8 shows one circuit for a resonant-inductor voltage regulator. Inductance L_1 is linear. Inductance L_2 and capacitor C_2 are parallel resonant at the supply line frequency and rated voltage. The pair draws very little current, so that the reactive voltage in L_1 is low. Output current flows through its secondary

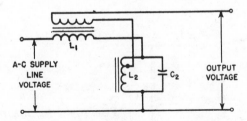

Figure 14.8. Resonant-circuit voltage regulator.

winding which is of such polarity as to maintain rated voltage. Inductance L_2 is partially saturated at this voltage. If line voltage falls below rated value, less current is drawn by L_2, and the L_2C_2 combination becomes untuned. Total current to the parallel circuit is then capacitive, and this capacitive current, drawn through L_1, raises the output voltage. Conversely, if line voltage rises above rated value, the L_2C_2 combination becomes untuned on the inductive side, and the output voltage falls below the line value. Output voltage variations of $\pm 1\%$ are obtained with $\pm 10\%$ line voltage variations in this manner, and with load changes from zero to full load.

Constant supply frequency is a condition for resonance at rated voltage; with the good frequency control of modern power systems this condition is generally fulfilled. Load power factor variations cause output voltage to change. Some regulators are provided with taps to minimize this effect. Output waveform contains a noticeable third harmonic, because the large magnetizing current of L_2 must flow through appreciable impedance in L_1. Owing to the partial saturation of inductor L_2, it tends to operate at a high temperature and requires good ventilation. Practical regulators are in use with ratings from 25 VA to several kVA.

14.6   CURRENT TRANSFORMERS

In the electric power industry instrument transformers are used to reduce the voltage or current being measured to a safe metering level; they are also used to provide the voltages and currents required for relay operation in control and protective circuits. In these applications the transformers must be accurate to a fraction of a percent from virtually no load to full load.

In electronic circuits high voltages are usually monitored through resistive or capacitive voltage dividers and consequently potential transformers are rarely used. Current-sensing transformers, on the other hand, are often used in electronic circuits to provide relatively gross indications of the magnitude of current or changes in current, and as a consequence they do not require the accuracy of those used in the electric power industry.

The current transformer is designed to have its primary winding connected in series with the circuit carrying the current to be measured or sensed. The most common type of current transformer used in electronic circuits is the window type, which may not have a primary winding. The transformer has an opening, or window, through which the conductor carrying the current to be sensed is passed while it is being assembled into the equipment as shown in Fig. 14.9.

In an ideal transformer, the transformer secondary current is equal to the primary current multiplied by the ratio of primary turns to secondary turns and the primary and secondary ampere turns are equal.

In practice the secondary current is always slightly less than that of the theoretical ideal because the vector difference of the primary and secondary ampere turns must sustain the magnetic flux.

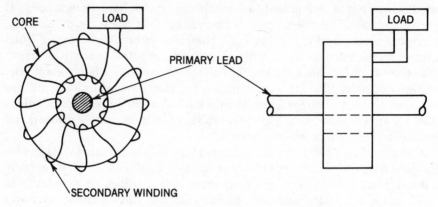

Figure 14.9. Window type of current transformer.

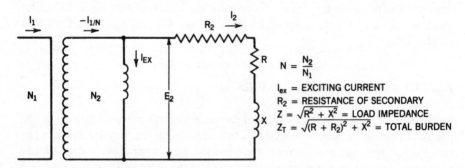

Figure 14.10. Current transformer equivalent circuit.

The equivalent circuit of the current transformer is shown in Fig. 14.10; it can be seen that the exciting current has an effect on both the current ratio and the phase angle. For that reason, in applications where currents are being sensed over a relatively wide range, the percent exciting current must be kept lower than is the case with power transformers. This is accomplished by using high-permeability core materials which are operated at relatively low flux densities to keep the exciting volt-amperes as low as practicable. Most often nickel-iron tape-wound toroids are used in current transformers.

14.7 BALUN TRANSFORMERS

The balun is a transformer used to connect a balanced load to an unbalanced (single-ended) source and is a special form of the wide-band transformer discussed in Chapter 7. The balun transformer has been used in a variety of forms for many years, some of which are bifilar, air-core devices used in the lower-frequency broadcast bands. The balun transformers used in electronic circuits, however, usually are wound on high-frequency low-loss high-permeability ferrite toroids which permits operation over very wide frequency ranges.

As with all wide-band transformers the low-frequency response of the balun is determined by the open-circuit inductance of the primary. Hence high-permeability core material is desired so that the primary turns are kept to a minimum and bandwidth is increased. However, bandwidth in the conventional

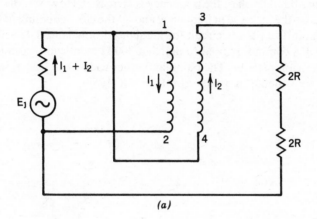

(a)

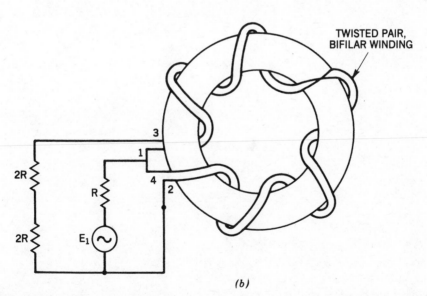

(b)

Figure 14.11. Balanced-unbalanced 4:1 impedance transformer. (a) Equivalent circuit; (b) wiring diagram.

transformer is limited by the resonance of the interwinding capacitance with the leakage inductance. To extend the high-frequency response, the balun is designed as a transmission line with the coils arranged so that the interwinding capacitance becomes a part of the characteristic impedance of the line. In this way the bandwidth is not limited by resonances. To obtain the close coupling required for good high-frequency response, the windings are normally wound bifilar, a twisted pair, usually of enameled magnet wire.

It was noted above that high-permeability core material was needed for good low-frequency response. The permeability of some ferrites is relatively high at low frequencies but falls off as the frequency is increased. However, the reactance is maintained by the increase in frequency and by the close coupling of the bifilar windings. Figure 14.11 is the circuit for a balanced-unbalanced 4:1 impedance transformer; it is commonly used to match a 300-Ω twin-lead transmission line to a 75-Ω television receiver. The balun and other wide-band transformers were discussed in some detail in a paper by Ruthroff (1959).

REFERENCES

Barnitkas, R., and E. J. McMahon, eds. 1979. *Engineering Dielectrics*, Vol. 1, *Corona Measurement and Interpretation*, STP-669. Philadelphia: American Society for Testing and Materials, pp.3–4.

Bode, H. W. 1945. *Network Analysis and Feedback Amplifier Design*, D. Van Nostrand and Company, New York.

British Electrical and Allied Industries Research Association. 1941. *Surge Phenomena*. London: BEAIRA, pp. 223–226.

Carter, C. E., and S. I. Rambo, 1969, "Derivation of Pulse Transformer Parameters on a Digital Computer." *IEEE Workshop on Applied Magnetics*.

Butterworth, S. 1926. "Effective Resistance of Inductance Coils at Radio Frequency." *Wireless Eng.* (April).

Dakin, T. W. 1986. "Insulation Reliability-5." *IEEE Electrical Insulation Magazine*, Vol. 2, No. 4, p. 54, July.

Dakin, T. W. 1948. "Electrical Insulation Deterioration Treated as a Chemical Rate Phenomenon." *Trans. AIEE*, **67**, 113.

Dakin, T. W., and Berg, O. 1962. "Theory of Gas Breakdown." Progress in Dielectrics-4, Heywood and Company Ltd., London.

Dishal, M. 1951. "Alignment and Adjustment of Synchronously Tuned Multiple Resonant Circuit Filters." *Proc. IRE*, **39**, 1448 (November).

Dornhoefer, W. J. 1949. "Self-Saturation in Magnetic Amplifiers." *Trans. AIEE*, **68**, 835.

Friend, A. W. 1947. "Television Deflection Circuits." *RCA Rev.*, 98 (March).

Glasoe, G. N., and J. V. Lebacqz. 1948. *Pulse Generators*, MIT Radiation Laboratory Series, Vol. 5. New York: McGraw-Hill Book Company, p. 567.

Gokhale, S. L. 1929. *J. AIEE*, **48**, 770 (October).

Gustafson, W. G. 1938. "Magnetic Shielding of Transformers at Audio Frequencies." *Bell Syst. Tech. J.*, **17**, 416 (July).

Hanna, C. R. 1927. "Design of Reactances and Transformers Which Carry Direct Current." *J. AIEE*, **46**, 128 (February).

Ingersoll, L. R., and O. J. Zobel. 1913. *The Mathematical Theory of Heat Conduction.* Boston: Ginn and Company, p.142.

Institute of Electrical & Electronics Engineers, Inc. 1977. *Standard Dictionary of Electrical and Electronic Terms*, IEEE Std. 100-1977. New York: IEEE.

Kantor, Myron. 1947. "Theory and Design of Progressive and Ordinary Universal Windings." *Proc. IRE*, 1563 (December).

Lee, R. 1954. "False Echoes in Line-Type Radar Pulsers." *Proc. IRE*, **42**, 1288 (August).

Lee, R., and D. S. Stephens. 1973. "Influence of Core Gaps in Design of Current-Limiting Transformers." *IEEE Transactions on Magnetics*, pp. 408–410 (September).

Lord, H. W. 1950. "The Design of Broad-Band Transformers for Linear Electronic Circuits." *Trans. AIEE*, **69**, 1005.

Lynn, G. E., T. J. Pula, J. F. Ringelman, and F. G. Timmel. 1960. *Self-saturating Magnetic Amplifiers.* McGraw-Hill Book Company, New York, 122–139.

Maurice, D., and R. H. Minns. 1947. "Very-Wide Band Radio-Frequency Transformers." *Wireless Eng.*, **24**, 168 (June).

McElroy, P. K., and R. F. Field. 1942. "How Good Is an Iron-Cored Coil?" *Gen. Radio Exp.*, *XVI* (March).

Melville, W. S. 1951. "The Use of Saturable Reactors as Discharge Devices for Pulse Generators." *JIEE (London)*, **98**, Part III, 185.

MIT Electrical Engineering Staff. 1943. *Magnetic Circuits and Transformers.* New York: John Wiley & Sons, Inc., p. 269.

Nyquist, H. 1932. "Regeneration Theory." *Bell Syst. Tech. J.*, **11**, 126 (January).

Partridge, G. F. 1936. *Phil. Mag.*, **22** (7th series), 675 (July–December).

Partridge, N. 1942. "Harmonic Distortion in Audio-Frequency Transformers." *Wireless Eng.*, **19** (September, October, and November).

Pen-Tung Sah, A. 1936. "Quasi Transients in Class B Audio-Frequency Push-Pull Amplifiers." *Proc. IRE*, **24**, 1522 (November).

Prince, D. C., and P. B. Vodges. 1927. *Mercury-Arc Rectifiers and Their Circuits.* McGraw-Hill Book Company, New York, p. 216.

Ramey, R. A. 1951. "On the Mechanics of Magnetic Amplifier Operation" and "On the Control of Magnetic Amplifiers." *Trans. AIEE*, **70**, 1214 and 2124, respectively.

Ruthroff, C. L. 1959. "Some Broad-Band Transformers." *Proc. IRE*, **47**, 1337–1342 (August).

Schade, O. H. 1943. "Analysis of Rectifier Operation." *Proc. IRE*, **31**, 341 (July).

Schade, O. H. 1947. "Magnetic Deflection Circuits for Cathode-Ray Tubes." *RCA Rev.*, 506 (September).

Shea, T. E. 1929. *Transmission Networks and Wave Filters.* New York: D. Van Nostrand Company, p. 187.

Snelling, E. C. 1969. *Soft Ferrites.* The Chemical Rubber Company, Cleveland, Ohio.

Storm, H. F. 1951. "Transient Response of Saturable Reactors with Resistive Load." *Trans. AIEE*, **70**, Pt. I, 99.

Tweedy, S. E. 1948. "Magnetic Amplifiers." *Electron. Eng.*, **38** (February).

Waidelich, D. L. 1941. "Diode Rectifying Circuits with Capacitances." *Trans. AIEE*, **61**, 1161 (December).

Waidelich, D. L., and H. A. Taskin. 1945. "Analysis of Voltage Tripling and Quadrupling Circuits." *Proc. IRE*, **33**, 449 (July).

Zverev, A. I. 1969. *Handbook of Filter Synthesis.* John Wiley and Sons, Inc. New York.

INDEX